HOW TO TEACH MATHEMATICS

SECOND EDITION

How to Teach Mathematics

Second Edition

Steven G. Krantz

American Mathematical Society
Providence, Rhode Island

2000 *Mathematics Subject Classification.* Primary 00A35, 00A05, 00A20.

Library of Congress Cataloging-in-Publication Data

Krantz, Steven G. (Steven George), 1951–
    How to teach mathematics : Second edition / Steven G. Krantz.
      p. cm.
    Includes bibliographical references and index.
    ISBN 0-8218-1398-6
    1. Mathematics—Study and teaching. I. Title.
QA11.K776 1998
510'.71'1—dc21
                                            98-31477
                                                CIP

---

10 9 8 7 6 5 4 3 2   04 03 02 01 00

To Robert L. Borrelli, teacher and friend.

Table of Contents

Preface to the Second Edition

"[When a mathematician speaks about teaching], colleagues smile tolerantly to one another in the same way family members do when grandpa dribbles his soup down his shirt." Herb Clemens wrote these words in 1988. They were right on point at the time. The amazing fact is that they are no longer true.

Indeed the greatest single achievement of the so-called "teaching reform" movement is that it has enabled, or compelled, all of us to be concerned about teaching. Never mind the shame that in the past we were *not* concerned about teaching. Now we are all concerned, and that is good.

Of course there are differing points of view. The "reform" school of thought favors discovery, cooperative and group learning, use of technology, higher-order skills, and it downplays rote learning and drill. The traditionalists, by contrast, want to continue giving lectures, want the students to do traditional exercises, want the students to take the initiative in the learning process, and want to continue to drill their students.[1] Clearly there are merits in both points of view. The good news is that the two sides are beginning to talk to each other. The evidence? **(1)** A conference held at MSRI in December, 1996 with the sole purpose of helping the two camps to communicate (see the Proceedings in [GKM]); **(2)** The observation that basic skills play a new role, and are positioned in a new way, in the reform curriculum; **(3)** The observation that standard lectures—the stock-in-trade of traditionalists—are not the final word on engaging students in the learning process; **(4)** The fact that studies indicate that neither method is more effective than the other, but that both have strengths; **(5)** The new wave of calculus books (see [STEW]) that attempt a marriage of the two points of view.

The reader of this book may as well know that I am a traditionalist, but one who sees many merits in the reform movement. For one thing, the reform movement has taught us to reassess our traditional methodologies. It has taught us that there is more than one way to get the job done. And it has also taught us something about the sociological infrastructure of twentieth-century mathematics. We see that our greatest pride is also our Achilles heel. In detail, the greatest achievement of twentieth-century mathematics is that we have (to the extent possible) fulfilled the Hilbert/Bourbaki program of putting everything on a rigorous footing; we have axiomatized our subject; we have precise definitions of everything. The bad news is that these accomplishments have shaped our world view, all the way down into the calculus classroom. Because we have taught ourselves to think strictly according to Occam's Razor, we also think that that should be the mode of discourse in the calculus classroom. This view is perhaps short-sighted.

---

[1] As I will say elsewhere in the book, the reformers constitute a heterogeneous group, just like the traditionalists. There is no official reform dogma, just as there is no official traditionalist dogma. Some reformers tell me that they strongly favor drill, but that drill should be built atop a bedrock of understanding. Many traditionalists seem to prefer to give the drill first—asking the students to take it on faith—and *then* to develop understanding. George Andrews has asked whether, if instead of calling it "mere rote learning" we called it "essential drill", would people view it differently? See the Appendix of Andrews at the end of this book.

First, students (and others, too!) do not generally learn *axiomatically* (from the top down). In many instances it is more natural for them to learn *inductively* (from the bottom up). Of course the question of how people learn has occupied educational theorists as far back as Beth and Piaget [BPA], and will continue to do so. But, as I say elsewhere in this book, the mathematics instructor must realize that a student cannot stare at a set of axioms and "see what is going on" in the same way that an experienced mathematician can. Often it is more natural for the student to first latch on to an example.

Second, we must realize that the notion of "proof" is a relative thing. Mathematical facts, or theorems, are freestanding entities. They have a life of their own. But a proof is largely a psychological device for convincing someone that something is true. A trained mathematician is taught a formalism for producing a proof that will be acceptable to his colleagues. But a freshman in college is not. What constitutes a believable proof for a freshman could easily be a good picture, or a plausibility argument. This insight alone can turn an ordinary teacher into a good one. What is the sense in showing a room full of freshmen a perfectly rigorous proof (of the fundamental theorem of calculus, say), secure in the knowledge that you have "done the right thing," but also knowing unconsciously that the students did not understand a word of it? Surely it is more gentle, as a didactic device, to replace "*Proof:*" with "Here is an idea about why this is true." In doing the latter, you have not been dishonest (i.e., you have not claimed that something was a strictly rigorous proof when in fact it was not). You have instead met the students half way. You have spoken to them in their own language. You have appealed to their collective intuition. Perhaps you have taught them something. Always keep in mind that persuasion has many faces.

I have witnessed discussions in which certain individuals were adamant that, if you give an explanation in a calculus class that is not strictly a proof, then you must say, "This is not a proof; it is an informal explanation." Of course such a position is a consequence of twentieth-century mathematical values, and I respect it. But I do not think that it constitutes good teaching. In the first place, such a mantra is both tiresome and discouraging for the students. The instructor can instead say, "Let's think about why this is true ..." or "Here is a picture that shows what is going on ..." and thereby convey the same message in a much friendlier fashion.

In my own mathematics department we have a "transitions" course, in which students are taught first-order logic, naive set theory, equivalence relations and classes, the constructions of the number systems, and the axiomatic method. They are also taught—at a very rudimentary level—how to construct their own proofs. Typically a student takes this course *after* calculus, linear algebra, and ordinary differential equations but *before* abstract algebra and real analysis. I think of the transitions course as a bellwether. Before that course, students are not ready for formal proofs. We should adapt our teaching methodology to *their* argot. After the transitions course, the students are more sophisticated. Now they are ready to learn *our* argot.

I have decided to write this new edition of *How to Teach Mathematics* in part because I have learned a lot about teaching in the five-year interval since this book first appeared. The teaching reform movement has matured, and so have the rest of us. I believe that I now know a lot more about what constitutes good teaching. I regularly teach our graduate student seminar to help prepare our Ph.D. candidates for a career in teaching, and I have an ever better understanding of how to conduct such training. I would like to share my new insights in this edition.

One of the best known mathematical errors, particularly in the study of an optimization problem, is to assume that the problem has a solution. Certainly Riemann's original proof of the Riemann mapping theorem is a dramatic example of this error, but the calculus of variations (for instance) is littered with other examples. Why can we not apply this hard-won knowledge to other aspects of our professional lives? Why do we assume that there is a "best" way to teach calculus? Or a "best" textbook? Teaching is a very personal activity, and different individuals will do it differently. Techniques that work for one person will not work for another. (Also, techniques that work in one class will not necessarily work in another.) I believe that we need, as a group, to acknowledge that there is a pool of worthwhile teaching techniques, and we should each choose those methods that work for us and for our students.

Ever since the first edition of this book appeared, mathematicians have approached me and asked, "OK, what's the secret? Students these days drive me crazy. I can't get through to them. They won't talk to me. How do you do it?" I wish that I had a simple answer. I would like to be able to say, "Take this little green pill." or "Say this prayer in the morning." or "Hold your mouth this way." But in fact there is no simple answer. Even so, I have invested considerable time analyzing the situation as well as talking to other successful teachers about how to make the teaching process work. I have come to the following conclusion.

Students are like dogs: They can smell fear. (I do not mean to say here that we should think of our students as attack dogs. Rather, they are sensitive to body language and to nuances of behavior. See also Section 2.9 on teaching evaluations.) When you walk into your classroom, the students can tell right away whether you really want to be there, whether you have something interesting to tell them, whether you respect them as people. If they sense instead that you are merely slogging through this dreary duty, just writing the theorems and proofs on the blackboard, refusing to answer questions for lack of time, then they will react to you in a correspondingly lackluster manner.

When I walk into my calculus class, I look forward to seeing the students perk up, with a look on their faces that says "Showtime!" In the few minutes before the formal class begins, I chat with them, joke around, find out what is going on in their lives. I relate to them as people. It will never happen that a student will go to the chairman or the dean and complain about me. Why? Because they know that they can come and talk to *me* about their concerns. If a student is not doing well in my class, that student is comfortable coming to me. And he knows that the fault for his poor performance is as likely *his* as it is mine, because he realizes that I am doing everything that I can. If you believe what I am describing here, then perhaps you can also understand why I enjoy teaching, and why I find the process both stimulating and fulfilling.

I recently taught a fairly rigorous course in multivariable calculus—a subject in which students usually have a lot of trouble. The main reason that they have so much trouble is that there are so many ideas—vectors, cross products, elements of surface area, orientation, conservative vector fields, line integrals, tangent planes, etc.—and they are all used together. Just understanding how to calculate both sides of the equation in Stokes's theorem, or the divergence theorem, requires a great deal of machinery. The way that I addressed their difficulties is that I worked the students hard. I gave long, tough homework assignments. A day or two before any given assignment was due, I would begin a class discussion of the homework. If necessary, I would work out the bulk of a problem on the board for them. But I would add that I expected each of them to write up the problem carefully and completely—with full details. And I would give them a few extra days so that they could complete the assignment. But I did not stop there. Next class, I would ask how the homework was going. If necessary, we would discuss it again. If necessary, I would give them another extension. The point here is that I made it absolutely clear to the students that the most important thing to me was that they would complete the assignment. I would give them whatever time, and whatever help, was needed to complete the work. During the long fifteen-week semester, attendance in the class was virtually constant, and always exceeded 95%. At the end, I gave them a long, tough final exam. And the average was 85%. I can only conclude that I set a standard for these students, and they rose to it. Both they and I came away from the course with a feeling of success. They had worked hard, and they had learned something.

You may be thinking, "Well, Krantz teaches at a fancy private school with fancy private students. I could never get away with this at Big State University." That is a defeatist attitude. If you expect your students to try, then you must try. I have taught at big state universities. I understand the limitations that teaching a large class of not particularly select students imposes. But you can adjust the techniques described in the last paragraph to most any situation. If you wonder how I can afford to spend valuable class time going over homework, my answer is this: I am an experienced teacher, and fourteen weeks is a long time. I can always adjust future classes, leave out a few examples, give short shrift to some ancillary topics. I never worry about running out of time.

I have gone on at some length in this Preface to give the uninitiated reader a glimpse of where I am coming from. I hope that on this basis you can decide whether you want to read the remainder of the book. This is a self-help book in the strongest sense of the word. It is a kit that will allow you to build your own teaching methodology and philosophy. I certainly cannot do it for you. What I *can* do is provide you with some tips, and advice, and the benefit of my own experience. Nothing that I say here is "correct" in any absolute sense. It is just what I know.

One of my disappointments pursuant to the first edition of this book is that nobody has taken it as an impetus to write his own book espousing his own teaching philosophy. There have been some reviews of this book—several of them rather strong and critical both of the book and of its author (see [MOO], [BRE]). I welcome such discussions, and would only like to see further discourse. I am delighted to be able to say that several distinguished scholars, who have been active in exploring and discussing teaching issues, have agreed to write

Appendices to this new edition of *How to Teach Mathematics*. Let me stress that these are not all people who agree with me. In fact some of us have had spirited public disagreements. But we all share some common values. We want to discover how best to teach our students. The new Appendices help to balance out the book, and to demonstrate that any teaching question has many valid answers.

When I teach the teaching seminar for our graduate students, the first thing I tell them is this: "In this course, I am *not* going to tell you how to teach. You have to decide that for yourselves. What I intend to do is to sensitize you to certain issues attendant to teaching. Then you will have the equipment so that you can build your own teaching philosophy and style." I would like to suggest that you read this book in the same spirit. You certainly need not agree with everything I say. But I hope you will agree that the issues I discuss are ones that we all must consider as we learn how to teach.

When I was a graduate student—in one of the best math graduate programs in the country—I never heard a single word about teaching. Actually, that's not true. Every once in a while we would be talking about mathematics and someone would look at his watch and say, "Damn! I have to go teach." But that was the extent of it. Six years after I received my Ph.D., I returned to that same Ivy League school as a visiting faculty member. Times had changed, and one of the senior faculty members gave a twenty-minute pep talk to all new instructors. He said, "These days, you can either prove the Riemann hypothesis or you can learn how to teach." He went on to tell us to speak up during lectures, and to write neatly on the blackboard. This was not the most profound advice on teaching that I have ever heard, but it certainly represented progress.

The truth is that, as a graduate student, I was so hellbent on learning to be a mathematician that I probably gave little thought to teaching. I would have felt quite foolish knocking on my thesis advisor's door and asking his advice on how to teach the chain rule. I shudder to think what he might have replied. But we have all evolved. It makes me happy that my own graduate students frequently consult me on **(i)** mathematics, **(ii)** teaching, and **(iii)** the profession. Though I secretly may relish **(i)** a bit more than **(ii)** or **(iii)**, I do enjoy all three.

Teaching is an important part of what we do. Because of economic stringencies, and new societal values, university administrations are monitoring every department on campus to ensure that the teaching is (better than) adequate and is working. My university is known nationwide for its good teaching. Yet an experienced administrator here said recently that 80% of the tenured faculty (campus-wide) could *not* get tenure today on the basis of their teaching.

We simply cannot get away with the carelessness that was our hallmark in the past. Thanks in part to the teaching reform movement, we have all come to understand this change in values, and we are beginning to embrace it. A book like [CAS], which offers advice to a fledgling instructor, could not have existed twenty years ago. Now it is a valuable part of our literature.

Teaching is a regimen that we spend our entire lives learning and revising and honing to a sharp skill. This book is designed to help you in that pursuit.

I am happy to acknowledge the advice and help that I have received from many friends and colleagues in the preparation of this new edition. I would like particularly to mention Joel Brawley, David Bressoud, Robert Burckel, John B. Conway, Ed Dubinsky, Len Gillman, David Hoffman, Gary Jensen, Meyer Jerison, Kristen Lampe, Vladimir Maşek, Chris Mahan, Deborah K. Nelson, Hrvoje Sikic, Nik Weaver, Stephen Zemyan, and Steven Zucker. Lynn Apfel was good enough to read several drafts of this manuscript with painstaking care, and to share with me her cogent insights about teaching; I am most grateful for her contributions. Jennifer Sharp of the American Mathematical Society gave me the benefit both of her editing skills and of her knowledge of language and meaning. Her help has been invaluable.

Last, but not least, Josephine S. Krantz is a constant wellspring of inspiration; her Mom, Randi Ruden, is a source of solace.

Of course the responsibility for all remaining errors or foolishness resides entirely with me.

Steven G. Krantz
St. Louis, Missouri

Preface to the First Edition

While most mathematics instructors prepare their lectures with care, and endeavor to do a creditable job at teaching, their ultimate effectiveness is shaped by their attitudes. As an instructor ages (and I speak here of myself as much as anyone), he finds that he is less in touch with his students, that a certain ennui has set in, and (alas) perhaps that teaching does not hold the allure and sparkle that it once had. Depending on the sort of department in which he works, he may also feel that hotshot researchers and book writers get all the perks and that "mere teachers" are viewed as drones.

As a result of this fatigue of enthusiasm, a professor will sometimes prepare for a lecture *not* by writing some notes or by browsing through the book but by lounging in the coffee room with his colleagues and bemoaning **(a)** the shortcomings of the students, **(b)** the shortcomings of the text, and **(c)** that professors are overqualified to teach calculus. Fortified by this yoga, the professor will then proceed to his class and give a lecture ranging from dreary to arrogant to boring to calamitous. The self-fulfilling prophecy having been fulfilled, the professor will finally join his cronies for lunch and be debriefed as to **(a)** the shortcomings of the students, **(b)** the shortcomings of the text, and **(c)** that professors are overqualified to teach calculus.

There is nothing new in this. The aging process seems to include a growing feeling that the world is going to hell on a Harley. A college teacher is in continual contact with young people; if he feels ineffectual or alienated as a teacher, then the unhappiness can snowball.

Unfortunately, the sort of tired, disillusioned instructors that I have just described exist in virtually every mathematics department. A college teacher who just doesn't care anymore is a poor role model for the novice instructor. Yet that novice must turn somewhere to learn how to teach. You cannot learn to play the piano or to ski by watching someone else do it. And the fact of having sat in a classroom for most of your life does not mean that you know how to teach.

The purpose of this book is to set down the traditional principles of good teaching in mathematics—as viewed by this author. While perhaps most experienced mathematics instructors would agree with much of what is in this book, in the final analysis this tract must be viewed as a personal polemic on how to teach.

Teaching is important. University administrations, from the top down, are today holding professors accountable for their teaching. Both in tenure and promotion decisions and in the hiring of new faculty, mathematics (and other) departments must make a case that the candidate is a capable and talented teacher. In some departments at Harvard, a job candidate must now present a "teaching dossier" as well as an academic dossier. It actually happens that good mathematicians who are really rotten teachers do not get that promotion or do not get tenure or do not get the job that they seek.

The good news is that it requires no more effort, no more preparation, and no more time to be a good teacher than to be a bad teacher. The proof is in this book. Put in other words, this book is not written by a true believer who is going to exhort you to dedicate every waking hour to learning your students' names and designing seating charts. On the contrary, this book is written by a pragmatist who values his time and his professional reputation, but is also considered to be rather a good teacher.

I intend this book primarily for the graduate student or novice instructor preparing to sally forth into the teaching world; but it also may be of some interest to those who have been teaching for a few or even for several years. As with any endeavor that is worth doing well, teaching is one that will improve if it is subjected to periodic re-examination.

Let me begin by drawing a simple analogy: By the time you are a functioning adult in society, the basic rules of etiquette are second nature to you. You know instinctively that to slam a door in someone's face is (i) rude, (ii) liable to invoke reprisals, and (iii) not likely to lead to the making of friends and the influencing of people. The keys to good teaching are at approximately the same level of obviousness and simplicity. But here is where the parallel stops. We are all *taught* (by our parents) the rules of behavior when we are children. Traditionally, we (as mathematicians) are not taught anything, when we are undergraduate or graduate students, about what constitutes sound teaching.

In the past we have assumed that either

(i) Teaching is unimportant.

or

(ii) The components of good teaching are obvious.

or

(iii) The budding professor has spent a lifetime sitting in front of professors and observing teaching, both good and bad; surely, therefore, this person has made inferences about what traits define an effective teacher.

I have already made a case that (i) is false. I agree wholeheartedly with (ii). The rub is (iii). If proof is required that at least some mathematicians have given little thought to exposition and to teaching, then think of the last several colloquia that you have heard. How many were good? How many were inspiring? This is supposed to be the stuff that matters—getting up in front of our peers and touting our theorems. Why is it that people who have been doing it for twenty or thirty years still cannot get it right? Again, the crux is item (iii) above. There are some things that we do not learn by osmosis. How to lecture and how to teach are among these.

Of course the issue that I am describing is not black and white. If there were tremendous peer support in graduate school and in the professorial ranks for great teaching, then we would force ourselves to figure out how to teach well. But often there is not. The way to make points in graduate school is to ace the qualifying exams and then to write an excellent thesis. It is unlikely that your thesis advisor wants to spend a lot of time with you chatting about how to teach the chain rule. After all, he has tenure and is probably more worried about where his next theorem or next grant or next raise is coming from than about such prosaic matters as calculus.

The purpose of this book is to prove that good teaching requires relatively little effort (when compared with the alternative), will make the teaching process a positive part of your life, and can earn you the respect of your colleagues. In large part I will be stating the obvious to people who, in theory, already know what I am about to say.

It is possible to argue that we are all wonderful teachers, simply by *fiat*, but that the students are too dumb to appreciate us. Saying this, or thinking it, is analogous to proposing to reduce crime in the streets by widening the sidewalks. It is doubletalk. If you are not transmitting knowledge, then you are not teaching. We are not hired to train the ideal platonic student. We are hired to train the particular students who attend our particular universities. It is our duty to learn how to do so.

This is a rather personal document. After all, teaching is a rather personal activity. But I am not going to advise you to tell jokes in your classes, or to tell anecdotes about mathematicians, or to dress like Gottfried Wilhelm von Leibniz when you teach the product rule. Many of these techniques only work for certain individuals, and only in a form suited to those individuals. Instead I wish to distill out, in this book, some universal truths about the teaching of mathematics. I also want to go beyond the platitudes that you will find in books about teaching *all* subjects (such as "type all your exams", "grade on a bell-shaped curve") and talk about issues that arise specifically in the teaching of mathematics. I want to talk about principles of teaching that will be valid for all of us.

My examples are drawn from the teaching of courses ranging from calculus to real analysis and beyond. Lower-division courses seem to be an ideal crucible in which to forge teaching skills, and I will spend most of my time commenting on those. Upper-division courses offer problems of their own, and I will say a few words about those. Graduate courses are dessert. You figure out how you want to teach your graduate courses.

There are certainly differences, and different issues, involved in teaching every different course; the points to be made in this book will tend to transcend the seams and variations among different courses. If you do not agree in every detail with what I say, then I hope that at least my remarks will give you pause for thought. In the end, you must decide for yourself what will take place in your classroom.

There is a great deal of discussion these days about developing new ways to teach mathematics. I'm all for it. So is our government, which is generously funding many "teaching reform" projects. However, the jury is still out regarding which of these new methods will prove to be of lasting value. It is not clear yet exactly how `Mathematica` notebooks or computer algebra systems or interactive computer simulations should be used in the lower-division mathematics classroom. Given that a large number of students need to master a substantial amount of calculus during the freshman year, and given the limitations on our resources, I wonder whether alternatives to the traditional lecture system—such as, for instance, Socratic dialogue—are the correct method for getting the material across. Every good new teaching idea should be tried. Perhaps in twenty years some really valuable new techniques will have evolved. They do not seem to have evolved yet.

In 1993 I must write about methods that I know and that I have found to be effective. Bear this in mind: Experimental classes are experimental. They usually lie outside the regular curriculum. It will be years before we know for sure whether students taught with the new techniques are understanding and retaining the material satisfactorily and are going on to successfully complete their training. Were I to write about some of the experimentation currently being performed then this book would of necessity be tentative and inconclusive.

There are those who will criticize this book for being reactionary. I welcome their remarks. I have taught successfully, using these methods, for twenty years. Using critical self examination, I find that my teaching gets better and better, my students appreciate it more, and (most importantly) it is more and more effective. I cannot in good conscience write of unproven methods that are still being developed and that have not stood the test of time. I leave that task for the advocates of those methods.

In fact I intend this book to be rather prescriptive. The techniques that I discuss here are ones that have been used for a long time. They work. Picasso's revolutionary techniques in painting were based on a solid classical foundation. By analogy, I think that before you consider new teaching techniques you should acquaint yourself with the traditional ones. Spending an hour or two with this book will enable you to do so.

I am grateful to the Fund for the Improvement of Post-Secondary Education for support during a part of the writing of this book. Randi D. Ruden read much of the manuscript critically and made decisive contributions to the clarity and precision of many passages. Josephine S. Krantz served as a valuable assistant in this process. Bruce Reznick generously allowed me to borrow some of the ideas from his book *Chalking It Up*. I also thank Dick Askey, Brian Blank, Bettye Anne Case, Joe Cima, John Ewing, Mark Feldman, Jerry Folland, Ron Freiwald, Paul Halmos, Gary Jensen, John McCarthy, Alec Norton, Mark Pinsky, Bruce Reznick, Richard Rochberg, Bill Thurston, and the students in our teaching seminar at Washington University for many incisive remarks on different versions of the manuscript. The publications committees of the Mathematics Association of America and of the American Mathematical Society have provided me with detailed reviews and valuable advice for the preparation of the final version of this book.

SGK

Chapter 1

Guiding Principles

I've got it all figured out: I pretend to teach and the students pretend to learn.
                                                                    Anon.

If logic is the hygiene of the mathematician, it is not his source of food.
                                                              André Weil

Only two things are infinite, the universe and human stupidity, and I'm not sure about the former.
                                                         Albert Einstein

Only professional mathematicians learn anything from proofs. Other people learn from explanations.
                                                       R. P. Boas [BOA]

A psychotic says that two and two are five. A neurotic says that two and two are four and hates it.
                                                     John D. MacDonald

I had so often in the past seen dumb domestic animals in Africa so aware of the secret intent of the people who had bred and reared them and earned their trust that they could hardly walk, knowing they were being led to a distant place of slaughter.
                                                   Laurens Van Der Post

I dreamed that I was in a large auditorium lecturing about calculus. Then I woke up, and I was in a large auditorium lecturing about calculus.
                                          An Anonymous Calculus Teacher

As long as there are math exams then there will continue to be prayer in school.
                                                                    Anon.

He not only overflowed with learning, he stood in the slop.
                                                Sydney Smith on Macaulay

1.0   Chapter Overview

Good teaching is a product of preparation, effort, and a good attitude on the part of the instructor. If you treat your teaching as analogous to having root canal treatments, then it's a certainty that you will not enjoy it. And your students probably will not either.

This chapter discusses broad issues connected with the teaching process: respect, preparation, time management, voice control, use of computers, and so forth. The ideas presented here are ones that you will want to internalize. You want them to be a natural part of your teaching habit.

1.1 Respect

You cannot be a good teacher if you do not respect yourself. If you are going to stand up in front of thirty people or three hundred people and try to teach them something, then you had better

- Believe that you are well qualified to do so.

- Want to do so.

- Be *prepared* to do so.

- Make sure that these characteristics are evident to your audience.

It is a privilege to stand before a group of people—whether they be young adults or your own peers—and to share your thoughts with them. You should acknowledge this privilege by **(a)** dressing appropriately for the occasion, **(b)** making an effort to communicate with your audience, **(c)** respecting the audience's point of view.

One completely obvious fact is that you should have your material *absolutely mastered* before you enter the classroom. If you *do* possess this mastery, then you can expend the majority of your effort and attention on conveying the material to the audience. If, instead, you have a proof or an example that is not quite right, and if you stand in front of the group trying to fix it, then you will lose all but the diehards quickly. (Note that this last statement is true both when you are giving a calculus lecture and when you are giving a colloquium talk at Harvard.)

It is easy to rationalize that if the students were more able then they could roll with the ups and downs of your lecture. This is strictly illogical. How do you behave when you are listening to a colloquium or seminar and the lecturer goes off into orbit—either to fix an incorrect argument or into a private conversation with his buddy in the front row or, worse, into a private conversation with himself? All right then, now that you have admitted honestly how *you* behave, then how can you expect unseasoned freshmen to be tolerant when you do not seem to be able to do the examples that *they* are expected to do?

One of the best arguments for even elementary college mathematics courses to be taught by people with advanced degrees is this: Because the material is all trivial and obvious to the professor, he can maintain a broad sense of perspective, he will not be thrown by questions, and he can concentrate on the act of *teaching*.

If you respect yourself then, it follows logically, you will respect your audience. You should prepare for your class. That way you will not be surprised by gaps in your thinking, you will not have to cast around for a necessary idea, and you will not lose your train of thought in class. You will be receptive to questions. You will sense immediately when the students are not understanding, and you will do something about it.

Throughout this book, I will repeatedly exhort you to prepare your classes. I do not necessarily have in mind that you should spend an inordinate amount of *time* preparing. Consider by analogy the psychology of sport. Weightlifters, for example, are taught to meditate in a certain fashion before a big lift. Likewise, preparing is a way to collect your thoughts and put yourself in the proper frame of mind to give a class. You might prepare by walking to the student union and buying a cup of coffee (thinking on the way about your lecture). Or you might prepare by browsing through some calculus books (or even—horror of horrors—browsing through the *actual text* for that particular class). I have sometimes prepared for a class by staring bleakly out the window pondering the notion that I have nothing to say. No matter—this is a yoga. It helps you to pull yourself together. It makes good sense. See also Section 1.3.

To me, preparation is the core of effective teaching. While this may sound like a tautology, and not worth developing, there is in fact more here than meets the eye. Just as being a bit organized relieves you of the stress and nuisance of spending hours looking for a postage stamp or a pair of scissors when you need them, so being the master of your subject gives you the ability to cope with the unexpected, to handle questions creatively, and to give proper stimulus to your students. An experienced and knowledgeable teacher who is comfortable with his craft is constantly adjusting the lecture, in real time, to suit the expressions on the students' faces, to suit their responses to queries and prods, to suit the rate and thoroughness with which they are absorbing the material. Just as a good driver is constantly (and unconsciously) making little adjustments in steering in response to road conditions, weather, and so forth, it is also the case that a good teacher (just as unconsciously) is engaging in a delicate give and take with the audience. Complete mastery is the unique tool that gives you the freedom to develop this skill.

You should treat questions with respect. I go into every class that I teach knowing full well that I am probably much smarter, and certainly much better informed, than most of the people in the room. But I do not need to use a room full of eighteen-year-olds as a vehicle for bolstering my ego. If a student asks a question, even a stupid one, then I treat it as an event. A wrong question can be turned into a good one with a simple turn of phrase from the instructor (see Section 1.7). If the question requires a lengthy answer, then give a short one and encourage the student to see you after class. If you insult—even gently—the questioner, then you not only offend that person but you perhaps offend everyone else in the room. Once the students have turned hostile it is difficult to win them back, both on that day and on subsequent days.

In fact one of the most dynamic and inspiring things that you can do for your students is to help them think that their questions are steering the class. Even though students know in their heart of hearts that you are the boss, and you are the person with all the knowledge, they deeply resent the notion that you are the high priest dispensing wisdom. It is so much more stimulating to respond to a student question by saying, "Yes, that was my next point ..." or "You are anticipating the ideas. That's very good ..." or "You have raised a crucial issue. That's great. Let me pursue it for a moment ..."

Sometimes you can actually develop an idea in dialogue with a student—right in front of the class. For example, you say, "This point is not a boundary point. Why not?" The student replies, "Some neighborhoods have no points from the complement." "Good," you say. "So, in a certain sense, this point is *away* from the complement of the set. Instead we might say it's ..." "Interior?"

interjects the student. How could a student *not* pay attention if he thinks that his thoughts, his questions, have shaped the discussion?

1.2   More on Respect

The last section laid some groundwork, in very broad strokes, concerning the issue of respect—that is, the respect that a teacher must show his audience. Now I shall pursue some more particular issues relating to respect.

If a student asks for permission to hand a paper in late, and you are tempted to say, "I've set the deadline and the deadline is the deadline and how dare you ask me for an extension?" then you should pause. You should ask yourself what you will have accomplished with this little speech. Does it really make any difference if this particular student hands in the homework paper tomorrow morning? If it does not, then is there any reason not to grant the extension? Is it possible that, given a little more time, this student will learn something extra? Couldn't you just wait until the next morning to give the stack of homework to the grader? This consideration is just showing the student the sort of respect that you would have liked to have been shown when you were a student.

If a student wants to discuss why an exam was graded the way that it was, or why so many points were deducted on Problem 5, or why he is not doing as well in the class as anticipated, then don't hide behind your rank. If you are not prepared to say a sentence or two about why you graded Problem 5 the way that you did, it may mean that you graded the problem very sloppily and are afraid that you will be made to look foolish. If you don't want to discuss how the student is doing in the class, it may mean that you don't know, or can't tell, or don't care. In fact the root of the problem may be that you are uncomfortable with confrontational scenes. But confrontational scenes may be avoided almost 100% of the time. If you are looking over Problem 5 with the student, and if it turns out that the problem was basically right but you only gave 3 points out of 10, then you might say, "I guess I read this problem too quickly. I get bleary-eyed late at night after reading sixty papers. Let me regrade it for you right now." The student is usually so grateful for the extra points that nothing more need be said.

To that (extremely) unusual student who says

> You are an incompetent boob. I am going to complain to the chairman of the department.

you might say, "That is your privilege. Let me phone his secretary and make an appointment for you." Most of the time the student will back down. In the rare instances when the student does not retreat, any good chairman will give you ample opportunity to clarify the situation and smooth things over.

I don't mean to suggest in these pages that teaching is by nature adversarial. On the contrary, it can and should be a nurturing activity. But the potential for conflict is there, and I shall avail myself of several opportunities to suggest how you can either avoid or ameliorate friction.

Another aspect of showing respect for your students is not springing nasty surprises on them. Do not say that you will grade a course one way and then grade it in a different fashion altogether. Do not act like an arrogant autocrat. (The truth is that, as a college teacher, you *are* an autocrat and a monarch and can do pretty much as you please. But there is no need to flaunt this fact before your students.) Do not create examinations that are full of dirty-trick questions. If you do, then you will be setting up a self-fulfilling prophecy in which everyone will do poorly. Then you can brag to your pals that the students are stupid. How does this profit you? It is easier to make up a straightforward exam (not necessarily a watered-down exam) that tests students *on the material that you taught them.* If, as a result, the average on the test is 80% (a not very likely eventuality), then you can tell the students how pleased you are that they have mastered the material so well.

If you give a miserable exam on which the average is 30% then you will have accomplished the following: For about 5% of the students (at the most) you will have set a standard that they can strive to meet. But you will have alienated the other 95%. Students simply do not have the maturity and experience to compensate for your inadequacies. If the average on the midterm is 30% and you tell the students, "Square your score and divide by 11 and that will give you a more realistic idea of how you are doing" then—truthfully—they won't know what you are talking about. It won't work and it will make the students feel angry and alienated. So avoid doing it. Students need not alienation but encouragement and (suitably tempered) challenge.

When talking among ourselves we instructors sometimes say that, "students today do not know what it means to rise to a challenge". I'm sorry, but if your students do not appear to be rising to a challenge then it is your fault as much as it is theirs. It is true that today's students are not accustomed, as perhaps was a student like Bertrand Russell or Ludwig Wittgenstein, to have ever higher hurdles set before them at a prodigious rate. Many American students went through a grade school and a high school program in which they played a somewhat passive role. In a number of troubled high schools, students receive a grade of "A" or "B" as a reward for nonviolence and overall cooperation. Many high schools, hellbent on preparing their students for the SAT,[1] tend to shortchange critical thinking skills and intellectual curiosity. This does not mean that students today are stupid. Rather, they may not have been stimulated to rise to their full potential. They may never have realized their capacity to be skilled thinkers. In spite of television and other ostensibly insidious forces in our society, it is still a part of human nature to want to excel. It takes some practice to learn how to bring out this attribute in students, but it can be done.

During the twenty-five years that I have been teaching, it has become common among mathematicians to observe that, "The students coming out of high school these days cannot add fractions, they don't know what a proof is. How am I supposed to teach people like that?" But I think that we are beginning to evolve. Among the people that I talk to (see also especially the Appendix by Steven Zucker), the attitude is now, "Let's figure out how to teach these eighteen

---

[1]The Scholastic Aptitude Test—a widely used test that measures college preparedness.

year olds." After all, these students are just as bright and just as able as you and I were when we were entering college. The difference is that these new students have never learned how to learn. So we had better step into the breach and show them how to do it. There *is* a difference between high school learning and university learning. Let's show the students what that difference is and then show them how to make the transition.

There are important philosophical and educational issues at play here. In America at the close of the millennium we attempt, as much as possible, to educate everyone. Whereas forty years ago this meant that "everyone" went to high school, now it means that "everyone" goes to college. It is vital in a free society that all citizens, regardless of financial resources, have an opportunity to pursue a college education. But society is set up in such a way now that a large percentage of young people go to college regardless of their interests or goals. For this we pay a price. We can rely less than we would like on the preparedness of our freshmen. We also can rely somewhat less on their attitudes and motivation.

What this means in practice is that, quite often, especially when teaching freshmen—and especially at a public institution—we are not necessarily teaching a very select group. Many public institutions these days have an open admissions policy. Anyone with a high school diploma has the right to attend the state university. From the taxpayers' point of view, such an admission policy makes perfectly good sense. If you are a professor at a state institution then you must make peace with the realities connected with such a policy. You must learn to adjust your expectations. You must learn a little patience, and learn to be flexible.

There are decided differences between the reality of what a high school education is and the dream of what a university education ought to be. The former is often a passive activity which does not challenge students to think creatively or incisively. The latter should be a stimulating and demanding rite that will turn students into productive citizens who will think creatively and profoundly and ask difficult questions of our leaders when appropriate (see Section 3.5 for more on the differences between high school and college). You, the college instructor, can and should play a critical role in helping students effect the transition from the first of these environments to the second.

I am not about to recommend that college math teachers spend their evenings reading position papers on motivational psychology. I *am* recommending that the college mathematics teacher exercise some tolerance. Students *will* rise to a challenge, provided that the teacher starts with small challenges and works up to big ones. If students stumble at the first few challenges then they need encouragement, not derision. Exercising patience requires no more effort than exercising your vocal chords with an insulting remark.

I have found persistence to be one of the most powerful pedagogical tools in my arsenal (see also Section 2.3). If my students aren't getting some idea I say, "OK, let's do it again." If that doesn't work, we do it some more. If I won't give up then the students are also unlikely to give up. They trust me; they know that I am trying to do something good for them.

The professor's attitude toward the class is apparent from his every word, every gesture, every action. If you are arrogant, if you despise your students, if you feel that you are above the task of teaching this course, then your students will get the message immediately. And what are you accomplishing by evincing these attitudes? Does it make you feel superior? More accomplished? More secure? More important? It should not. Teaching is an essential part of what we do. Society sees us as teachers and, increasingly, the university sees us as teachers. Why not see yourself as a teacher? It will increase your self-esteem. And it will increase the esteem that others have for you. Being a good teacher does not detract from your mathematical attributes. It augments them.

1.3 Prepare

Some people rationalize not teaching well by saying (either to themselves or to others), "My time is too valuable. I am not going to spend it preparing my calculus lecture. I am so smart that I can just walk into the classroom and wing it. And the students will benefit from watching a mathematician think on his feet." (As a student, I actually had professors who announced this nonsense to the class on a regular basis. And, as you can imagine, these were professors who royally botched up their lectures on a regular basis.)

It is true that most of us can walk into the room most of the time and mostly wing it. But most of us will not be very successful if we do so. Thirty minutes can be sufficient time for an experienced instructor to prepare a calculus lecture. A novice instructor, especially one teaching an unfamiliar subject for the first time, may need considerably more preparation time. Make sure that you have the definitions and theorems straight. Read through the examples to make sure that there are no unpleasant surprises. It is a good idea to have a single page of notes containing the key points. To write out every word that you will say, write out a separate page of anticipated questions, have auxiliary pages of extra examples, have inspirational quotes drawn from the works of Thomas Carlyle, make up a new notational system, make up your own exotic examples, and so forth, is primarily an exercise in self-abuse. Over-preparation can actually stultify a lecture or a class. But you've got to know your stuff.

I cannot emphasize too strongly the fact that preparation is of utmost importance if you are going to deliver a stimulating class. However it is also true that the more you prepare the more you lose your spontaneity. You must strike a balance between (**i**) knowing the material cold and (**ii**) being able to "talk things through" with your audience.

My own experience is that there is a "right amount" of preparation that is suitable for each type of course. I want to be confident that I'm not going to screw up in the middle of lecture. But I also want to be actually thinking the ideas through as I present them. I want to feel that my lecture or class has an edge. It *is* possible to over-prepare. To continue to prepare after you have already prepared sufficiently is a bit like hitting yourself in the head with a hammer because it feels so good when you stop.

You must be sufficiently confident that you can field questions on the fly, can modify your lecture (again on the fly) to suit circumstances, can tolerate a diversion to address a point that has been raised. The ability to do these things well is largely a product of experience. But you can *cultivate* this ability too. You cannot learn to play the piano by accident. And you will not learn to teach well by accident. You must be aware—in detail—of what it is that you are trying to master and then consciously hone that skill.

If you do not prepare—I mean *really* do not prepare—and louse up two or three classes in a row, then you will experience one or more of the following consequences: **(i)** Students will take up your time after class and during your office hour (in order to complain and ask questions), **(ii)** Students will stop coming to class, **(iii)** Students will complain to the undergraduate director and to the chairman, **(iv)** Students will (if you are really bad) complain to the dean and write letters to the student newspaper, **(v)** Students will write bad teaching evaluations for your course.

Now student teaching evaluations are not gospel (see Section 2.9). They contain some remarks that are of value and some that are not. Getting bad teaching evaluations does not necessarily mean that you did a bad job. And I know that the dean will only slap me on the wrist if he gets a complaint about my teaching (however, if there are ten complaints, then I had better look out). Finally, I know that the chairman will give me the benefit of the doubt and allow me every opportunity to put any difficult situation in perspective. But if I spend thirty minutes preparing each of my classes then I will avoid all this grief and, in general, find the teaching experience pleasurable rather than painful. What could be simpler?

As well as preparing for a class, you would be wise to debrief yourself after class. Ask yourself how it went. Were you sufficiently well prepared? Did you handle questions well? Did you present that difficult concept as clearly as you had hoped? Was there room for improvement? Be as tough on yourself as you would be after any exercise that you genuinely care about—from playing the piano to engaging in a tennis match. It will result in real improvement in your teaching.

Read your teaching evaluations (Section 2.9). Many are insipid. Others are puerile. Most, however, are thoughtful and well-meant. If ten of your students say that your writing is unclear, or that you talk too quickly, or that you are impatient with questions, then maybe there is a problem that you should address. Teaching is a yoga. Your mantra is "Am I getting through to them?"

It is a good idea to try to anticipate questions that students will ask. But you cannot do this artificially, as a platonic exercise late at night over a cup of coffee. It comes with experience. Assuming that you have adopted the attitude that you actually care whether your students learn something, then after several years of teaching you will know by instinct what points are confusing and why. This instinct enables you to prepare a cogent lecture—to know what to emphasize, where to slow down, where to provide extra examples. It helps you to be receptive to student questions. It helps you to have a good attitude in the classroom.

An easy way to cut down on your preparation time for a class is to present examples straight out of the book. The weak students will appreciate this repetition. Most students will not, and you will probably be criticized for this policy. On the other hand, it is rather tricky to make up good examples of maximum-minimum problems or graphing problems or applications of Stokes's theorem.

It can be time-consuming as well. If you need more examples for your calculus class, then pick up another calculus book and borrow some. Develop a file of examples that you can dip into each time you teach calculus. You will learn quickly that making up your own examples is hard work. Do you ever wonder why most calculus books are so disappointing? All right, *you* try to make up eight good examples to illustrate the divergence theorem.

1.4 Clarity

When you teach a mathematics class, clarity (or lack thereof) manifests itself in many forms. If you are the most brilliant, and even the most well prepared, mathematics teacher in the world, but you stand facing the blackboard and mumbling to yourself, then you are not being clear. If instead you shout at the top of your lungs so that all can hear, but your handwriting is cryptic, then you are not being clear. If your voice is clear, your handwriting clear, but your blackboard technique nonexistent, then you are not being clear. If your voice is beautiful, your handwriting artistic, your blackboard technique flawless, but you are completely disorganized, then you are not being clear. If you speak clearly, write clearly, have good blackboard technique, are well organized, but speak with a foreign accent, then don't worry. You are being clear.

Here is the point: Mathematics is hard. Do not make it harder by putting artificial barriers between yourself and your students. If you are shy and simply cannot face your audience, then perhaps you chose the wrong profession. More seriously, be extremely well prepared. *Make* yourself confident. Calculus is one of the most powerful analytic tools that has ever been created. It is a privilege to be able to pass it along to the next generation. Be proud of what you are doing. It is no less an event for you to teach the fundamental theorem of calculus to a group of freshmen in the 1990s than it was for Archimedes to teach his students how to calculate the area inside a circle.

I have atrocious handwriting. When my departmental librarian got her first written message from me she thought it had been written in traditional Chinese characters. But when I lecture I slow down. I write deliberately and clearly. I *want* my audience to understand me and to respect me and I take steps to see that this actually happens.

Suppose that you are in the middle of a lecture and you are making a very important point. How can you drive it home? How can you get the students' attention? We all know that students drift into a malaise in which they are copying and not thinking (after all, we were once students and did the same).[2] How do you wake them up? It's easy. Pause. State the point clearly and simply. Write it clearly and simply. Say "This is important." Repeat the point. One of Mozart's most decisive tools in his compositions was repetition of a particularly beautiful passage. We can benefit from his example.

Ask whether there are any questions. Repeat the point again. Assure students that this point will be on the exam, and that it will come up over and over

---

[2]Surely you have heard the old saw about the concepts passing from the professor's lecture to the students' notes without ever entering the students' brains.

again in the course. Smile wickedly. Tell them that if they do not understand this point, then they will be hampered later in the course. Knowing how to make certain that students know when you are making an important point is a big—and infrequently mentioned—aspect of the "clarity" issue.

If your teaching evaluations say that "the exam didn't cover the points stressed in class," it may mean that you don't know how to write a good exam (Section 2.10). But it may also mean that you don't know how to put your point across, or how to tell the students what is important—in other words, how to make yourself clear.

Let me expand on this last point. We faculty are trained to think as mathematicians. A corollary of this observation is that, when we write an exam, we tend to suppose that the key ideas in the course are fairly obvious. Therefore we are inclined to fill the exam with *questions that would amuse a mathematician*—by which I mean questions about material that is secondary or tertiary. Now let us look through the other end of the telescope. In the students' eyes, such a test is *not* about the main ideas in the course. The students feel misled by such an exam, since they studied (what they perceived to be) the main ideas but they were tested on something else. Thus, in writing such an exam, you definitely have not made yourself clear to the students (about what is *expected* of them).

Another possible scenario is that each of your lectures—indeed your entire course—comes across as an uninflected monotone. Thus the students have *absolutely no notion* of what is important and what is not. It all sounded the same! From the students' point of view, the course looks absolutely flat, like the EKG of a person whose heart has stopped. Thus the students do not know what is expected of them. The trouble with a textbook is that all the paragraphs and sentences look the same. The instructor is supposed to provide context and to prioritize the ideas. If he has not done so, then he has instead succeeded in misleading the students. He has not been clear. No matter what such an instructor puts on the exam, the students feel cheated because they cannot recognize the questions as being about the main ideas in the class, or about anything in particular that has caught their attention.

The upshot of the considerations in the last two paragraphs is that you must be conscious—all semester long—of what are the key ideas you are trying to communicate to the students and what you intend to test them on. It certainly should be no secret what will be on the tests. Don't be afraid to say—from time to time—"This is important. You *will* be examined on this." This type of activity is the single most critical manner in which you can and must make yourself clear.

1.5   Speak Up

If you are going to be a successful teacher, then you have to find a way to fill the room with yourself. If you stand in front of the class (be it a class of ten or a class of a hundred) and just mumble and look at the floor, then you will not successfully convey the information. Even the most dedicated students will have trouble paying attention. You will not have stimulated anyone to think critically. In fact you will have lost your audience.

You do not need to be a showoff or a ham or a joke teller to fill a room with your presence. You can be dignified and reserved and old-fashioned and still be a successful instructor with today's students. But you must let the students know that you are there. You must establish eye contact. You must let them know that you are *talking to them*.

Before I start a class session, especially with a large group of students, I engage some of them in informal conversation. I get them to talk about themselves. I ask them how they are doing on the homework assignment. I comment about the weather. Then I make a smooth transition into the more formal class activities. That way I already have half a dozen people on my side. The others soon follow.

Some new instructors—especially those who are naturally soft-spoken or shy—may need some practice with voice modulation and projection. If you are such a person, then get together a group of friends and give a practice lecture for them. Ask for their criticism. Make a tape recording of your practice lecture and listen to it critically.

If there is any doubt in your mind as to whether you are reaching your audience during a particular class, then *ask* about it. Say "Can you hear me? Am I talking loud enough? Are there any questions?" This is one of many simple devices for changing the pace of a class, giving note-takers a break, allowing students to wake up.

Think of a good movie that you've seen recently. Now remove the changes of scenery; remove the voice modulation and changes of emotion; remove the changes in focal length; remove the skillful use of silence as a counterpoint to sound; remove the musical soundtrack. What would remain? Could you stay awake during a showing of what is left of this movie? Now think about your class in these terms.

1.6 Lectures

In an empty room sits a violin.

One person walks in, picks it up, draws the bow across the strings, and a horrible screeching results. He leaves in bewilderment.

A second person walks in, attempts to play, and the notes are all off key.

A third player picks up the instrument and produces heavenly sounds that bring tears to the eyes. He is Isaac Stern and the instrument is a Stradivarius.

Wouldn't it have been foolish to say, after hearing the first two players, that this instrument is outmoded, that it doesn't work? That it should be abandoned to the scrap heap? Yet this is what many are saying today about the method of teaching mathematics with lectures. Citing statistics that students are not learning calculus sufficiently well, or in sufficiently large numbers, government-sponsored projects nationwide assert that the lecture doesn't work, that we need new teaching techniques.

Whether you like lectures or not, we have to face some facts here. Most of us don't lecture very well. After all, when did you and I learn to lecture, and in what forum? How many excellent lecturers do you know? Where did they learn to lecture?

OK, so we admit it: The lecture doesn't work very well because most of us aren't very good at it. The trouble is that most of us aren't very good at any other method of teaching either. We are not skilled at any of these methods because we have received no training in them, and because we have not given them careful thought. While one possible solution to the problem is "The lecture is dead so let's move on to something else," another possible solution is "Let's learn how to lecture."

Those who say that "the use of the lecture as an educational device is out-moded" rationalize their stance, at least in part, by noting that we are dealing with a generation raised on television and computers. They argue that today's students are too ready to fall into the passive mode when confronted with a television-like environment. It follows that we must teach them interactively, or in groups, or using cooperative learning. Perhaps we should use computers and software to bring students to life.

Lectures have been used to good effect for more than 3000 years. I am hesitant to abandon them in favor of a technology (personal computers, videos) that has existed for just ten years. In spite of popular rumors to the contrary, a lecture does not need to be a bone-dry desultory disquisition. It can have wit, erudition, and sparkle. It can arouse curiosity, inform, and amuse. It is a powerful teaching device that has stood the test of time. The ability to give a good lecture is a valuable art, and one that you should cultivate.

However, you really have to work at making your lectures reach your students. It is true that mathematics teaching in this country is not, overall, a great success. The reason is not that the lecture method is "broken". Rather, we tend not to put a lot of effort into our teaching because the value-reward system is often not set up to encourage putting a lot of effort into it. Many of us spend semester after semester facing down rooms full of calculus students who are there only because the course is required for their major. And it is not required for their major because their department wants them to know calculus. Rather, it is required for their major because their department hopes that half of them will flunk. It is a sorry situation. And it is easy for us, the underpaid and overworked faculty, to become demoralized. Certainly "The lecture is dead." is one way to rationalize an already dreary reality.

Of course many of us content ourselves with internal rewards, with the sense of satisfaction in doing a good job. No matter what rewards you seek, you must identify and learn to use the tools that will make you an effective teacher. You must learn to develop eye contact with your audience, to fill the room with your voice and your presence, and to present your ideas with enthusiasm and clarity. Other sections of this book discuss these techniques in detail.

Turn on your television and watch a self-help program, or a television evange-list, or a get-rich-quick real estate huckster. These people are not using overhead projectors, or computer simulations, or `Mathematica`. (Incidentally, they are also not using group learning or self-discovery!) In their own way they are lecturing, and *very powerfully*. They can persuade people to donate money, to change religions, or to join their cause. Of course your calculus lecture should not literally emulate the methods of any of these television personalities. But these people and their methods are living proof that the lecture is not dead, and that the traditional techniques of Aristotelian rhetoric are as relevant and effective as ever. By watching even a charlatan do his stuff, you can learn something about how to engage an audience, how to answer questions, how to interact with people.

I like to shock people by telling them that I have refined my teaching technique by watching David Letterman (of late night television fame). This is not an exaggeration. Letterman is a master at communication, at dealing with many different types of people, at taking a situation that is going sour and turning it around. *I am not talking here about telling jokes.* I am instead talking about skill at handling people.

My friend Glenn Schober was once teaching a class to help train graduate students to teach. For the first day he carefully crafted a lecture on elementary mathematics in which he purposely made 25 cardinal teaching errors. He walked into class on the first day, told the students he was going to give a sample lecture, and did so. At the end he said, "There were a number of important teaching errors that I intentionally committed in this lecture. See how many you can identify." The students found 32 errors.

This story is amusing, but it is also an important object lesson. There are two kinds of "important" in the world—your kind and my kind. Any classroom situation has a promulgator and an audience. If the lecturer and the audience work together, and share the same goals, then they will reach the same place at the same time, with mutual satisfaction as the result. If not, then they will be working at cross purposes, with mutual frustration as an outcome. These statements apply no matter *what* teaching method you are using. The key to success is that you must *communicate* with your audience.

There are other useful teaching environments besides lectures. Although less common in mathematics than in some humanities courses, group discussions can be useful. If you want to get students interested in what the boundary of a set in a metric space ought to be, then you can begin with a discussion in which students offer various suggestions. Before you define what a finite set is, ask the students to suggest a definition. See also Section 2.5 on the benefits of group activity.

It is not difficult to see that putting a student with a group of four or five peers, so that they can conduct an intimate, one-on-one discussion of a mathematical topic, is bound to generate student interest. It also may tend to help timid students to open up, and to engage in communication. But we must acknowledge that educational activities that, in effect, make the classroom the venue where all learning takes place use time differently than learning activities that make the student's dorm room (or the library) the place where learning takes place. To be specific, the old-fashioned paradigm for student learning was that the student would sit in class for an hour listening to a lecture and taking notes. Then he would go home and spend three to five hours deciphering the notes, filling in the gaps, and doing the homework. The new paradigm has the student learning *right now*, either before a computer screen or interacting with a group of other students. The reform teacher makes sure that students are engaged in the learning process because he actually *manages the engagement*.

The traditional teacher leaves more of the responsibility for engagement with the student himself.

To repeat, many of the reform methods use time in a different, and less familiar, manner than do the traditional (lecture) methods. When you are giving a lecture class, you know just what is going on during the allotted class time, and you also know what is *supposed* to be going on during out-of-class study time. If you instead teach with group work, self-discovery, computer labs, `Mathematica` notebooks, and other new devices, then you must be retrained as an instructor. You must learn anew how to monitor what is going on in your class and how to evaluate student progress.

Reformers, among them Ed Dubinsky, assert that they work their students much harder than do the traditionalists—and that the students are so fulfilled by the learning process that they are glad to do the work. Since reform teachers have many more student contact hours than do traditional teachers, they are probably well qualified to make this assertion (see [ASI] for more on this idea). The jury is still out on the question of whether students taught with reform methods or students taught with traditional (lecture) methods derive the most from their education. Which students learn more? Which retain more? Which have greater self-esteem? Which feel more empowered? Which have greater interest in the learning process? Which teaching method encourages more students to become math majors? Frankly, we don't know.

Of course it is as unfair to group "reformers" together as it is to group "liberals" or "anti-vivisectionists" together. I have heard reformers say in a public venue that, in the reform environment, only half as much material can be presented to the students (as is/was presented in the traditional lecture environment). At the same time, they assert that that's about as much as the students ever learned anyway. So nothing is really lost. Traditionalists, remembering the lectures they attended and the way that they worked when they were students, might bridle at this reasoning. A traditionalist will watch a reform class, perceive unmanaged use of time, and conclude that less learning is taking place. The reformer will argue that his students are working harder and internalizing more.

Clearly we must weigh any teaching method according to both its merits and its demerits. Traditional techniques (lectures) can have the effect on today's students of making learning and erudition seem to be dry, dusty, uninteresting, and irrelevant. However, they are efficient at promulgating a great deal of information. Some of the new techniques are terrific at getting students involved in the material, at making the ideas come alive, and perhaps at aiding student retention. But these methods use time in new ways, and it seems possible that less material will be taught when they are used.

It has been natural, in this section on the lecture, to digress about reform. For one of the battle cries of the reform movement has been that "The lecture is dead." Let us now conclude this section by mentioning some teaching devices that should be of interest both to reformers and to traditionalists.

It can be instructive to have students volunteer to do problems at the blackboard (see Section 3.12). Once in a great while—when I am lecturing—if a student offers an alternative proof of a proposition or another point of view, I hand him the chalk. Everyone is usually quite surprised, but the results are generally pleasing and it provides a nice change of pace.

Computer labs can also be a useful instructional device (see Section 1.10). The subject of sophomore-level differential equations lends itself well to helping students explore the interface between what we can do by hand and what the machine can teach us. Let's be frank. We do not know how to solve most differential equations explicitly, or in closed form. Thus it is important for students to see how much analysis one can do with traditional methodologies and then to see how the machine can use phase plane analysis, numerical methods, and graphing to provide further concrete data.

We should continue to seek new and better methods and technologies for teaching. This author, and this book, has a built-in bias toward traditional methods, such as lectures. That is because he has watched them work and used them successfully for more than twenty-five years. I hope that other writings will describe and explore some of the new teaching techniques. The new Appendices to this second edition of *How to Teach Mathematics* provide descriptions of some of the new teaching techniques—by the people who have developed them.

1.7   Questions

In a programmed learning environment, whether the interface is with a PC or with `Mathematica` notebooks or with a MAC, the student cannot ask questions. The give and take of questions and answers is a critical aspect of the human part of the teaching process. Teachers are *supposed* to answer questions.

There is more to this than meets the eye. When I say that a teacher answers questions I do not envision the student saying, "What is the area of a circle?" and the teacher saying " πr^2". I instead envision the student struggling to articulate some confusion and the experienced teacher turning this angst into a cogent question and then answering it. To do this well requires experience and practice. I frequently find myself responding to a student by saying, "let's set your question aside for a minute and consider the following." I then put the student at ease by quickly running through something that I know the student knows cold, and that serves as a setup for answering the original question. With the student on my side, I can answer the primary problem successfully. The point is that some questions are so ill-posed that they literally cannot be answered. It is the teacher's job to make the question an answerable one and then to answer it. See also Section 3.12 on asking and answering questions.

A similar, but alternative, scenario is one in which the student asks a rather garbled question and I respond by saying, "Let me play the question back for you in my own words and then try to answer it ...." The point is that the responses "Your question makes no sense" or "I don't know what you mean" are both insulting and a cop-out. To be sure, it is the easy answer. But you will pay for it later. It takes some courage for the student to ask a question in class. By treating questions with respect, you are both acknowledging this fact and helping someone to learn. If instead you stare at the student as though he has weevils in his eyebrows, then you will gain no allies and will most likely lose several friends and make a few enemies.

Yet another encouraging response to a student question is to say, "Thank you. That question leads naturally to our next topic ..." Of course you must be quick on your feet in order to be able to pull this off. It is worth the trouble. Students respond well when they are treated as equals—see the discussion of this technique at the end of Section 1.1.

There are complex issues involved here. A teacher does not just lecture and answer questions. A good teacher helps students to discover the ideas. There are few things more stimulating and rewarding than a class in which the students are anticipating the ideas because of seeds that you have planted. The way that you construct your lecture and your course is one device for planting those seeds. The way that you answer questions is another.

When I discuss teaching with a colleague who has become thoroughly disenchanted with the process, I frequently hear complaints of the following sort: "Students these days are impossible. The questions that they pose are unanswerable. Suppose, for example, that I am doing a problem with three components. I end up writing certain fractions with the number 3 in the denominator. Some student will ask 'Do we always put a 3 in the denominator when doing a problem from this section?' How am I supposed to answer a question like that?" (See also Section 4.5 on frustration.)

Agreed, it is not obvious how to answer such a question, since the person asking it either **(i)** has not understood the discussion, **(ii)** has not been listening, or **(iii)** has no aptitude for the subject matter. It is tempting to vent your spleen against the student asking such a question. Do not do so. The student asking this question probably needs some real help with analytical thinking, and you cannot give the required private tutorial in the middle of a class hour. But you can provide guidance. Say something like "When a problem has three components it is logical that factors of $1/3$ will come up. This can happen with certain problems in this section, or in any section. But it would be wrong to make generalizations and to say that this is what we do in all problems. If you would like to discuss this further, please see me after class." In a way, you are making the best of a bad situation. But at least you are doing something constructive, and providing an avenue for further help if the student needs it.

The Dalai Lama once visited the headquarters of *Time* magazine in Chicago. He was given the chef's tour, and then there was a grand formal lunch at which the various executives of the enterprise pontificated *ad nauseum*. The Dalai Lama—an elfin man—sat swathed in his saffron robe, an inscrutable smile on his face, saying nothing. After about an hour, the CEO of *Time* turned to the Dalai Lama and said, "Do you have any questions about *Time*, the nation's premiere news magazine? Go ahead, ask us anything at all." The Dalai Lama bowed his head for a moment, apparently deep in thought. Then he looked up and said, "Why do you publish it?"

We mathematicians are very much like the executives of *Time* in that story. We are wrapped up in our own world, we all speak the same language, and we suffer intruders with pained resignation. Sadly, our students are like intruders. They come to us with questions we would never have dreamed of, and they expect answers. They do not speak our language, and they do not necessarily respect our mores. Yet it is our job to talk to our students, to engage them in discourse, to answer their questions. We must exercise patience in order to gain their trust. And we must try to speak to them in their language, rather than in our own.

Let us consider some further illustrations of the principle of making a silk purse from a sow's ear—that is, answering unexpected or awkward questions in a constructive manner. The first example is a simple one.

Q: Why isn't the product rule $(f \cdot g)' = f' \cdot g'$?

The answer is *not* "Here is the correct statement of the product rule and here is the proof." Consider instead how much more receptive students will be to this answer:

> **A:** Leibniz, one of the fathers of calculus, thought that this is what the product rule should be. He recorded this thought in his diary. Ten days later, he gave the correct form—with a proof and the cryptic statement that he had known this to be the correct form "for some time". Because we have the language of functions, we can see quickly that Leibniz's first idea for the product rule could not be correct. If we set $f(x) = x^2$ and $g(x) = x$ then we can see rather quickly that $(f \cdot g)'$ and $f' \cdot g'$ are unequal. So the simple answer to your question is that the product rule that you suggest gives the wrong answer. Instead, the rule $(f \cdot g)' = f' \cdot g + g' \cdot f$ gives the *right* answer and can be verified mathematically.

The second example is more subtle.

Q: Why don't we divide vectors in three-space?

The *wrong* answer is to tell about Stiefel-Whitney classes and that the only Euclidean spaces with a division ring structure are $\mathbb{R}^1, \mathbb{R}^2, \mathbb{R}^4$, and $\mathbb{R}^8$. A better answer is as follows.

> **A:** J. Willard Gibbs invented vectors to model physical forces. There is no sensible physical interpretation of "division" of physical forces. The nearest thing would be the operations of projection and cross product, which we will learn about later.

Notice that in both illustrations an attempt is made to turn the question into more than what it is—to make the questioner feel that he has made a contribution to the discussion.

Q: Why isn't the concept of velocity in two and three dimensions a number, just like it is in one dimension?

If you are in a bad mood, you will be tempted to think that this person has been dreaming for the past hour (or the past week!) and has understood absolutely nothing that you have been saying. Bear up. Resist the temptation to voice your frustrations. Instead try this:

A: Let me rephrase your question. Instead let's ask, "Why don't we use vectors in one dimension to represent velocity just as we do in two and three dimensions?" One of the most important features of vector language is that a vector has *direction* as well as magnitude. In one dimension there are only two directions—right and left. We can represent those two directions rather easily with either a plus sign or a minus sign. Thus positive velocity represents motion from left to right and negative velocity represents motion from right to left. The vector language is *implicit* in the way that we do calculus in one dimension, but we need not articulate it because positivity and negativity are adequate to express the directions of motion.

In dimensions two and higher there are infinitely many different directions and we therefore require the explicit use of vectors to express velocity.

As the author of this book, I have the luxury of being able to sit back and drink coffee and think carefully about how to formulate these "ideal" answers to poor questions. When you are actually teaching you must be able to do this on your feet, either during your office hour or in front of a class. At first you will not be so articulate. This is an acquired skill. But it is one *worth acquiring.* It is a device for showing respect for your audience, and in turn winning *its* respect.

Large lectures pose special problems with the issue of student questions. Obviously you cannot let each student ask his little question. You cannot let your lecture get bogged down with questions like "How do you do problem 6?" or "Will this be on the test?" See Section 2.14 for a discursive discussion of questions in the large lecture context.

A final note about questions. Even though you are an authority in your field, there are certainly things that you do not know. Occasionally these lacunae in your knowledge will be showcased by a question asked in class or during your office hour (it does not happen often, so don't get chills). The sure and important attribute of an intelligent, educated individual is an ability to say, "I don't know the answer to that question. Let me think about it and tell you next time." On the (rare) occasions when you have to say this, be sure to follow through. If the item that you don't know is an integral part of the class—and this had better not be the case very often—get it down cold because the question is liable to come up again in a different guise later in the course. If it is not an integral part of the course, then you have no reason to feel bad. Just get it straight and report back.

The main point is that you should never, under any circumstances, try to fake it. If you do, then you will look bad, your interlocutor will be frustrated and annoyed, and you will have served no good purpose. If there is any circumstance in which honesty is the best policy, this is it. Professor of Economic History Jonathan R. T. Hughes was wise to observe that "There is no substitute for knowing what you are talking about."

1.8 Advanced Courses

The teaching problems that arise in an advanced course are rather different from those in a lower division course. You are dealing with a more mature audience and, in at least some advanced courses, many of your students will be math majors. The main message for a new teacher is: Don't get carried away. Don't try to tell them about your Ph.D. thesis in the first week of class. Try to remember the troubles you had learning about uniform continuity and absolute convergence and the descending chain condition. Give lots of examples. Prepare your lectures well and *slow down*. Be receptive to questions and sympathetic to awkward struggles with sophisticated new ideas. Be willing to repeat yourself.

It is probably best in an advanced undergraduate course to cover less material but to cover it in depth—to endeavor to give the students a real feel for the subject—rather than to race through a lot of material. Again I shall repeat an implicit theme of this book. Most undergraduate students do not have either the maturity or the experience to put the shortcomings of your teaching into perspective. Good teaching is your responsibility.

Even in these courses, be sure to use the inductive method instead of the deductive method as a vehicle for conveying ideas (refer to Section 3.8). Any hard theorem should be suitably motivated. Do even more examples than seem necessary. Refer to the Beresford Parlett quote near the beginning of Section 3.8 for inspiration.

It is tempting in an upper-division course to assume, at least subliminally, that your students are little mathematicians. They are not. This course may be their first exposure to rigorous thinking, to ϵ's and δ's, to Theorem–Proof–Theorem–Proof, to careful use of "for all" and "there exists", to quashing a possible theorem with a single counterexample. In short, you are not just teaching these students some advanced mathematics. You are also teaching them how to think. This is an important opportunity for you, the instructor; and it is an important juncture in the students' education. You must use it wisely.

Today's undergraduate students do not have the background and experience in rigorous thinking that we all fancy we had when we were students. They are unaccustomed to proofs and to the strict rules of logic. It is often a good idea to have a whirlwind review of logic at the beginning of an upper-division mathematics course—especially in real analysis or algebra where *modus ponendo ponens*, contrapositive, proof by contradiction, induction, and so forth are used frequently. It would not be out of place to present some material on set theory and number systems as well. Some mathematics departments have a "transitions" course (see the discussion of such a course in the Preface to this edition) designed to bridge the gap in methodology between lower-division courses and upper-division courses. If yours is such a department, then you may be able to skip the review that I am advocating here.

If you do not make some extra effort to help the students in your advanced courses over the "hump" that separates math enthusiasts from mathematicians, then you are missing an opportunity to contribute to the pool of mathematical talent in this country. It's your decision, but if you decide not to participate, then you have no right to complain when your department's graduate program is reduced for lack of students, or your undergraduate program curtailed for lack of majors.

1.9 Time

There are several aspects of teaching that require time management skills. When you are giving a lecture, you must cover a certain amount of material in the allotted time—and at a reasonable rate. When you give a course, you must cover a certain amount of material in one semester or term. When you give an exam, it must be doable by an average student in the given time slot. When you answer a question, the length of the answer should suit the occasion.

All of these topics will be touched upon in other parts of this book. They require some thought, and some practice and experience, so that they become second nature to you.

Nobody can design a class so that (as in the movies) the last 'QED' is being written on the blackboard just as the bell rings. There are certain precepts to follow in this regard:

- Have some extra material prepared to fill up extra time.

- If you finish your lesson with five minutes to spare, don't rocket into a new topic. You will have to repeat it all next time anyway, and students find this practice confusing.

- If the clock shows that just five minutes remain, and you have ten or fifteen minutes of material left to present, then you will have to find a comfortable place to quit. Don't race to fit all the material into the remaining time. At the same time—if possible—don't just stop abruptly and in mid-thought, thinking that you can pick up a calculation cold in the following class.

  An experienced teacher will know which will be the last example or topic in the hour, and that he might get caught for time. Therefore the instructor will plan in advance for this eventuality and think of several junctures at which he might bring the hour to a graceful close. With enough experience, you will know intuitively how to identify the comfortable places to stop; thus end-of-the-hour time management problems can be handled on the fly. In particular, if five minutes remain then *do not begin* a ten-minute example!

- If you prepare (the last part of) your lesson in units of five minutes' duration, and if you are on the ball, then you should never have to run over by more than two minutes nor finish more than two minutes ahead of time. (The idea here is if there are three minutes remaining, then you can include another five-minute chunk without running over by more than two minutes. If there are just two minutes remaining, then you should stop.)

- If you *run out of time*, do not keep lecturing past the end of the period—at least not by more than a minute or two. Students have other classes to attend, and they will not be listening. If the time is gone, then just quit. Make up for your lapse in the next class (this will require some careful planning on your part). Best is to plan your class so that you do not get caught short of time.

A special note about buzzers: Some math buildings have a loud buzzer or bell that sounds at the end of the class period. Once that buzzer sounds, all is lost. Most students will instantly start packing up their books and heading for the exit. If there is a clock on the wall, then you will know when the buzzer is going to sound, and you can be adjusting your lecture and zeroing in on a suitable conclusion as you go along. At a school without a buzzer (especially one that also has no clocks in its classrooms!), you have a bit of slack since no two wrist watches are in agreement. You may want to interpret the advice in this section according to the physical environment in which you are teaching.

Some students have the annoying habit of setting the alarms on their electronic watches to chime precisely on each hour. Worse, some students set their alarms for five minutes before the hour or three minutes before the hour or two minutes after the hour. Since no two watches agree, what you experience at the end of class is a cacophony of electronic beeping. This distracts the students, it disrupts your class, and it is a damned nuisance. Tell the students in advance—on the first day of class—that you want all electronic alarms turned off during your class. Be stern about it, and nail those who fail to comply.

- If a student asks a question that requires a long answer, don't let your answer eat up valuable class time. Tell the student that the question is ancillary to the main subject matter of the course (it had better be, or else you evidently forgot to cover an important topic) and that the question can best be treated after class. However, do not let the student get the impression that the question is being given the brush-off.

- On the other hand, if a student asks a question for which a brief answer is appropriate (such as "Shouldn't that 2 be a 3?" or "When is the next homework assignment due?") then do give a suitably brief answer. Anecdotes about your childhood in Shropshire are probably out of place.

By the way, this last is more than a frivolous remark. As we slide into our golden years, we seem to be irretrievably moved to share with our students various remembrances of things past—"It seems to me that twenty years ago students worked much harder than you people are willing to work." or "When I was a student, we put in five hours of study for each hour of class time." or "I used to walk six miles barefoot through the snow to attend calculus class, and it was uphill both ways." Trust me: Students hate this sort of emotional slobbering. You will defeat all the other good things that you do by giving in to this temptation to prattle.

If you have the time problem under control at the level of individual class meetings, then you will have the ability to pace your course in the large as well. You should have a good idea how much material you want to cover. And when you plan the course you should allot a certain number of class periods for each topic. If you are teaching undergraduates, they depend on your course for learning a certain body of material (that may be prerequisite for a later course). Don't shortchange them.

A test should be designed for the allotted time slot. You can rationalize giving a two-hour exam in a one-hour time slot by saying to yourself that there is so much material in the course that you simply *had* to make the test this long. This is nonsense. The point of the exam is not to *actually test* the students on every single point in the course, but to make the students *think* that they are being tested on every point in the course. Ideally, the students will study everything—but your test amounts to a spot check. Even if you had a four-hour time slot in which to give the exam, you couldn't really test them on everything, now could you? See Section 2.10 on exams.

If you give a two-hour exam in a one-hour time slot, then you run several risks: that students will become angry, demoralized, alienated, or all three. Telling a student not to worry about his grade of $37/100$ because the average was $32/100$ does not work (see also Section 1.2). Students are unable to put such information into perspective.

1.10   Computers

Computers are everywhere, especially in mathematics departments. Software for teaching mathematics is also everywhere. Most publishers of basic mathematics texts are absolutely convinced that they cannot market their products without making extensive software resources available to mathematics instructors. It is not clear, as of this writing, that much of this software is actually being used in the classroom. This is so in part because most instructors are not conversant with what is available, are not comfortable with these products, or simply cannot be bothered. There is little objective information available as to which of these products, if any, is a useful teaching tool.

One form of software for the classroom that is being heavily touted these days is the `Mathematica` notebook. Briefly, a `Mathematica` notebook is a self-contained environment by means of which students can interact with the computer over various mathematical issues *without* knowing anything about computing or computer languages. What the student sees when he boots up a `Mathematica` notebook is plain English, and mathematics, on the screen. The computer poses questions to the student and offers guidance in helping the student to find the correct answer(s). When the student errs, the machine provides hints (and encouragement!). (`Maple` has a similar environment called the "Worksheet".)

All of this hardware and software raises fundamental issues about the way that mathematics is taught and the way that it ought to be learned. I drive my car every day and am perfectly comfortable in doing so while not knowing in intimate detail how it works. Ditto for my computer, my telephone, my television, and so forth. Some would say that a measure of how civilized we are is how many black boxes we are willing to use unquestioningly. To what extent should this point of view be allowed in the mathematics classroom?

Most of us were trained with the idea that the whole point of studying mathematics is to understand precisely why the ideas work. To make the point more strongly, this attitude is what sets us apart from laboratory scientists. We make no statement unless we can prove it. We use no technique unless we fully grasp its inner workings. We would not dream of using the quadratic formula unless we knew its genesis. We would not be comfortable using the fundamental theorem of calculus unless we had seen its proof.

Yet in many colleges these days the linear programming course consists primarily of learning to use `LINDO` or some other canned computer package for applying the simplex method and its variants. The statistics course consists primarily in learning to use `SAS` or `MINITAB` or another commercial product. The undergraduate numerical analysis course consists in learning to use `IMSL`.

It has been argued that the use of programmable calculators and software such as `Mathematica` notebooks in the calculus classroom will free students from the drudgery of calculation and will allow us to teach them how to analyze multi-step word problems (see, for instance, [STE2]). Thus, it is hoped, our lower-division calculus classes will be more closely tied to the way that mathematics is used in the real world. I see this as double talk.

It is certainly the case that the less able students are so hampered by their inability to take the derivative and set it equal to zero or to find the roots of a quadratic equation or to calculate the partial fraction decomposition of a rational expression that they have little hope of successfully analyzing a multi-step word problem whose solution includes one or more of these basic techniques. But there is no evidence to support the (apparent) contention that a person who is unable to use the quadratic formula will somehow, if these technical difficulties are handled for him by a machine, be able instead to analyze conceptual problems. Using the quadratic formula is easy. Analyzing word problems is hard. A person who cannot do the first will also probably not be able to do the second—with or without the aid of a machine.

Consider the following analogy. A software wizard approaches a frustrated piano student and says "For \$100 I'll give you some software that will play scales and emulate hand positions for you. That way you can go straightaway to playing Tchaikovsky." Few people would lend credibility to such a story, yet in the context of mathematics education we sometimes delude ourselves into using analogous teaching techniques.

I think, as I have already indicated, that the computer can be used successfully to provide helpful graphical analyses. If you want to draw a surface in space and then move it around to analyze it from all sides, then there is nothing to beat `Mathematica`. If you want to illustrate Newton's method, or use the Runge-Kutta method, or implement Simpson's rule, then a computer is almost indispensable. But the use of the computer should be based on a firm foundation of conceptual and technical understanding.

Let me stress that most reformers concur with the statements in the last paragraph. It is incorrect—and certain passages in the first edition of this book could be construed as making this claim—to suggest that all the reformers want to do is abandon lectures and let the kids spend their time horsing around with computers (see J. Jerry Uhl's Appendix, where he spells out in detail how his teaching methodology uses computers). First, there is no single reform dogma. Second, there are many different opinions—both among reformers and among traditionalists—about how best to use computers in instruction. My few experiences using computers in instruction have been rather positive. I would encourage others to be open to at least trying some of the different ways that they can be used in the classroom.

My own view about computers is that one of the highest and best uses of the computer in mathematics instruction is as the basis for *laboratory* work. I would like to see a lower-division mathematics class still consist largely of classroom instruction—either lecture, or group work, or discovery activities—but I would like to see the classroom work re-enforced by well-thought-out computer labs. Chemists have understood this notion for many years, and have made laboratory work an integral part of basic instruction. We should do the same. Mathematical ideas are abstract. Elementary mathematics students feel as though they are staring into a crystal ball, with nary a place to gain a foothold in the subject. We should use laboratory activities to make the subject more tactile for them.

For example, if the class is treating level sets for functions of two variables, then there should be a lab activity in which the student (with the aid of a machine) constructs level sets for various functions, and then sees how they are assembled to form the graph in three-dimensional space. The student should be able to "walk around" on the graph, and verify that the gradient of the function is perpendicular to the level sets. The student should be able to do barehands confirmation of the method of Lagrange multipliers. Properly viewed, this is fertile and largely unexplored territory.

There is an important point to be kept in mind. The computer lab activities should be tightly integrated into what is happening in the classroom—reinforcing what happened in the previous class and anticipating what will happen in the next. The lab activity might involve graphics, or numerical calculations, or computer algebra. But it should not be computer activity for its own sake—it should be part of the learning process.

We had better realize that making the highest and best use of computers in teaching our lower-division courses might entail rethinking our traditional classroom activities. I don't have all the answers here, and neither does anyone else. But if you try tacking some silly little computer activities onto your lectures, and if the end result is not particularly satisfying, then that is not necessarily the end of the story. You may have to rethink the way that the entire course is taught. Just as the computer activities are supposed to reinforce the ideas introduced in class, so the activities in class should reinforce the computer activities. It's a symbiotic process, and one which will require careful planning if it is to be successful.

We have had catastrophic experiences, at my own university, trying to introduce `Maple` or `Mathematica` or some other software into isolated courses. Students are not stupid. They catch on right away to the fact that this software is specific to the particular course, and as soon as the semester is over they are unlikely to see it again. They resent having to learn a whole new language—

especially since it will be used in just one course for only one semester.

What these life experiences tell me is this: If we are going to introduce serious software into the lower-division curriculum then we should do it globally instead of locally. Of course this is tough. We'll really have to plan. But it makes sense that we should tell students from day one that their entire lower-division mathematics curriculum will depend on `Maple` (or `Mathematica`, or another alternative) and that they need to master it right away. Having understood this dictum, they will comply (if only out of an instinct for survival) and the software will become a part of their *lingua franca*. They will (we hope) carry it (or the analytical skills attendant to it) on to the rest of their education, and their lives.

At my own university, when we teach sophomore-level ordinary differential equations, we commonly include a "computer unit" in the course. This means that there are four or five computer-oriented homework assignments that the students must do—in addition to the more traditional handwritten assignments (with separation of variables, first-order linear, second-order, constant coefficient, and so forth)—that are geared to the text and to the lectures. A typical computer-oriented homework assignment sets a specific ODE for the student to consider, and guides him through several steps of hand calculation—to a point where it becomes clear that further hand calculation is infeasible. And then it says, "Now let's do a phase plane analysis," or "Now let's look at Runge-Kutta," or "Now let's do a numerical or modeling experiment." My experience is that students have gone into this computer-oriented material kicking and screaming. But, at the end of the semester, they have come out of it convinced that they have learned a new set of tools, and a new view of scientific analysis. I take no credit for the construction of these computer-oriented problem sets, nor for their efficacy (they were created by my colleagues). But I think they are marvelous, and I would like to see this approach used in more undergraduate courses.

Of course what I am saying here is not news to those—both of the traditional and of the reform turn of mind—who have been studying mathematical pedagogy professionally. The articles [BOU] and [HEI] discuss many aspects of computer use in mathematics teaching.

Certainly some of the most innovative uses of the computer in the lower division classroom are due to Ed Dubinsky and his coworkers [DUB]. Dubinsky uses the programming language `ISETL`, which has the attractive feature that its code is very much like the mathematical notation to which we are all accustomed. Dubinsky is a firm believer in the notion of Beth/Piaget [BPA] that each student must build the mathematical ideas for himself in his own mind. He argues that `ISETL` is a powerful tool in helping students to do so. In particular, the programming language cuts through the mental inertia that we have all experienced. Instead of sitting and staring at a new idea in bewilderment, the student turns to `ISETL` and *begins to try things*. There is no arguing with success. Dubinsky and his collaborators have calculus texts [DSM] and [C4L], an abstract algebra text [DUL], and a discrete mathematics text [DUF] that implement these ideas.

There are some instructional uses of the computer that I don't understand. Let me begin with an analogy. Imagine a chemistry instructor facing a classroom of 100 students, each with chemistry apparatus set up before him. The instructor says, "OK, today we are going to make nitroglycerine. Take your flask, and put these chemicals in it. Now heat it over your Bunsen burner ..." Just imagine! Some student on the left, in an effort to keep up with the instructor's pace, fumbles and spills a caustic substance in his lap; a student on the right gets excited, shakes the nitroglycerine, and is blown to smithereens.

You chuckle. But what better sense does it make to have a mathematics classroom with a computer before each student and the instructor delivering commands to the students? Many students are not conversant with the computer environment. Perhaps their hand-eye coordination is not fully up to snuff. Many will not be able to keep up. Hit the wrong key and you get a screen full of chaotic gibberish, or you lose your work, or you accidentally reboot, or you erase your boot sector.

People need to perform laboratory activities in their own time and at their own pace. They need to have the leisure to make errors and to repeat steps. The chemists know what I am talking about. They let the students schedule their own labs, and do things (under mild supervision) in their own time. The proof is in the pudding: It works. It makes sense to me that, as we rethink the way that we teach mathematics, we should take into account what other sciences have learned from their experiences.

Now let me summarize my point. I am all in favor of computer labs to accompany mathematics classes. But I can make no sense of "electronic classrooms"—in which each student, or each pair of students, has a computer before him. How much could you accomplish in a 55-minute class if 45 of those minutes are spent fumbling around on computers?

Of course the computer also raises epistemological issues, and I hinted at some of these at the start of the section. Do we want our students to understand how to integrate by parts, or do we want them to know how to tell **Mathematica** to do it for them? Do we want students to understand the simplex method, or do we want them to know how to tell **LINDO** to do it for them?

I think that there are several correct answers to these questions, and I think that the appropriate answer for any given context depends on who is in the class we are teaching. Are we teaching future mathematicians, or future nurses, or future businessmen, or future chemists? See Section 3.9 for a discussion of the issue of suiting your teaching to your audience.

Peter W. Jones once made the following cogent observation. After being in school for a number of years, students learn that their instructor is fallible, and they learn that even the text is fallible. But if the calculator or computer says that something is so, then it must certainly be correct. I have had the experience of giving students assignments to complete in the computer lab, with the following result. After spending many hours, they would bring to me pages of gobbledegook—long streams of meaningless ASCII code. They assumed that this is what I wanted, because it is what the computer produced.

Perhaps the discussion in the last paragraph illustrates the following point. A live teacher does things in real time. But a computer does things so fast that it trivializes them. When something is trivialized, then understanding is lost. These statements are not made to downplay the potential value of the computer in the classroom, but rather to put it in perspective.

It is probably the case that students learning calculus from **Mathematica** notebooks respond positively to the extra attention they are receiving, to the novelty of the teaching environment, and to the fact that making a mistake when interacting with the computer is less heinous than asking a dumb question in class. A good `Mathematica` notebook will never produce streams of nonsense symbols. So in that sense the environment is controlled and the flow of errors kept in check. In the `Mathematica` notebook environment, the student is probably more willing to try things and to experiment. Such an environment could serve as a catalyst for creativity. We should think carefully about how to capitalize on the special features that computers can contribute to the learning experience. However it is clear that we do not yet know all the answers, nor have we realized all the potential.

Today there is much discussion of self-discovery, and of students making their own conjectures, of students doing group work, and of students building the ideas in their own minds. All well and good; but how can a student discover the fundamental theorem of calculus for himself when it took great historical geniuses like Newton and Leibniz to do it the first time around? How can a person with no specialized training make conjectures? How can students carry on cogent group discussions if they don't know what they are talking about? How can a student with no experience in critical thought build ideas in his own mind?

These are serious questions, and members of the reform movement are finding ways to answer them. The computer can be a powerful tool in this quest. If two or three students are sitting together in front of a computer screen, going through the paces of a well-constructed lab, then they definitely have something to talk about. They have the catalyst for some group work. The computer lab can certainly give them grist for formulating conjectures. And the computer can pace them through a self-discovery process. Ed Dubinsky's use of the programming language `ISETL` ([BDDT], [DUB], [DUF], [DSM], [C4L]) provides evidence that the computer can be used to help students to reconstruct ideas in their own minds. The computer is a powerful new tool that is at our disposal. We must train ourselves how best to use it.

But let us bear in mind the techniques that we already know, and that are of proven value. Don't forget: If a student spends an hour with a pencil—graphing functions just as you and I learned—then there are certain specific and *verifiable* skills that will be gained in the process. By learning how to *create* a graph, the student will also learn how to *read* a graph. By contrast, `Mathematica` or `Maple` can create a flawless graph much more quickly and easily than can a human being, but these software products *will not teach the student how to read the graph*. If we keep these guiding principles in mind as we develop methods for using computers in the classroom, then I think we will end up with more valuable teaching tools.

S. Hildebrandt used to remark that a student who solves a problem two different ways will learn a great deal more than the student who solves two different problems, each in only one way. There is wisdom in this remark, and it is salient to the issue of teaching using computers. As already noted, computers do things too quickly, and tend to trivialize them in the mind of the user. The student who can solve a problem swiftly and easily at the keyboard is not likely to want to engage in the conceit of solving it twice.

A wise man once told me that the computer is a solution looking for a problem to solve. It is obviously a powerful tool in the right circumstances. The generation of minimal surfaces of arbitrary genus by Hoffman, Hoffman, and Meeks ([HOF]) is a dazzling use of computers. The simulations that these three mathematicians performed would have been inconceivable without this advanced computer graphics technology. `Mathematica` is a powerful tool in the hands of the right user, as are `AXIOM`, `MACSYMA` and `Maple`.

Calculus is perhaps the most powerful body of analytic tools ever devised. All young scientists should learn calculus in essentially the traditional fashion so that they have these tools at their disposal. *Graphing a function* is one of the most basic processes of analytical thinking—analogous to finger exercises for the piano. The partial fractions technique is also one of the most far-reaching algebraic devices that we have. Integration by parts is perhaps the most ubiquitous and powerful tool in all of mathematical analysis. Letting a computer do these processes for the students abrogates much of what we have learned in the past three hundred years.

Nobody would be foolish enough to assert that one can learn to spell by using a spell-checker or to write by using a grammar-checker or to play the piano by listening to someone else do it. I sometimes wonder why people think that students will learn mathematics by letting `Mathematica` do the thinking for them.

Let me assure you that I have discussed this point with the presidents of large high tech corporations and they agree with me absolutely. Use of the new technology should be layered atop a traditional foundation. And, if you'll pardon my being a bit droll, that traditional foundation should consist of "readin', writin', and 'rithmetic". That is what works in the classroom and that is what works in the real world. In fact in one memorable conversation that I had at a board meeting of H. R. B. Singer (a company that produces products for electronic warfare), I said to the CEO, "There are a lot of new ideas in the air about teaching mathematics. We are wondering which ones to adopt. Are you content with the traditional mathematical training of the college graduates that you hire?" He said, "Their mathematical training is fine. I just wish that they could write English."

1.11   Applications

One of the most chilling things that can happen to an unprepared, unseasoned faculty member is to have a belligerent (engineering) student raise his hand and say, "What is all this stuff good for?" And one of the most irresponsible things that a faculty member can say in response is, "I don't know. That is not my problem." If you do not have an answer for this student question, then you are not doing your job.

I have found it useful in all of my undergraduate classes to tell the students about applications of the techniques being presented *before* the aforesaid chilling question ever comes up. This requires a little imagination. If I am lecturing about matrix theory, then I tell the students about Markov processes, or I say a little bit about image processing and data compression. Or I tell them about eigenvalue asymptotics for clamped beams and applications to the building of a space station (not coincidentally, this is a problem that I have worked on in my research). If I am lecturing about surfaces then I tell them about the many applications of surface design problems. If I am lecturing about uniform continuity or uniform convergence, then I tell them about some of the applications of Fourier analysis. The pedagogical technique that I am describing in effect defuses any potential belligerence from engineering or other students who have no patience for mathematical abstraction.

Carrying out this teaching technique requires a little forethought and a little practice. After a while it becomes second nature, and you will find yourself thinking of potential answers while taking a shower or walking to class. If it suits your style, keep a file of clever applications of elementary mathematics. It is not true that the concept of uniform convergence is used on a daily basis by civil engineers to construct bridges. Do *not* use this facile line of reasoning to talk yourself into abandoning the effort to acquaint lower-division students with the applications of mathematics. Instead, reason that uniform convergence is a bulwark of the theory behind the practical applications of mathematics. It is important. Act as though you believe it.

Try to be flexible and to reach out for up-to-date and striking applications. Uniform convergence is a basic idea in the convergence of series of functions. One of the most interesting uses of series is in Fourier analysis. And what is Fourier analysis good for? Mention the hot new theory of wavelets and some of its uses. Wavelets were used to clean up a recording of Brahms; they are used every day by the FBI to analyze and compress fingerprint files; they are used to analyze electrocardiogram readings; they are used to analyze biological neural systems. Don't you think this is impressive? Wouldn't your students agree?

It is a matter of personal taste (and much debate) as to how much should be done with applications in, say, a calculus class. For most of us the problem is solved by the very nature of the undergraduate curriculum. There is little opportunity to do any but the most routine applications. But times are changing and many mathematics departments are re-evaluating their curricula. There is considerable enthusiasm for infusing the freshman-sophomore curriculum with more applied material. Examples of new approaches to the calculus, by way of applications and technology, can be seen in the Amherst project materials [CAL] and the Harvard materials [HAL]. Other "reform" projects include [ANG], [C4L], [DIP], [DSM], [OSZ], [SMM], [STEW], [WAT]. The project [KOB], available for free on the Worldwide Web, does not fit comfortably into the "reform" rubric. But it is a new approach to the calculus text, and it is noteworthy for its applications.

If you decide to work applications into your class then consider this: Mathematical modeling is complicated and difficult. If you take an already complex mathematical idea and spend an hour applying it to analyze a predator-prey problem, or to derive Kepler's third law, or to design the Wankel engine, then you are likely to lose all but the most capable students in the room. How will you test them on this material? Can you ask the students to do homework problems if their understanding is based on such a presentation?

Think of what it is like to teach the Divergence Theorem. There are almost too many ideas, layered one atop the other, for a freshman or sophomore to handle. Students must simultaneously keep in mind the ideas of vector field, gradient, surface, surface integral, curl, orientation, and so forth. Most cannot do it. The same phenomenon occurs when one is attempting to get students to understand a really meaty application. I am not advising you against doing these applications. But if you should choose to do one, go into it with your eyes open. If, after fifteen minutes, the students' eyes glaze over then you will have to shift gears. Be prepared with a physical experiment, or a film strip, or an overhead slide, or a transition into another topic.

A stimulating presentation of an application should be broken into segments: a little analysis, a little calculation, a little demonstration. Lower-division students cannot follow a one-hour analysis. Always bear in mind that, no matter how satisfying you find a particular application, your audience of freshmen may be somewhat less enthusiastic and may require help and encouragement.

On the other side of the coin, don't get sucked into doing just trivial, artificial applications. This cheapens our mission in the students' eyes and makes us seem disingenuous. The calculus and its applications are among the great achievements of western civilization. Be proud to share with your class the analytical power of calculus. Do so by presenting some profound applications, but put some effort into making the presentation palatable.

My experience is that, for freshmen, brief (*not* trivial) applications are the best. They can be modern, they can be interesting, but if the analysis entails layering too many levels of ideas on top of each other, then most students will be lost. What you should do, in preparing to present an application to your class, is to lay the whole thing out on a piece of paper and then examine each step. Consider which steps you can actually discuss in detail with the class, and which steps you should summarize. It is perfectly all right, in the middle of your presentation, to say, "Now I'm going to skip some difficult calculations. I have them available on a handout. The calculations produce the following formula." Or you could say, "A delicate physical analysis, which you can view at thus and such a Web site, yields the following mathematical model." It is part of our training to want to show *every single dirty rotten detail*. When teaching your freshmen, learn to resist this temptation. Make your presentation fit into a fifteen minute window, and make it accessible to your audience.

No matter how you package your wares, there is always the danger that afterward some student will ask the question, "Will this be on the test?" What can you say? Will a long, elaborate application be on the test? No, but some of the analytical techniques that you used in the example could be. You had better have an answer prepared for questions such as these. See also Section 4.6 on answering difficult questions.

Do not give applications that consist of a lot of terminology with a negligible mathematical kernel buried in it. For example, a recent calculus text has a problem about the destruction of trees in a tropical rain forest. After wading through dozens of pieces of unfamiliar terminology, the student finds that all he is required to do is to take the derivative of the given function and set it equal to zero. Doing such an example gives the students the message that applications are nothing but mindless drivel. And that is the *wrong* message. Another calculus text has a problem that begins (here I am paraphrasing) "The function $y = x^2$ is important in relativity theory and cosmology ..." The author is of course alluding to Einstein's equation $E = mc^2$. It is easy to see that such a problem is thoroughly disingenuous, and that any student who knows a little physics will be insulted and turned off. This is *not* the way to gain credibility with your students.

Learn not to treat applications as something "apart" from the subject proper of the course. In fact applications can be a device for helping students to learn key mathematical ideas. As a simple instance, the notion of velocity helps students to understand what the derivative means. As a more sophisticated example, the notion of centrifugal force helps students to understand what curvature means. Applications should not be a tiresome appendage that is tacked onto the course. They should be an essential part of the course.

Recently I taught a rather sophisticated course in multi-variable calculus. I say that it was sophisticated because it was preceded by a mandatory course in linear algebra, and linear algebra was used throughout the multi-variable calculus. Advanced ideas were considered frequently. When we got to Green's theorem, I took great pleasure in telling students something of the life of Green, and that one can visit his mill today and purchase (for fifteen pence) a pencil with Green's theorem emblazoned on the side. Then I told them, as an application of Green's theorem, that you can calculate the area inside a planar region Ω by calculating the following boundary integral:

$$\oint_{\partial\Omega} \frac{1}{2}x\,dy - \frac{1}{2}y\,dx. \qquad (*)$$

Most sophomores are not sufficiently experienced to appreciate what this statement says, so I told them about the "planimeter", which is a hardware device—commonly used by draftsmen—that actually implements equation $(*)$. When I said to them, "You take this little doohickey and run it around the edge of a figure on a piece of paper and it tells you the area inside," they just went wild. *This* practical piece of information they really understood. We looked at several examples of pairs of domains with equal perimeter but different areas and discussed how a mechanical device tracing the boundary might distinguish the two. The students had a great many questions. When class was over, several of them stormed over to our rather distinguished School of Architecture and demanded to see a planimeter. The rejoinder from the architects was, "What's that?" When told, the architects said, "That's impossible. You'd need a computer to do something like that." Well, there was nothing for it. I had to go out and buy a planimeter.

This cost me a pretty penny, but it was a terrific object lesson for the students. I cared enough to follow up on what I had told them. I stood my ground against the architects. And I triumphed!

William Thurston, in his article [THU] on the teaching of mathematics, points out that math is a "tall" subject and that math is a "wide" subject. The tallness articulates the fact that mathematics builds up and up, each new topic taking

advantage of previous ones. It is wide in the sense that it is a highly diverse and
interactive melange. It interfaces with all of the other sciences, with engineering,
and with many other disciplines as well. It is our job as teachers of mathematics
to introduce students to this exciting field, and to motivate students to want to
study mathematics and to major in it. Applications are a device for achieving this
end. Using them wisely and well in the classroom is a non-trivial matter. Talk
to experienced faculty in your department about what resources are available to
help you to present meaningful applications in your classes.

An unpublished biography [TBJ] of R. L. Moore (see Section 1.12) tells of
Moore's rough-and-ready childhood schooling in a one-room schoolhouse in late
nineteenth century East Texas. One Fall, the headmaster was overheard on the
first day of the school year saying to a parent "Do you want me to teach the
earth round or flat?—I can do it either way." This is an approach to applications
that we should all best avoid. When you give applications, tell the truth. Or
else don't give any.

1.12   The Moore Method

R. L. Moore was an important and influential faculty member at the University
of Texas from the 1920s to the 1970s. Possessed of a strong and single-minded
personality, he developed a special method of teaching that has been successfully
practiced and disseminated by his students and followers.

The Moore method worked like this—and it was practiced in all his classes,
from the most elementary undergraduate to the most advanced graduate. Students would show up on the first day, usually with no notion of what Moore
was like or what he expected. He would hand out a single mimeographed sheet
with some definitions and some axioms and some theorems stated on it. And he
would say, "Who would like to go to the board and prove the first theorem?" Of
course Moore was usually greeted by bewildered silence. Many of his students
did not know the meaning of the words "definition" and "axiom" and "theorem"
and "proof", nor did they know the basic rules of logic. But Moore could take it.
He sat and he waited. Often the first hour would go by without a word spoken.
And so would the second. And many times the third as well. Eventually, some
brave soul would go to the blackboard and attempt a "proof" of Theorem 1.
Moore would rip that person apart. And that set the tone for the class.

In each session of R. L. Moore's class, some student would go to the board
with a proof he had concocted, and everyone else in the class—student and
professor alike—would examine it critically. There were no holds barred. Moore
would not *assist* the students in creating a proof, but he would *respond* to
whatever they said.

The good thing about Moore's classes is that his students came to *possess*
the mathematics. If you put it incorrectly on the blackboard, then you were
subjected to merciless criticism. If you got it right, then you received quiet but
sincere approbation.

Moore did not allow his students to read books or papers (this included even
his graduate students!). Indeed Moore saw to it that the mathematics library

got no funds for new books! As legend had it, if he found a student of his in the library he would throw that person (bodily) out the door. Students were to create the mathematics themselves—from whole cloth. The class worked best if the students began with approximately equal abilities. They were not allowed to read, and they were not allowed to collaborate outside of class. Moore was merciless in weeding out those students who did not cooperate or did not fit.

I leave it to you, the reader, to determine the good and the bad points of the Moore method. Mary Ellen Rudin, one of Moore's most prominent and successful students, does not recommend it. She says, in part, "All you get [from this method] is a (probably overconfident) uneducated ignoramus." By contrast Paul Halmos, an eloquent avatar of good teaching, says, "The Moore method is, I am convinced, the right way to teach anything and everything—it produces students who can understand and use what they have learned." Moore's teaching method—and his techniques for selecting the students whom he wanted in *his classes*, and whom he would mentor—caused emotions to run high in the University of Texas mathematics department. It is said that "...the department was literally divided into two camps: the Moore and the anti-Moore. The Moores had the fourth floor, and it was armed: guns in the drawers up there." See [BAP], [TBJ], [HALM], [WIL] for my sources, and for more on the Moore method.

The Moore method is obviously a powerful way to teach any subject. Its success was as much a product of Moore's strong personality as it was of its essentially combative nature. And Moore's students—especially his graduate students—often became strong proponents of the method. It spread far and wide, and it was difficult to grow up in American mathematics in the 1960s and 1970s without hearing of, and experiencing, the Moore method.

I took one class that was taught by Moore's method. Fittingly, it was point-set topology (Moore's pet subject). I both loved this class and hated it. I loved it because I loved the subject matter and I loved learning it; I hated it because I thought the professor was a no-good, lazy bum. Moore would have smiled in approbation.

Chapter 2

Practical Matters

> What a waste it is to lose one's mind. Or not to have a mind as being truly wasteful. How true that is.
>
> > Dan Quayle, former Vice President of the U.S.
>
> In God we trust; all others bring data.
>
> > A Statistician
>
> All innovations succeed in the hands of the innovator, and none succeed in other hands.
>
> > David S. Moore
>
> You may as well put the sidewalk where the students walk, because they'll walk there anyway.
>
> > William James Lewis
>
> The pleasure of learning, and knowing, though not the keenest, is yet the least perishable of pleasures; the least subject to external things, and the play of chance, and the wear of time. And as a prudent man puts money by to serve as a provision for the material wants of his old age, so too he needs to lay up against the end of his days provision for the intellect.
>
> > A. E. Houseman
>
> People who become legends in their own time usually have very little time left.
>
> > John D. MacDonald
>
> The most irresponsible thing that a civilized adult can do is to stand up in a crowded room and shout "Proof!"
>
> > James Carlson
>
> Everybody will be famous for fifteen minutes.
>
> > Andy Warhol
>
> An expert is just some guy from out of town.
>
> > Mark Twain

2.0   Chapter Overview

Like many activities in life, fine teaching is composed of many technical components. When everything works properly, then the whole is considerably greater than the sum of its parts. However, if some of the crucial parts are rusty or, worse, non-functional, then the whole will creak and drag and not do a good job of it.

The novice instructor should probably read every section in this chapter. The more experienced instructor may wish to pick out particular sections for concentrated effort.

2.1   Voice

There is nothing more stultifying than a lecture in a reasonably large classroom on a hot day delivered by an oblivious professor mumbling to himself at the front of the room. We are not all actors or comedians or even great public speakers. But we are teachers, and we must convey a body of material. We must capture the class's attention. We must *fill the room.*

If you are unlucky, you may be assigned to teach in a classroom that works against you. Perhaps visibility is poor for the students in the back, or the acoustics are bad, or the blackboard is substandard. If you are burdened with such a teaching environment, try your best to get it changed. If you cannot, then think hard about how to get the best out of this classroom. If the blackboard is unusable, then consider lecturing with an overhead projector. If the acoustics are bad, then consider using a microphone. If visibility is poor, then think about changing the seating arrangement. No matter what the liabilities, you must take charge.

Your voice is one of your primary tools. And you must use that tool in part to control the environment in your classroom. The most important presence in the room is not the blackboard, nor the desk, nor the text. It is *you.* You want the students' attention focused on you.

I am not saying that you must lose your dignity, or act silly, or show off. You must learn to use your voice and your eyes and your body and your presence as a tool. If you are going to say something important, then make a meaningful pause beforehand. *Say* that it is important. Repeat the point. Write it down. Give an example. Repeat it again.

You can gain the attention of a large group by lowering your voice. Or by raising it. Or by pausing. One thing is certain: You will not gain the audience's attention by rolling along in an uninflected monotone. Again, I am not suggesting that you undergo a personality change in order to be a sound teacher. What I am suggesting is that you find ways to talk to them as a *person interacting with people*, rather than as an ill-at-ease, out-to-lunch egghead.

At a well-known university (of good quality!) in southern California they once tried bringing in actors from Hollywood to help professors spice up their delivery. One instructor of an Abnormal Psychology course was advised to use the line, "I never teach about any mental illness until I try it out myself." Such pandering is inappropriate, offensive, and childish. Can you imagine yourself using such patter? Who would want to?

What I am suggesting here is that you take just a little time and contemplate your lecture/classroom style. A lecture or class should be a controlled conversation between you and your audience. It is a trifle one-sided, of course. But there must be cerebral interaction between the teacher and the students. That means that you, the instructor, must grab and maintain the attention of the class. Your behavior in front of the group is a primary tool for keeping the lines of communication open.

When you are talking about a subject that you perceive to be trivial, and when you are nervous, you tend to talk too fast. Novice instructors find themselves barreling through their lectures. You must resist this tendency. If you are

really new at the business of teaching, then practice your lectures. Get a friend
to listen. In calculus, a fifty minute lecture with three or four good examples
and some intermediate explanatory material is probably just about right (I'm
thinking here of a lecture on max-min problems, for example). Try to make each
class consist of about that much material, and make it fill the hour. If you finish
early, that is fine (but it may mean that you talked too fast). You can quit early
for that day, or do an extra example, or use the extra time to answer questions.

Don't give the students the impression that you are in a rush. It puts them
off, and reflects a bad attitude toward the teaching process. If on Wednesday
you plan to explain the chain rule, then do just that. If the chosen topic does
not fill the hour, then do an extra example or field questions. Do not race on
to the next topic. One idea per class, at the lower-division level, is about right.
(Of course if you are teaching a multi-section class at a big university, then it
is important to keep pace with the other instructors. This is yet another reason
for keeping careful track of your use of time. See also Section 1.9.)

It is something of an oversimplification, but still true, that a portion of the
teacher's role is as a cheerleader. You are, by example, trying to persuade the
students that this ostensibly difficult material is doable. Part of the secret to
success in this process is to have a controlled, relaxed voice, to appear to be at
ease, and to be organized. Don't let a small error fluster you. Make it seem as
though such a slip can happen to anyone, and that fixing it is akin to tying your
shoelaces or pulling up your socks.

But, as with all advice in this book, you must temper the thoughts in the
last paragraph with a dose of realism. If you make the material look easy, then
students will infer that it *is* easy. The psychological processes at play here are
not completely straightforward. Nobody would be foolish enough to go to an
Isaac Stern concert and come away with the impression that playing the violin
is trivial. Yet students attend my calculus classes, watch me solve problems,
conclude that the material is easy and that they have it down cold. They decide
that in fact they *don't* need to do any homework problems or read the book, and
then they flunk the midterm.

These are the same students who come to me after the exam and say, "I
understand all the ideas. The material is absolutely clear when you talk about
it in class. But I couldn't do the problems on the exam." How many times have
you heard such a statement from your students? I like to tease my students by
reminding them that this is like saying, "I really understand how to swim, but
every time I get in the water I drown."

On the one hand, you don't want to make straightforward material look
difficult. After 300 years, we've got calculus sewn up. There is no topic in the
course that is intrinsically difficult. We merely need to train our students to do
it. So *do* make each technique look straightforward. But remind the students
that *they themselves need to practice*. Do this by telling them so, by giving
quizzes, by varying the examples and introducing little surprises. Ask the class
questions to make the students turn the ideas over in their own minds. Use your
voice to encourage, to wheedle, to cajole, to question, to stimulate.

Mathematics instructors in general, whether they are "reformers" or "traditionalists" or "high techers" or "plug-and-chuggers", agree that each student

must take each idea in the course and rebuild it in his own mind. This is nothing new. Go read Beth and Piaget [BPA]—they discuss this notion in detail. An awareness of this concept will help you to shape your teaching methods. If the students cannot understand what you are talking about, then it is unlikely that they will take the ideas home and think about them. If the students watch you state and prove a lot of abstruse theorems, and in the process become terminally depressed, then it is unlikely that they will take the ideas home and analyze them and internalize them. If the students watch you flounder around, unable to complete an example coherently or explain a concept neatly, then it is unlikely that they will take the ideas home and rebuild them in their own minds.

If instead you kindle the students' curiosity, plant in them a desire to learn, show them something they have never seen before and *make them realize that it is something they have never seen before*—and certainly never understood before—then there is a real likelihood that they will leave class turning the new thoughts over in their minds, talk among themselves about it, ask questions, and come back to you with their own ideas. *That* is teaching.

Even if you know how to use your voice well with a small audience, and to capture their attention and get them excited about learning, there are special problems with the large classes that are used in the teaching of calculus (for instance) at many universities. Refer to Section 2.14 for more on this matter.

2.2   Eye Contact

We all know certain people who invariably emerge as the leader of any group conversation. Such people seem to sparkle with wit, erudition, and presence. They usually pick the topic and they usually aim the discussion. They have a sense of humor, and they are intelligent. What is their secret?

It is partly a matter of attention and awareness. The sort of person I am describing has an inborn curiosity; he is aware of you, and interested in you, and genuinely eager to learn about your opinions and experiences and interests. When you ask yourself what makes another person interesting, an honest answer would have to entail that such a person is outer directed, and cares about others.[1]

This is obviously a talent that is partly inborn and partly cultivated. Some of the trick is to show genuine interest in what other people have to say before bounding ahead with what you have to say. Another part is to talk about subjects, and to tell anecdotes, that you know will interest other people. Being charming and witty helps too, but in this section I want to concentrate on more mechanical features of repartee.

Many of the devices that make for an engaging conversationalist also make for an engaging teacher. A review of the last paragraphs, and of the rest of this book, will bear out this assertion. In this section I will discuss the importance of *eye contact*.

Telling a good joke while staring at the floor with your thumb in your ear will not have the same effect as telling the joke while looking at your listener,

---

[1]A boring person is one who talks about himself. An interesting person is one who talks about you.

engaging his attention, and reacting to the listener while the listener is reacting to you. A good joke teller has his audience starting to chuckle half way through the joke and just dying for the punch line. Getting a good laugh is then a foregone conclusion.

Giving a good lecture or class is serious business, and is not the same as telling a joke. But many of the moves are the same. If you want to hold your audience's attention then you must look at your audience. You must engage not one person but all. You must learn to use your body as a tool. Step forward and back. Force the eyes of the audience to follow you. A good lecturer speaks to individuals in the audience, to grouplets in the audience, and to the whole audience. Like a movie camera, you must zoom in and zoom out to get the effects that you wish to achieve. A ninety minute movie filmed at the same constant focal length would be dreadfully boring. Ditto for a lecture.

Some people are very shy about establishing eye contact. It is a device that you must consciously cultivate. The end result is worth it. The teacher who can establish eye contact is also the teacher who is confident, who is well prepared, and who conducts a good class.

2.3   Blackboard Technique

Write neatly. Write either in very plain longhand or print. Be sure that your handwriting is large enough. Be sure that it is dark enough. Endeavor to write straight across the blackboard in a horizontal line. Proceed in a linear fashion. Don't have a lot of insertions, arrows, and diagonally written asides.

Don't put too much material on each board. The ideas stand out more vividly if they are not hemmed in by a lot of adjacent material. In particular, it is difficult for students to pay attention when the teacher fills the board with long line after long line of print. An excellent guitarist once said that the silences in his music were at least as important as the notes. When you are laying material out on a blackboard, the same can be said of the blank spaces.

Some people find it useful to divide the blackboard into boxes. This practice makes it easier for the lecturer to organize what he is writing, and also makes it *much, much easier* for the students in the audience to organize the material in their minds and in their notes.

Label your equations so that you can refer to them verbally. Draw sketches neatly. Use horizontal and vertical lines to set off related bodies of material.

You can control your output more accurately by keeping the length of each line short. Think of the blackboard as being divided into several boxes and write your lecture by putting one idea in each box. To repeat: If necessary, *actually divide the blackboard into boxes.*

If the classroom has sliding blackboards, think ahead about how to use them so that the most (and most recent) material is visible at one time. For those combinatorial theorists among you, or those experts on the game of NIM, this should be fun.

If you are right-handed, consider writing first on the right-hand blackboard and then working left. The reason? That way you are never standing in front of what you've written. Good teaching consists in large part of a lot of little details like this. You shouldn't be pathological about these details, but if you are aware that they are there then you will pick up on them.

Every now and then during your presentation, you should stand aside and pause. Don't say *anything*. This gives you an opportunity to collect your thoughts and catch your breath. You can verify the accuracy of what you have written. It gives the students an opportunity to catch up and to ponder what they've heard. They may decide to formulate questions. If instead you are barging ahead at full speed for the entire hour, then students never have a moment to think about what they are hearing. They cannot interact with you because you are not interacting with them.

Try to think ahead. Material that needs to be kept—and not erased—should be written (probably in a box) on a blackboard to the far left or far right where it is out of the way but can be referred to easily. You may wish to reserve another box on the blackboard for asides or remarks. Some instructors put material that needs to be seen for the entire hour on an overhead slide. This frees up all the blackboards, and keeps those important equations or definitions front and center.

These ideas are another facet of the precept that you know the material cold so that you can concentrate on your delivery. Just as an actor knows his lines cold so that he can make bold entrances and exits, and not trip over his feet, so you must be able to focus a significant portion of your brain on the *conveying of the information*.

If your lesson will involve one or more difficult figures then practice them on a sheet of paper in advance. Remember that you are a mathematical role model for the students. If you make it appear that it is difficult for *you* to draw a hyperboloid of one sheet, then how are the students supposed to be able to do it? Of course you can prepare the figure ahead of time on an overhead slide (or even photocopy it straight out of a book, or straight from `Mathematica` or `Maple`, onto a transparency). This solves the problem of having a nice figure to show the students. It does not solve the problem of *showing the students how to draw the figure*. As a result, it puts a barrier between them and the ultimate goal of learning how to *read* the graph. If necessary, consult a colleague who is artistically adept for tips on how to draw difficult figures.

You will find students quite resistive to learning to graph—especially in three dimensions. I learned a useful teaching device when last I taught multi-variable calculus, and it became clear to me when I was showing my students how to graph. That device is *persistence*. I made it clear to them that anything that gave them a pain in the neck was going to appear repeatedly in subsequent work. For example, almost all of them did rather poorly on the graphing portion of the first midterm. So I gave them several followup quizzes to help them hone their graphing skills. After each quiz they all gave me their "Is that finally the end of graphing?" look. But after examining their work I said, "Nope; not good enough yet." And on we went. Over and over again I graphed functions in three dimensions. I went through every step. And I did it *at the blackboard*, just as I expected them to do it with pencil and paper. Graphing appeared again on the final. And I *told* them that it would so appear. In fact I told them that the best way to study for the final was to find everything on the first two midterms that they hated, and that this material would certainly be on the final. They believed me, and it worked.

If you cannot organize the steps of a maximum-minimum problem, then can you really expect the students to do so? In the best of all possible worlds, the students' work is but a pale shadow of your own. So your work should be the platonic ideal. Sometimes, in presenting an example or solving a problem at the blackboard, you may inadvertently gloss from one step to another. Or you might make a straightforward presentation look like a bag of tricks. This practice is very confusing for students, especially the ones who lack confidence. By organizing the solution in a step-by-step format you can avoid these slips.

After you have filled a board, it should be neat enough and clear enough that you could snap a Polaroid$^®$ snapshot and read the presentation from the Polaroid$^®$. In particular, you should not lecture by writing a few words, erasing those, and then writing some more words on top of the erased old words. Students cannot follow such a presentation. I cannot emphasize this point too strongly: Write from left to right and from top to bottom. *Do not erase.* When the first box is filled, proceed to the second. *Do not erase.* Only when all blackboards are full should you go back and begin erasing. Students must be given time to stare at what they've just seen as well as what is currently being written. Keep as much material as possible visible at all times.

BUT: When it is time to erase, be sure to erase thoroughly. It is well worth spending a few extra moments being sure that the blackboard is sparkling clean before you begin a new block of material. For if you endeavor to write over a sloppily erased blackboard then your writing will be obscure at best. This is really a psychological issue. Of course the students can squint and strain and figure out what you are writing (even if it is a virtually unreadable palimpsest), but it bums them out to have to do so. Try to make their job as easy as possible.

Do not stand in front of what you are writing. Either stretch out your arm and write to the side or step aside frequently. Read aloud to the class as you write. Make the mathematics happen before their eyes and *be sure that they can see everything.* Every once in a while, pause and step aside to catch your breath and to let them catch up.

Here is a common error that is made even by the most seasoned professionals. Imagine that you do an example that begins with the phrase "Find the local maxima and minima of the function ... " And so forth. Say that you've worked the example. Now suppose that the next example begins with the same phrase. It is a dreadful mistake to erase all but that first phrase and begin the new example on the fly, as it were.

Why is this a mistake?—it *seems* perfectly logical. But the students are taking notes! How can they keep up when you pull a stunt like this? *Slow yourself down.* Write the words again. If a student gets two sentences behind then he may as well be two paragraphs behind. Give frequent respites for catch up.

And now a coda: How much of what you are saying should you write? In my experience, the answer is "As much as possible." When you are transmitting sophisticated technical ideas verbally, students have trouble keeping up. Many of them are not native English speakers. They need a little help. Write down everything except asides (actually, some asides are worth recording as well). Say the words as you write them. This is also a device for slowing yourself down. Most of us tend to talk far too fast—at least about mathematics. Slowing yourself down and writing deliberately will help you to keep your handwriting clear and will make the lesson as a whole appear to be neat and clean. If the *appearance* is neat and clean then perhaps the *ideas* are neat and clean—at least that's what you want the students to think.

The flip side of the last paragraph is that the tendency to talk too rapidly may cause you to write too rapidly (and therefore sloppily). Thus periodically checking the quality of your handwriting on the blackboard can serve as a means of telling whether either your verbal or written delivery is too speedy.

Let me reiterate one of my most fundamental precepts. There is a real psychological barrier for the instructor to overcome when learning blackboard technique, and voice control. When we understand very deeply what we are talking about, then it all seems quite trivial. We can convince ourselves rather easily— at least at a subconscious level—that it is embarrassing to stand in front of a group and enunciate whatever mundane material is the topic of the day. Thus we are inclined to race through it, both verbally and in the way that we render it on the blackboard. *Be conscious of this trap and do not fall into it.* I have never been criticized for being too clear, whether I was giving a calculus lecture to freshmen or a seminar lecture at the Mathematisches Forschungsinstitut Oberwolfach. *Slow down.* Be deliberate. Enunciate. Explain.

Many of us, at the beginning of the class, rattle on verbally at some length before we finally persuade ourselves that we had better start writing something on the board. Please don't do this. Start writing the material from the very outset. If you want the students to notice it, and write it down, and get it straight, then you had better set the example by writing it.

Writing material neatly and slowly is a subtle way of telling the students that this material is important. If you are taking the trouble to write it down deliberately, then it must be worth writing deliberately. Conversely, if you scribble some incoherent gibberish, or scribble nothing at all, then what signal are you sending to the students?

2.4   Body Language

If you skulk into your classroom, stand slouching in front of the class with a furtive and disreputable expression, wear slovenly clothing, and give off a ripe odor that permeates the room well into the fifth row, then you are sending numerous negative signals to your class. It sounds trite to say it, but dress neatly and attractively when you go to teach. Stand erect and look dignified. Master the basics of personal grooming. Attending to these mundane matters really does make a difference. If you want the class to behave as though *you* are in charge, then look and act the part. It is true that Grigory Rasputin broke all these precepts and still managed to rule a country, but he had many other talents that are foreign to most mathematicians.

Think about the last time you attended a colloquium party or other mathematics social function. In my experience, there is always a group of eight or ten people who don't have the good sense to go home at the propitious time (obviously the only reason I know this is that sometimes I've been one of those people). These stragglers usually pull chairs into a ring and sit and stare at each other for an hour or so before the host finally says a cheery, "Well gang, time to go!" This "circle of death" is a strange and embarrassing artifact of mathematical—and perhaps academic—social life. One cannot help but suppose that people who would display this sort of social ineptitude might be equally gauche in other circumstances, such as when they teach. *Do* endeavor to be aware of the world around you. Endeavor to get to know your students, and to get to know who they are. Try to understand their needs, and their goals. Use your *body* (as well as your wit) to show them that you are there to help them learn the material, to be their guide in this new territory.

A friend of mine, fresh from Romania and with minimal English skills, once sat in a lunch room in Chicago endeavoring to formulate his order. He stared forlornly at "Ruben Sandwich", "Chile Size", "Three Pigs in a Blanket", "Montecristo Sandwich", and other solecisms on the menu trying to determine what in the world they could be. His dictionary was of little help. Finally he found a menu item that he could translate: "bacon, lettuce, and tomato sandwich". He happily signaled the server and painstakingly ordered "a bacon, lettuce, and tomato sandwich and a Coke." The server scribbled down the order and drawled "Ya want mayo on your BLT?"

Has it ever occurred to you that we treat our calculus students in just the way that that server treated my Romanian friend? The students are struggling with the chain rule and the mean value theorem and related rates in just the way that my friend struggled with the concept of Ruben sandwich, and we often answer (because we are so accustomed to the ideas) "Ya want mayo on your BLT?" It is important that we develop some sensitivity to the student's point of view, his values, and his vocabulary. We need not *admire* these attributes, but we must *deal with them*. That is the teacher's job.

Not long ago I asked a young assistant professor how he taught the Mean Value Theorem to freshmen. He began by saying, "First I tell them about the Axiom of Choice ..." I could not resist interrupting and asking "Why?" He seemed to have some vague idea that the Axiom of Choice had something to do with the completeness of the real numbers (this turns out to be a misapprehension). But so what if it does? I wouldn't say a word about the completeness of the reals when explaining the Mean Value Theorem to eighteen year olds, much less about the Axiom of Choice. In fact I'm quite sure that I wouldn't even prove the Mean Value Theorem. I would concentrate on getting them to understand what the theorem *says*. In short, while my friend would be asking the freshmen "Ya want mayo on your BLT?" I would be serving them a nice hot sandwich on a platter.

It is best—really—if your students can think of you as a human being, part of the same species to which they belong. Intellectually you may be on an entirely different plane. Nevertheless, when you are not running bulls in Pamplona or swapping gestalts with the Maharishi, you are in fact just the calculus instructor.

It is an important job, and you should endeavor to carry it out with dignity and professionalism.

2.5   Homework

In most lower-division courses, and many upper-division ones, it is by way of the homework that you have the greatest direct interaction with your students. When students waylay you after class or come to your office hour, it is usually to ask you about a homework problem. This is why the exercise sets in a textbook are often the most important part of the book (textbook authors do not seem to have caught on to this observation yet) *and* why it is critical that homework assignments be sensibly constructed.

Let me stress again that I am not trying to sell you a time-consuming attitude or habit. If you take twenty minutes to compose a homework assignment then you are probably taking too much time. But consider the following precepts:

- Do not make the homework assignment too long.

- Do not make the homework assignment too short.

- Check over the problems you assign to confirm that there are no notational or obvious typographical errors. (Students can waste great amounts of time trying to fathom typos that are trivial to you and me. As a result, they become quite frustrated and angry. Doing this sort of checking shows them that you are on their side.)

- Be sure that the assignment touches on all of the most important topics.

- Be sure that the homework assignment drills the students on the material that you want them to learn and the material that you will be testing them on.

- *Make sure that at least some of the homework problems are graded.*

- Plan ahead. The exams that you give should be based only on material that the students have seen in the classes and in the homework.

If homework does not count and is not graded, then students will not do it. That is a fact. I realize that many of us have neither the time nor the inclination to spend long hours each evening grading homework. Many universities and colleges these days simply do not have the resources to provide enough graders for lower-division courses. But there are compromises that you can make. For example, you can tell the students that, of ten problems on the homework assignment, just three will be graded. But don't tell them which three. This device will force most of the serious students to do *all* the homework problems, but it requires much less grader time to get the grading done.

If the last suggestion will not work for you, then you can give weekly quizzes that you yourself will grade. The amount of your time involved will be little, and it is a device to force students to keep up with the work. Incidentally, this device also gives you a gentle way to keep your finger on the pulse of the class.

Consider the implementing following policy to help get your students more interested in doing the homework. Students can and do benefit from collaboration, just as we mathematicians do in our research. While you probably do not want to encourage collaboration on exams, you may wish to encourage it on homework. Of course I'm not talking about "I'll copy yours this week and you can copy mine next week." Instead, I'm talking about an intelligent exchange of information among equals.

Some studies have shown that one reason that Oriental students in this country tend to do very well in their mathematics classes (and there are surely many reasons) is that they work in groups. More precisely, they first work hard as individuals. Then they get together and compare results. In short, they collaborate in much the same way that mature mathematicians collaborate. They are willing to say, "I can do this but I cannot do that. What can you contribute?" At the same time, the studies indicate that certain other elements of the student population are either loath to work in groups or are unaware of the benefits of this activity. These strata tend to do poorly in mathematics classes. See [TRE] for details.

Some of the more interesting teaching reform projects, including those from Harvard and Duke, are specifically designed to encourage students to learn mathematics through group activities. Reports on these experiments are encouraging.

If you do decide to encourage group work in your classes, then you will have to make peace between said collaboration and your grading policies. If homework is not collected, then there is no problem and you can separate the good students from the bad through exams and quizzes. If instead homework is collected, then you will have to consider carefully how to tell whose work is whose, or at least how to divide up the credit.

2.6 Office Hours

At most universities the instructor is required to hold three or more office hours per week. Choose three hours that are convenient for you or convenient for the students or both. Monday/Wednesday/Friday at 11:00 A.M. is, on most campuses, one of the most popular times for classes. If you schedule your office hour at that time then many students will not be able to attend. One good strategy is to stagger your office hours, so that they are at different times on different days. Another is to make an office hour from half past the hour to half past the hour, so that a student's class is likely to overlap only half of it rather than all of it.

Of course you cannot select a time for office hours that will please *everyone*, so don't even attempt to do so. Set your office hours, and announce them, and explain to the students that you can make appointments for those who cannot attend the regularly scheduled hours. Such an announcement will not appreciably increase the number of visitations from your students, and it is just good business to set such a policy.

Promise students that you will be there during your office hour. And be there. Students should be made to understand that they need not wait for a natural or personal disaster in order to come to your office hour. It is perfectly all right for a student to come to your office hour and say "I don't get problem 6." or "The chain rule makes no sense to me."

During your office hour, you will usually not be overwhelmed with students (except perhaps just before an exam). In fact it is a general rule of thumb that, the larger the class, the smaller the percentage of students who will come to your office hour. But those who do show up will appreciate your attentions. Of the hours that you have designated, you can spend some of them catching up on your correspondence, making up the next homework assignment, or reading the *Notices* or the *Monthly* or the *Mathematical Intelligencer*.

If you have sufficient space in your office, it is a good idea to have a table and a couple of chairs set up in a special part of the office—*away from your desk and your papers and books and personal artifacts*—where you will consult with students. What is the reason for this affectation? First of all, you don't want students inadvertently walking away with your papers or your correspondence. Second, you don't want them spilling coffee on your latest manuscript or your new book that you purchased at great expense from Marcel Dekker. Third, students are by nature careless. They may put their feet on your desk or use your telephone or grab your fountain pen. Rather than appear to be an old fuddy-duddy and constantly be scolding, it is so much easier to have a special venue in which to "hold court".

When a student comes to my office expressing befuddlement over a particular type of problem, I have a powerful and decisive weapon that I unleash. I begin by asking, "Do you have a half hour or so?" If the answer is "Yes", then I sit the student down and say, "Try a problem of the kind you are having trouble with. When you get stuck, tell me." Of course the student invariably gets stuck, and I give him a little help. I might need to intervene three or four times during the first problem that the student does for me. But the second problem may require only two interventions, and the third only one. By the time the fourth problem rolls around, the student's newfound confidence is irrepressible, and the transaction is a great success. The student goes away pleased and happy that he has now mastered a heretofore mystifying mathematical idea. Of course I always tell the student, "If you get home, and you find that you are still confused, then come back and we'll do this again."

On days when your office hour is not crowded, and you only have a couple of customers, I highly recommend that you try this teaching technique. It's good business, and it always produces satisfied customers. Word gets around in the class that the professor is not such a bad guy after all. Perhaps, as a result, a few other students drop by for help.

You want to convey to students that the office hour is a particular time that you have set aside for them. If you consult with students while sitting at your desk and glancing at your mail, or scribbling notes for an upcoming seminar, or answering phone calls, then you in fact will *not* convince them of your dedication. Instead, if you hold court in the special part of your office that you have set aside for consultations, then your students will understand that this time is theirs. If you really want to do it right, then let your voice mail pick up on your phone calls during your office hour. For those sixty minutes, give yourself to your students.

The office hour is your opportunity to get to know at least some of your students personally. Of course I do not mean by this that you should get involved in their *personal lives*. Problems about their love lives or their parents or their social diseases should be referred to the professional counselors that are on the staff of every college. What I mean is that you should take the opportunity to get to know some of your students as people, and to let them get to know you as well.

This activity has several beneficial side effects, both for you and for them. When you are lecturing, you can have certain individuals in the room in mind as you formulate your remarks. You can make reference (*without* mentioning any names) to questions that came up during office hour. It is reassuring to the average student (the type that *does not* go to office hour) to know that conscientious students (the type that *do* go to office hour) have some of the same questions that they have.

This point is in fact worth developing. Some components of teaching may be compared with certain components of psychotherapy. One big aspect of therapy—certainly an aspect that is exploited by popular psychology and self-help books—is to assure the patient that he is not alone. There are thousands of people with exactly the same problems, suffering in just the same ways. And they have been treated successfully.

Just so, when you teach you must give both tacit and explicit reassurances to students that their questions and confusions are not theirs alone. An eighteen year old is scared to death that he is the only person in the room who doesn't understand why the numerator in the quotient rule has the form that it has—or why it does not seem to be symmetric in its arguments. Such a student would not dare ask about it in front of a room full of his peers. The student may not even be sure how to articulate the question, so surely will not want to flounder about in front of the entire class. At the same time the student may be afraid to come to your office hour and, alone but in *your* august presence, ask for a clarification.

Thus you must signal to students that questions are a good thing. When a student asks a question in class that might be of general interest, I not only repeat it but I often state that I am glad this question was raised. I carefully record the question on the blackboard. Several people have visited me privately, I add, and asked variants of the same question. If there is a question that should be asked but has not been, then I ask it myself. I say that if this point is unclear to them (the students) then they should come see me in my office hour and get it straightened out. You don't need to give away door prizes to drum up business at your office hour. However, it is psychologically important for students to know that you are available, whether they actually come to see you or not.

I have said repeatedly in this book that persistence is an important attribute for the successful teacher. Another such attribute is patience. If a student finds the nerve to ask a question in class or during office hour—*even if it is a question that I have answered before*, in fact *even if I have answered it several times before*—then I treat the questioner with respect and I pay due homage to the question and I answer it. I *never* say, "I've already answered that question. Go home and read your notes." Such a rejoinder would be counterproductive, and would discourage further question-asking in class.

I often announce to my classes that students may drop by my office even when it is not my office hour. If I am not busy, I'll be happy to talk to them. In practice, this charity does not appreciably increase the flow of business. There are always students who strictly respect your designated office hour and there are always those who drop by whenever they please. But making an announcement of this nature is one of those little details that contribute to a favorable student attitude. For it sends a signal to the class that you care, and that you truly want to help them learn mathematics. If you do make such an announcement, be courteous to those who take you up on it. If you are busy and must send the student away, do so with respect and suggest another time for the student to return.

I once had a colleague who, whenever a student would show up at his door, would crawl under his desk until the student went away. This is certainly a memorable way to deal with students, but not one that I would recommend. When a student comes to your office, make him feel welcome. Act as though you are happy that he dropped by. Endeavor to adopt the same cheery tone that you would assume if a good friend paid a surprise visit. Such an action on your part will put the student at ease, and will make the transaction go smoothly and productively.

The office hour is a way to step out of your role as instructor and let the students know that you are a person. It is a way to become acquainted with some of your students. Any good public speaker "works the audience" before his speech. Holding your office hour is one way to work the audience. You will also get a feeling during the office hour for how the class is doing, what problems and concerns have arisen, how the pace is working. It is wrong, and self-defeating, to view your office hour as a dreary duty. It is a teaching tool that you should use wisely.

2.7   Designing a Course

Many of us never have the privilege of actually designing a course. Instead, we are assigned to teach prepackaged courses that the department has already assembled. This will especially be true if your teaching load is primarily "service courses": precalculus, calculus, linear algebra, ordinary differential equations. In non-service courses—upper-division courses or courses that are taught for majors—you may in fact have considerable discretion as to what you will include in the course, and how you will organize it. In what follows, I shall draw a sketch of what input you may have into the structure of a course, and also what input you may not have.

If you are teaching one of the prepackaged service courses, then certainly the content will be pre-specified. And, like it or not, you had better stick to the syllabus or outline that the department provides for you. Your students are taking this course *only because* it is required for their major. If you are an instructor who louses up this course for students (so that they must repeat it), then before you know it other departments will be designing their own course to substitute for this one. Such a consequence of your actions will not endear you to your department head. Funding and hiring at many universities is linked rather directly to the number of courses offered and number of students taught. You do not want to be known as the instructor who killed Math 117.

Even if the course you are teaching seems to be good old-fashioned pure math, and is taught almost exclusively to mathematics majors, you may find that your department has rather rigid ideas about what should be included in the course. The department may also have a committee that selects the text.[2] If you are not sure what text to use, or what course outline to follow, don't be afraid to approach a more experienced faculty member for advice on this matter (see also Sections 2.13, 3.9).

If indeed you are teaching a course—perhaps an advanced topics course— where you will have a free hand, then the main precept is to *slow down*. It is too easy to fall into the trap of preparing the course for your fellow faculty (don't worry—they probably have other things to do). This is another good opportunity for obtaining advice from a more experienced colleague. Draw up a tentative outline for the course and show it to a friend. Design the course in such a way that there are several natural places to quit for the semester. Don't drive students away with a syllabus that reads like "Everything that I know, or wish that I knew, about mathematics." Generally it is better to slow down, to give more examples, and to achieve more depth, than to set up a situation that allows you to go to the next conference and tell of all the topics that you covered and all the others that you wanted to cover but could not.

If you have the opportunity to design a course, then approach the task logically. First write down the key ideas in the subject. Then think about how you want to flesh out each one. How will you illustrate the important techniques? What links will you construct between the big ideas? How will you examine the students on this material? What you are doing, in effect, is living through the course once in your head, so that when you do it with the students it will go smoothly. It is too easy for a mathematician to spend a couple of weeks of class time going down a (rather foolish) primrose path that is dictated by his momentary fancy. A little planning can help you to avoid such a reckless teaching error.

Draw up a syllabus (Section 2.12) that is reasonable, and that accurately reflects what you intend to cover in the course. Most students do not have the maturity or experience to look at a completely impossible course syllabus and reason that "The instructor will never cover all this. Of course he will slow down." You have to provide the leadership, both in the small and in the large.

Your reading list also sends important signals to your class. If your required reading is *Linear Operators*[3] by Dunford and Schwartz and the complete works

---

[2]The only "letter of reprimand" that is in my personnel file at my current university stems from my having selected my own text for a course for which an official text had already been selected by a committee.

[3]An important historical work comprising 2500 pages and three volumes.

of Leonhard Euler,[4] then (no matter what the merits of these books, or of your course), you will have no audience.

Use a little practical sense when selecting a text. Also remember that, although most of your students love knowledge, they will not spend $250 for a book. Section 2.13 offers detailed advice on how to select a text.

2.8 Handouts

It is tempting to write up a lot of handouts for your class. If you give a class on Stokes's theorem and feel that you have not made matters clear, then you might be inclined to draw up a handout to help students along. You also might suspect that this extra effort on your part will improve your teaching evaluations and, in particular, that students will appreciate all this additional work that you have put in. Well, it won't and they don't. Only prepare a handout when it will really make a difference. Students feel that they already have enough to read. Inundating them with handouts will only confuse them.

Some professors prepare handouts to expose the students to more meaty problems than those in the text, or to explain ancillary material, or to give summaries of key ideas. Sometimes a handout can supplement what is in the text. I cannot argue that in every circumstance at every college this is a bad thing to do.

What I can do is examine my own conscience and tell you what I see. If I give a lesson that is not up to snuff, or if I do a poor job explaining what curvature is, or if I goof up a proof in class, then I can salve my conscience by writing up a handout. It takes about an hour, it is a way of doing penance, and it is a way of working past the guilt of having screwed up in class. In my heart of hearts, however, I know that what I *should* do is strive to give better classes.

Let me temper these remarks with one important exception. At many universities, it is common to distribute prepared lecture notes. At some, such as UCLA, the student association hires graduate students to prepare careful lecture notes of key courses (such as calculus) and sells them in the student store. This can be a real boon to the students. First, many a student is unable to take good notes and listen to the lecture (and think!) at the same time. Knowing that good notes are available for a modest price gives such a student the freedom to sit back and really listen. Second, having prepared notes available makes missing class a less onerous inconvenience than it would be otherwise.

Having an institutionalized lecture notes system is akin to providing students with a textbook. It does not really fly in the face of what I said in the first few paragraphs of this section. You will have to use my advice here in the context of what resources are available at your institution.

---

[4]The complete works comprise 29 volumes.

2.9 Teaching Evaluations

Mention teaching evaluations to most faculty members and you will get reactions ranging from horror to nausea: "Teaching evaluations are just a popularity contest." or "Kids these days don't want to be taught; they want to be entertained." or "I know that I'm a good/competent teacher, and I certainly don't need to be evaluated. I only distribute the damned forms because the administration makes me do it."

In this section I am going to tell you that, yes, individual student teaching evaluations can be irritating. They can reflect nothing more than the students' immaturity, or pique, or the sorry state of their digestion. But, taken as an aggregate, student teaching evaluations contain valuable information that will help you to become a better teacher.

The simple fact is that most people in most lines of work are regularly evaluated. The evaluation of physicians and lawyers is perhaps rather distant and painless. But it is done. It is a relatively recent development that tenured professors are evaluated—*by their students*—for teaching competence. How did this come about?

Part of the root of the teaching evaluation process comes from the student unrest—especially at U. C. Berkeley—in the 1960s. Students were partly distressed by free speech issues, and partly because they could never find their professors outside the classroom. In those days, professors were not required to hold office hours! The then chancellor at Berkeley, Clark Kerr, recalled wistfully that in the old days the duties of a chancellor were to provide "parking for the faculty, football for the alumni, and sex for the students." He could see the tides of change coming, however, and he helped to usher them in.

Today, there are manifold reasons for teaching evaluations. University administrations are taking a hard look at tenured university faculty and demanding accountability (in some contexts, this sort of examination operates under the rubric "Total Quality Management" or TQM—see Section 2.14 for more on TQM); a system of post-tenure review is being put in place at more and more universities; taxpayers are pressuring universities to hire faculty who can teach (and who can speak English); if a department chair wants to get a faculty member tenured or promoted, then he must provide ample evidence that the candidate is a good teacher.

You may find that reading your teaching evaluations is a gut-wrenching experience. As an instance, James R. Martino of Johns Hopkins University compiled the following pairs of quotations (as cited in [MAR]). Each pair comes from different students evaluating the same instructor for the same course.

> **1.** Does a good job of explaining new concepts and doing examples to demonstrate how it's actually done.
> **2.** Hopefully, he can explain the concepts at a more fundamental level; he assumes that some of the concepts are too trivial to be explained fully in class.

1. Great in all aspects of communication . . . is extremely intelligent when it comes to knowing [the] course.
2. The man is a horrible teacher. I'm sorry to say that he reflects the math department.

1. [The professor] is a giant. The man is humorous and very intelligent. He actually makes math interesting.
2. I don't go to lecture anymore because I never learn anything from the instructor. He seems to be talking about irrelevant topics and phrases questions in ways which are hard to understand. I see no benefit and reason to go to lecture except to obtain the homework assignment.

Not logically consistent, are they?[5] Do these examples prove that teaching evaluations are worthless? That they are the puerile rantings of unformed minds? I don't think so. The extremes of opinion that you see here are no more bizarre than those that you would see in any public opinion survey. Any statistician will tell you that a sample of two is far too small for obtaining useful information.

When you read your teaching evaluations, don't let the outliers upset you. On the one hand, the isolated opinions may be those of thoughtful iconoclasts who really have something to say. You may indeed learn something from the musings of those independent thinkers. On the other hand, the outliers could be people who have been struggling all semester, who are too timid to get help, who are having personal problems, who have been ill, or who just like to complain. As award-winning teacher Tom Banchoff [BAN] says, "good teaching is not identical with perfect ratings." Instead, if you want to learn something about your teaching and improve it, look for trends in the student evaluations. If ten students say that you talk too softly or too rapidly, or if many of the evaluations say that you are a poor communicator, or if a plurality indicate that you cannot explain ideas at the students' level, then you should consider these criticisms carefully. Examine your conscience and determine whether you can learn something from your teaching evaluations and improve your technique.

If I have succeeded in convincing you that there may be something of value in teaching evaluations (taken as a group, not necessarily individually), now let me send chills up your spine. The article [AMR] describes the following experiment (which I present here in slightly simplified form), that was performed at an American college. Ten instructors were chosen from ten different departments. A random group of thirty undergraduates was assembled. Over the course of several days, each instructor gave a fifty minute lecture to this same group. At the end of each lecture, each student completed a standardized written teaching evaluation form for that instructor. The results were tabulated. Now each lecture was also filmed, and a ten second slice was taken from the beginning, from the middle, and from the end of each lecture. The sound was removed, and the thirty excerpts were spliced together in random order. The result was a five

---

[5]Note that we are unlikely to see pairs of teaching evaluations that differ over whether the professor wore plaid socks. What they will differ on is *subjective matters*, like clarity or level.

minute silent film showing randomly arranged ten second clips of each of the ten instructors.

Next, a hand-picked group of thirty sophomore women was assembled, and they viewed the five minute silent movie. Each young woman completed the same standardized written teaching evaluation form for each instructor—based only on what she saw in the silent movie. The results were tabulated. The startling fact is that the data from the first group of students—who saw the original lectures—correlated extraordinarily well with the data from the second group—who only saw the silent film. What conclusions may we draw from these observations?

It is easy to focus on the flaws of this experiment. You may well ask, "How can a ten second slice show what a professor can really do?" If the professor is in the middle of the proof of the fundamental theorem of calculus, or explaining the more technical aspects of Marx's theory of commodity valuation, then he will not show well in a ten second slice. Perhaps, as in the review of a good restaurant, one should view the professor over several different days. Keeping these limitations in mind, let us see what we might learn from the study.

I think there is a nugget of insight buried in this exercise, and it may be this: Young people do not generally have the intellectual equipment to determine whether their calculus teacher or their genetics teacher really knows his stuff, or is doing an optimal job, or knows how to communicate the material most cogently. But young people have a lifetime of experience evaluating body language and nonverbal communication. The second group of students (the sophomore women) *only had* body language and nonverbal communication on which to base their evaluations.[6] Since their evaluations correlated very closely with those of the first group (the students who experienced *all* aspects of the lectures), one is tempted to conclude that the first group based its assessments also on the body language and nonverbal communication aspects of the lectures.

Cynics like to conclude from this study that students simply cannot evaluate your teaching. They are just reacting to the "vibes" that you give off. If you are hellbent on writing off teaching evaluations as worthless, this reasoning may appeal to you.

An informal study conducted at Cornell University [CEW] performed the following experiment. A single professor taught the same class, to similar audiences, two semesters in a row. The first time around the lecturer evinced a monotone, staid teaching style. The second time around, he showed great enthusiasm for his task, using many hand gestures and vocal diversity. The students reacted far more positively in the second semester than in the first; in fact in the second class they even gave the textbook more enthusiastic praise. What may we conclude from this information? One arguable conclusion is that it is almost impossible for a person to evaluate his own teaching. While the Cornell professor was being monotone and staid, he may also have discouraged student questions and done a poor job of answering those that he did receive. In the second semester, while being enthusiastic, he also may have interacted much more vigorously with the

---

[6]Sophomore women were chosen because, in the opinion of the experimenters, young women are more sensitive to nonverbal communication than are young men.

students. Even so, the experiment is worth considering. Studies show that students react more to the instructor's enthusiasm than to anything else. Perhaps the Cornell experiment only serves to affirm that finding.

Students are not scholars. But they are people. They probably react to human input more than they react to intellectual input. That is just a given, and it is not likely to change. Even so, what they have to say has value. If a freshman says, "This instructor doesn't know what he is talking about." then you have reason to be skeptical. If instead that freshman says, "This instructor cannot present the material in a manner that is accessible to freshmen." then perhaps he is giving you information that you could not obtain in other ways. A student certainly has something of value to say about whether he feels he is learning something; about whether the instructor can explain the material. You should give the matter careful thought so that you can put your teaching evaluations into context and learn something from them.

There are objective studies that suggest that student evaluations of a given instructor are consistent over time, that they correlate well with administrative and peer review, that (taken as a whole) they are independent of extraneous student characteristics, and that they correlate significantly with how much students actually learn (see [CEN], [COH], [FEL1], [FEL2], [HCM], and further references in [NAR]). The paper [KULM] offers statistical evidence that student teaching evaluations measure (*accurately*) (**1**) instructional skill, (**2**) respect and rapport, and (**3**) instructional organization (remember that these are *statistics*; they do not speak of any one particular teaching evaluation form, but of the overall content of the aggregate of teaching evaluation forms). This information may or may not appeal to you. But there it is.

I repeat: It is a noteworthy observation that student teaching evaluations correlate well with peer evaluation (see [DAV]). They do *not*, however, correlate well with self-evaluation. On days when you are sitting around badmouthing your teaching evaluations as immature and worthless, bear this thought in mind.

In my experience, teaching evaluations (again taken as a group) have empirical value. Obviously, what information you glean from your teaching evaluations is up to you. If a significant number of your students say that you make too many mistakes in class, or you ridicule people who ask questions, or you don't prepare, or your tests are unfair, then I think it's a cop-out to just claim that they are reacting to your vibes. After all, this is your audience giving you feedback. Why not try to learn from it?

On the other hand, if a large number of your students say that you are inspiring, well-informed, creative, and an excellent pedant, then don't just say, "Aw shucks" and forget about it. This is positive reinforcement. You must be doing something right!

Bear in mind that the evaluation of teaching need not take place only at the end of the semester. Mid-semester evaluations can be extremely useful. They tell you how the course is going, whether the students perceive that they are learning anything, what problems may have arisen.[7] It is one thing to get your

---

[7] Note that, at many institutions, midterm teaching evaluations can be just between you and the students. Nobody else need see them after you have collected them.

evaluations at the end of the term and say to yourself, "Boy, there's something I did wrong. I'll try to fix it next time." It is quite another to realize after only half the course has gone by that corrections must be made *while the course is in progress.* For then you have an opportunity to make things right, and your end-of-term evaluations should improve as a result. The book [GOL] offers a number of insights on course evaluation techniques.

Some math faculty are uncomfortable with the standardized teaching evaluation forms that the university provides. Usually these forms ask the student to answer twenty or so questions with a rating of "1" to "5" to indicate "Poor", "Fair", "Good", "Very Good", "Excellent". While some of the questions are about reasonable issues (such as "Was the text useful?", "Was the instructor prepared?", etc.), others are rather vague ("Evaluation of the Instructor Overall" and "Evaluation of the Course Overall"). It is a fact that the dean is a busy person and wants a quick and dirty estimation of the teaching of any given faculty member. Often the numerical responses to these last two questions can give such a gauge, but it is not a gauge with which the faculty member being examined is comfortable.

You may feel, with some justification, that the evaluation process described in the last paragraph does not give the mathematics professor an adequate chance to show what he can do. If this describes you, then I encourage you to discuss the matter with your colleagues, your chairman, and your dean. The deans I have met are not averse to individual departments developing their own methods of teaching evaluation. Possibilities to consider are

- Videotaping of lectures

- Peer review

- Self-evaluation

- Consultation with other faculty experts

- Exit interviews by a third party professional

Let me give a brief explanation of each of these techniques.

(1) Few things are more honest, and sometimes demoralizing, than a videotape of a lecture. The videotape will show all your awkward mannerisms, your squeaky voice, your dandruff, your strange pauses and facial expressions. A lecture that you thought was ethereal will come across as peculiar when first viewed on videotape.

If you agree to have yourself videotaped, then view the tape with an experienced faculty member who can point out both what is good and what needs improvement. You will need some help in keeping the matter in perspective.

(2) Peer review consists of having some of one's colleagues attend some class lectures, and perhaps also review some class materials that were prepared, by the person being reviewed. Most intelligent people feel comfortable receiving cogent remarks from someone whom they respect and admire. Why not learn about teaching from a fellow mathematician?

(**3**) Self-evaluation might consist of the candidate preparing a "Teaching Portfolio". This portfolio would contain an enunciation of the candidate's overall teaching goals, plus a list of particular goals connected with particular courses that the candidate is going to teach. A mentor would review the portfolio regularly with the candidate, and help him assess whether he is achieving his objectives.

(**4**) Some people are shy, or uncomfortable in front of groups, or are poor speakers. Notable success has been had by having such individuals work with experts from the Communications Studies or Speech Department. While it may be uncomfortable to have a fellow mathematician tell the candidate that his teaching is inadequate, it may be more natural (akin to going to a fitness trainer or a podiatrist) to have the candidate consult with a well-meaning faculty member from another department. And I can tell you that this is a method that works in practice.

(**5**) I rather like this last method for evaluating teaching, but it is expensive both in terms of time and in terms of money. The idea is that a professional interviewer, perhaps someone with a background in psychology, will interview each student at the end of the course. The interview can be brief—perhaps ten minutes. But the interviewer can ask questions that will draw out the student's concerns. He can also zero in on important points that the student is trying to articulate and help him to develop them. At the end of the interview process, the interviewer will write up an in-depth report on the class, and the instructor, in question.

Once you start thinking creatively about ways to evaluate teaching, you will certainly develop ideas of your own. Bear in mind, as you do so, that the dean has an affection for "Evaluation of the Course Overall" and "Evaluation of the Instructor Overall" because these simple questions give him two numbers. He can quickly assess whether a given candidate cuts the mustard or not. When you devise alternative assessment techniques, be sure that each one results in useful and accurate advice for the teacher being examined and also in a *quick and incisive* take on the candidate's teaching abilities. The dean does not have the time to view videos, or read long position papers. He may not insist on a *number*, but he needs the evaluation to be of the nutshell variety.

2.10   Exams

In this section I will discuss how to compose an exam, how to formulate questions on an exam, how to judge the length of an exam, how to grade the exam, and so forth.

I will also discuss larger issues: **(i)** How much should you tell your students about what is on the exam?, **(ii)** How should you handle student questions about how the exam was graded?, **(iii)** How comprehensive should you make your exam? I will also discuss, in some detail, the question of whether exams should be multiple choice or of the (more traditional) written-out variety.

Let me state my thesis quite plainly. Handwritten exams, in which students write out complete solutions to stated problems, are good. Multiple choice, machine-graded exams are not so good. Of course nothing is black and white. Handwritten exams have their down side and multiple choice exams have their up side. The relevant issues will be developed as the section unfolds.

In most elementary math classes (and many advanced ones) the principal device for determining grades is the examination. These are usually (but not always) given in class, or during a special time slot in the evening. There are a number of points of view about what constitutes a good exam.

Some professors attempt to put together elaborate exam problems, each of which synthesizes several of the concepts introduced in the course. This practice causes me to pose some questions which you should ask yourself frequently when you teach or write: "Who is my audience? Am I trying to teach eighteen year olds or am I trying to impress myself? Am I trying to effect an educational experience? Or am I trying to put together an exam that I can show to my cronies while crowing about how dumb it proves the students to be?"

By contrast, there is the "minimalist" exam. A famous old exam from MIT consisted of the single problem

> You have a pile of warm metal shavings in the shape of a cone. Discuss.

There's a conversation stopper. On the other hand, a notable instructor at that same venerable institution for many years formulated final exams as follows

> There are fifteen important concepts in this course. Discuss any thirteen of them, outlining key ideas and providing proofs as time permits.

These types of exams may be suitable for certain students at MIT some of the time. They are not appropriate at most universities today most of the time.

My practice is extreme in yet a third direction. I usually tell my students what will be on the exam. No, I don't write each exam problem on the blackboard during a review session. But if a student asks, "Will we be tested on the chain rule?", I give him an honest answer (with the understanding that if I say "yes" then this should be construed as "maybe"). If the student says "How many problems are on the exam?" then I tell. If a student wants to know how many questions are multiple choice and how many not, I give. To deny this information is just power tripping. It serves no good purpose.

To be honest, 95% of my exam questions (in an elementary course) are straightforward. They offer no surprises. They are similar, but not isomorphic to, homework exercises. With the other 5% I am more fast and loose. I use these as a vehicle to identify the really bright and able students in the class.

I know good teachers at first class universities who take the straightforward approach one step further. They have a blanket policy in all elementary classes (calculus and linear algebra and ODE, let's say) that *all* exam questions come directly from the homework. Literally. And they announce this on the first day of class and repeatedly throughout the course. It's an interesting policy. They tell the students exactly what will be on the test (in a sense), but on the other hand they really don't. This policy leaves students little room for complaining about the content of exams. On the other hand, it does not challenge them. And it encourages them to memorize (and perhaps to cheat!). Use this policy with caution.

Exam time is when you really have the students' attention. Get as much from it as you can. Drive home the important ideas of the course. Give a thumbnail sketch of the evolution of these ideas a few days before the exam. Such a review helps students to organize their thoughts.

Your exams are one of your most important tools for communicating with your class. The students may be at only half mast during some of your classes. But at exam time they are giving you their full and rapt attention. This is your big chance to tell them what this course is about, and how they are doing in it. There is no sense to use your exams as a device for alienating the class, and there are so many ways in which you can do so. If you are consciously going to give your students a killer exam then you should ask yourself *why* you are doing so. What are you trying to accomplish? Whom are you trying to impress? Consider carefully before you give such an exam. If the class is already dead then giving a hairy exam will pound the final nail into the coffin's lid. If the class is instead on your side, then why make a conscious effort to drive the students away?

Put another way, the purpose of a class is to transmit knowledge and information. Any given class has a dozen or more key ideas in it. That is what the tests should be about. A midterm or final exam in a basic course should not be a repository for ancillary theorems. It should not be a forum for obscure results not covered in class, or touched upon only in passing. An exam should be about the *principal topics* in the course—ones that you have emphasized and illustrated and repeated (*ad nauseum* if necessary). Topics covered on the exam should be ones that the students have heard about in class and seen in the homework (see also Section 1.4 on clarity).

Make sure that the questions you ask elicit the basic information that you seek. If your question about the chain rule turns into an algebraic morass, then it does not test the students about the essential material that they are to have mastered. If your maximum-minimum problem involves arithmetically or algebraically complex expressions that obscure what is going on, then you are not really testing the students as you wish to do. Thus it is important that you, the instructor, work the test problems through in advance. This takes some time, but less time than all the aggravation that ensues if you give a poorly formulated or carelessly prepared exam.

Multiple choice or show the full solution? There are arguments for and against both systems. From the professor's point of view, one argument for multiple choice is that the grading of the exams requires no effort (in many cases it can be done by machine). And the exam is completely objective. But these reasons are a bit self-serving, and there is another more interesting consideration.

If you give traditional exams on which students write out solutions to the problems, then you usually fall into the malaise when grading of giving a lot of partial credit. Since you are human, you may tend to give even more partial credit on the 75th examination paper than on the 5th. The upshot is that it is actually possible for a student to get through the entire calculus sequence, with a grade of "C" or better, not knowing any particular calculus technique in its entirety. By contrast, it can be argued, the multiple choice exam has the advantage of requiring the student to actually *get to the correct answer* on a number of problems. But there is more to mathematics than just getting the correct answer. So you must consider to what extent your multiple choice exam is exposing students to the wrong value system.

On the other side, it can be argued that multiple choice exams involve a lot of gamesmanship. A student who has not studied, but who is clever, can sometimes get a reasonable grade on such an exam just by guessing shrewdly. (Of course you can offset this feature with negative scores for wrong answers. Also, if you give about ten possible choices for each question, and if the exam is otherwise well constructed, then you can make this eventuality unlikely.) It can also be argued that it is easier for students to cheat on a multiple choice exam.

I think that a more serious point about multiple choice exams is similar to the liability of large lectures. They don't do a good job of engaging the student in the learning process (see Section 3.5). A handwritten exam is a form of discourse between the student and the teacher (Section 3.14). The student writes his thoughts, the teacher evaluates those thoughts, and the student (ideally) learns from the exchange. A multiple choice exam is more like getting money from an Automatic Teller Machine. The job gets done, but no nurturing or growth occurs.

You also have to ask yourself which type of exam really tests the students on what they should be learning. Are they learning problem-solving skills? Are they learning the key ideas? Can they state the theorems? Can they prove them? Do they understand the definitions? Can they reproduce, with comprehension, the important examples? It seems patently clear that a written-out exam can do all of these. A multiple choice exam would be less informative in almost all instances.

A common student complaint about multiple choice exams, and one which I find difficult to gainsay, is that the student can do a problem almost completely correctly but have a small arithmetic slip, with the result that he cannot find the correct choice among those given. If, instead, the exam had allowed the student to submit his full answer for reading by the instructor, then the student would no doubt have received substantial credit. Instructors will argue that students should learn to be accurate. A small arithmetic slip will cause the bridge to fall down or the brain surgery to go awry. I imagine that the same professors would not expect their next scholarly papers to be refereed with this thought in mind.

Perhaps especially critical these days is that multiple choice exams do not appear to be a good vehicle for training students to do multi-step word problems. This is one aspect of mathematical training in which American students lag behind students in Japan and other countries. A well-crafted written-out exam can walk the student through six or more steps, beginning at square one and ending with the solution of some really interesting problem or phenomenon. This can be done with a multiple choice exam too, but it is much trickier to pull it off.

If I am teaching a large class (200 students or more) in which a hand-graded exam is infeasible, then I find it useful to compose my exams as follows: If there are twelve problems on the exam then ten of them are multiple choice and two are "short answer". The short answer problems are of the sort that I can grade instantly—just by glancing at them.

The students in large classes that I have taught are comfortable with an exam that is primarily multiple choice. But they appreciate the personal touch suggested by a couple of short-answer problems that are graded by hand.

It seems to me that in a small class (60 students or fewer) the professor can write a traditional exam requiring full answers to questions and then spend some time grading the papers carefully. In this context you can not only attend to the grading yourself but you can make constructive comments. These comments can be brief, and they can be encouraging. The serious students do read them, and do benefit from them.

I have presented arguments in favor of machine-graded multiple choice exams and also arguments against them. Once again, I shall be prescriptive: Hand-graded exams are better. They keep you in touch with how the class is doing as a whole, and also with individuals in the class. They give you the opportunity to discern what topics require additional coverage in class. Your comments on the exam are a useful part of the teaching process. If it is at all feasible, even in a class of eighty or more students, endeavor to give traditional hand-graded exams (or at least an exam that has a hand-graded component).

It is tempting, especially for new instructors, to hold review sessions for exams. This is a way of making yourself feel generous, it is easier than doing something more productive, and it will make the students grateful. But it also makes exams seem more onerous than they really are. (If you do decide to hold a review session anyway, then read Section 2.16 on problem sessions.) And it makes the students who cannot attend the review session feel as though they are at a serious disadvantage. I find it more useful to write a practice exam that I distribute a week in advance of the real test. About two days before the test I post solutions to the practice problems (either on the class Web page or on a bulletin board or both). Of course there is always the danger that students will think that first reading the practice problems and then reading your solutions will constitute studying for the exam. I always caution the students strenuously against this trap. No system is perfect.

Tests that are too long, or too involved, do not work. Your exam should contain a reasonable number of questions of reasonable length, and they should not be inter-linked. If problems are interconnected, and if a student makes a critical error in one of these, then all of the related problems are affected. If test problems are too involved then students can panic, mismanage their time, and turn in a performance that does not at all reflect their true abilities.

Master teacher Tom Banchoff [BAN] recommends the following technique for dealing with student panic on exams. He gives regular, 50 minute, in-class exams—as we all do. But each student has the option of going home and re-working the exam at leisure to show what he really knows. Banchoff takes both performances into account when he does the end-of-term grading.

By the same token, it is sometimes appropriate to give a "take-home" exam. You will have to decide whether the particular class you are teaching can be trusted with such an exam. And then you will have to lay down some ground rules. Open book or closed? Timed or not? Can consult other people or not? A take-home exam gives the students an opportunity to really show what they can do. But it has many unmanaged aspects that can lead to trouble.

It is very easy to misjudge a test that you write. A problem that seems trivial at first blush may have complex arithmetic or algebra hidden in it. Thus *you must personally work the test out completely before you give it to your class.* An exam that you can do in twenty minutes—with all solutions written out neatly—is probably about right for a 50 minute exam for a class of freshmen. If it takes you 40 minutes, and you find yourself laboring over the algebra or arithmetic, then obviously this is not a suitable 50 minute exam for freshmen.

The point value of each exam question should be clearly exhibited on the exam. The total number of possible points on the exam should be displayed. It is tempting to make difficult problems worth a lot of points and trivial problems worth very few. But of course the end result, since many students will not do well on the hard problems, is that the class average is pushed down. On the other hand, you don't want to make the easy problems worth a lot of points and the hard ones worth just a few—this sends entirely the wrong message to the class about what is important. So you must strike a balance.

It is a useful device to break difficult exam questions up into steps. This practice helps the weaker students to get started, and to display what portion of the material they actually know. It also makes the exam easier to grade, and increases the consistency of your grading.

When you are grading exams, it is important to be as consistent as you can be. Begin by writing out the solution to each problem. Break the solution into pieces and assign a point value to each part. Thus, in a maximum-minimum problem, setting it up might be worth 3 points, doing the calculations another 3, and enunciating the answer another 3. One spare point for overall analysis makes a total of ten. Some instructors like to be even more precise than this. Refer to Section 2.14 for the concept of "horizontal" grading for insuring uniformity.

Remember that some students, the day the test is returned to them, will come to you with questions about how their individual exams were graded. In some cases, they will come with a friend and ask why two similar solutions were graded differently. If you are systematic, then you can handle such transactions with dispatch.

Should you write your exams out in longhand (with a pen), or should you word-process or TeX the exams? The obvious advantage of word-processing or TeXing an exam is that it is then lovely to view, all the characters are legible, there is virtually no chance that a student will misconstrue a problem for having misread what you wrote, and the exam has a professional appearance. Many mathematics departments keep an archive of all exams (especially finals, but sometimes even midterms) in order to handle student complaints, to give

guidance to future new instructors, to keep tabs on the faculty, and generally to
have a paper trail of what is going on in the department. Clearly a file full of
carefully typeset or word-processed exams gives a much better impression than
a file full of hastily scrawled longhand exams. But there is a flip side to this
picture.

Writing an exam on a computer entails special formatting problems. It is
difficult with a word-processor, and agonizing with TEX, to render an exam
in the usual format—with a lot of indenting, alignment, vertical spaces, hang-
indented material for point values and numeration, running heads, and so forth.
It is also extremely easy to introduce inadvertent mathematical (and other)
errors when keyboarding an exam. If figures or annotations or other pixilations
are needed then you will likely be tempted to throw your keyboard out the
window. Whereas writing an exam by hand could take an hour, it often happens
that writing an exam on a computer could take several hours. Thus many an
instructor—including yours truly—finds himself writing his exam with a *pen* by
hand. When I do so, I make strenuous efforts to write neatly and clearly. I
proofread meticulously. And I save a lot of time.

In this instance I am not going to make a crisp recommendation. Word-
processed or typewritten exams are formally more attractive, and look more
professional. But they are time-consuming to write and prone to error. Proceed
with caution.

Now back to the trenches. When teaching a big class, it is best to generate
some statistics about each exam that you give. When you hand an exam back
to 200 people and someone asks, "What is the average?" or "What is the cutoff
for an 'A'?" then you had better have an answer ready. The alternative is chaos.
Therefore consider calculating the mean, the mode, and the median (if you don't
know these words then look them up). Calculate the standard deviation and use
it as a guide in setting up your grading curve. Draw a histogram. When you
are explaining to a student how the exam was graded, such statistics are a great
help.

Incidentally, hand exams back at the *end* of the class period. For if you
return them at the beginning of the hour then students will spend the period
reading the exam and comparing grades rather than listening to your lecture.

Hand back exams just as soon as possible after the exam is given—in the
next class period if possible. If you procrastinate, and do not return the exams
to students for a couple of weeks or, worse, until right before the next exam,
then much of the didactic value of the exam will be lost. Students will have put
that material on the back burner while they are learning new material, and their
overall interest in the exam and its contents will have waned. If you can return
exams in the very next class, then you can bring that portion of the course to
closure and move confidently into the next body of material that you must teach
your students.

It seems natural to spend the class period following an exam actually working
the exam at the board. Let me tell you decisively that this is not a good use of
time. First, students resent your implicit statement that "Look at me—unlike
you, I can do the exam quickly and easily." Second, what each student really
cares about is how *he* performed on the exam. If a student did a problem

incorrectly, it is possible that he will want to see you give the correct solution. Otherwise interest is minimal. Best is to make the written-out exam solutions available to all class members (in a posted copy, or in a hard copy that you distribute in class, or on the Web). Always insist that, before a student comes to you to complain about how a problem is graded, the student must first read your solution. You cannot imagine how much time and travail this practice will save.

You also need to get on with the course you are teaching. Make it clear to the students that you welcome their comments about the exam. Make yourself available during office hours, and even have extra office hours for discussing the exams if appropriate. But don't waste valuable class time hashing over an exam that has been taken and graded. Such an exam is basically a dead issue, and it is time to move on.

Generally speaking, it is best to try to deal with specific student questions about the way a specific person's exam was graded in a private, one-on-one fashion. You should never handle such complaints in front of a class. It is also a bad idea to handle them in front of a group of six after class. This is a personal matter. Treat it like an appointment with a physician. See also Section 2.11 on grading.

2.11 Grading

The pot of gold at the end of the rainbow, from the student's point of view, is the grade at the end of the course. Grading is a multi-parameter problem. The students want to be treated fairly, yet they want to to feel that the course has substance. They want to be enlightened, yet they want (to some degree) to be delighted, to be entertained. They want to respect you (the instructor), but they want to be your friend. There are a variety of devices for making your grading scheme palatable (without being essentially more lenient) to students. What is the most evenhanded and efficient way to determine grades?

I have used a number of grading schemes successfully, and some unsuccessfully. I would like to record a few of the former here—merely for the reader's delectation. My main goal in formulating my grading policies is to make the greatest number of students feel that they have been treated fairly (and, not incidentally, to reduce student complaints). This does not mean that I am a lenient grader, nor that I give away grades for no special reason.

Always tell students on the first day of class, and in your syllabus (see Section 2.12), how you will grade the course. You want this to be a matter of public record. If students complain about your grading practices, and there will occasionally be some who do, then you have your public statements to fall back on. And don't lie. If you say that you will grade according to a certain scheme—with exams worth so much and homework worth so much and so forth—then do so. If you say in your syllabus that you will grade on a curve, then do so. If you say in the syllabus that you have an absolute grading method (90% is an "A", 80% is a "B", etc.), then stick to that.

You may wish to consider in advance how you will handle students who are distraught about their midterm grades. One method that usually works for me in borderline cases is to say to the complaining student, "If these few points really make a difference in your course grade at the end of the semester then let's discuss it at that time." This arrangement usually makes everyone happy, and very few students will take you up on the offer.

If a student comes to complain about a grade, then show the student some courtesy. If you cannot come up with a cogent reason for the way that you graded an exam or a problem, then that is *your fault*. Rethink your grading practices. Never fall back on your august position as your first line of defense. You show the student absolutely no respect by saying, "that's the way I graded your test and I'm the boss." That is not how you would wish to be treated. You can always turn a session of "Why was this test graded this way?" into a favorable transaction. What does it cost you to give the student a few extra points if the points are merited?

However *never* penalize a student for being honest. If the student comes to you and says, "You added up my points incorrectly. I should have received an 87 instead of a 90." then just send the student home with a little praise for being so perceptive. Tell the student that if you ever inadvertently give him too few points then he should feel free to approach you at that time as well.

It is tempting, especially when you are a new instructor, to endeavor to take an "organic" approach toward grading. Students are very receptive to the instructor who says, "I try to grade on a subjective system. If your strong grades are on the midterm exams, then I downplay the homework and the final. If your work shows improvement, then I take that into account when I determine your grade. I try to emphasize everyone's strengths. I am your friend." This approach works well in the short run. It is a good opiate—for you as well as for the students.

One weakness of the organic grading method is that it is intrinsically unfair. A student who does his homework with friends (and who therefore, despite his own personal lack of understanding, gets good homework grades), but who takes his exams alone (and gets poor exam grades because of his lack of true understanding) could still garner an "A" or "B" in the course.

Perhaps a more pragmatic and immediate liability of the "organic" grading method is that, if a student complains about his course grade, then you have nothing to fall back on. You cannot show him your calculations and you cannot show where his score fits on a histogram. The trouble with an intuitive methodology is that you cannot explain it or defend it. Even though it sounds a trifle cold, you are much better off with an objective system of grading. In the end, everyone is more comfortable with a dispassionate approach. And, in the rare event that you have to defend yourself to the chairman, or to an angry parent, or to the dean, or even to a colleague, you will be prepared.

One device that I have used in large calculus classes (see also Section 2.14 for additional thoughts on grading in large classes) is the following: I tell students that I determine their grade by weighting their midterms as 50% of the grade, the final as 30% of the grade, and the homework as 20% of the grade (for example). But the *caveat* that I throw in is that anyone who gets an "A" on the final gets an "A" in the course. This assumes, of course, that the final exam is comprehensive. Thus if a student comes to me during the term and is distraught about his homework grade or his midterm grade, then I can simply enjoin that student to do well on the final. In fact not many students pull their grade up with the final exam (never more than 5%) and this simple device helps to keep morale high.

As with many items in this book, I offer the last tentatively. I have had some of the brighter students complain about this policy: "Why should some jerk who didn't work all semester be able to pull off an 'A' at the last minute just by cramming?" I patiently explain that it is virtually impossible for such a "jerk" to pull off such a miracle, that the purpose of the policy is to help and to offer encouragement to students who have been struggling, or for whom this is the first difficult math class. It also sends the important message to students that what is important is that, by the end of the semester, they ultimately master the material. Some faculty have told me that the skewed value system that the just-described policy implies is sending a counterproductive message to the students. My primary motive in formulating the policy was to give the students hope, and to quell their misgivings and their fears; and not least I wanted to minimize their complaints. In a substantial course—say real analysis—in which it takes time for the students to internalize the ideas, this policy helps students to show (integrated over time) what they have ultimately learned.

Sometimes you must change a student's grade—either on an exam or in a course. Perhaps you made a clerical error in recording the grade, or you made an error in grading a problem, or you were the victim of any number of other human frailties. Do not be afraid to change a grade when it is merited. *However:* You do not want to develop the reputation among students as an instructor with whom grades can be negotiated. I've had this rep, and I don't know how I got it. But there was a time when, the day after an exam, 85% of my students lined up outside my office to take turns slugging it out with me—point by point and problem by problem—over their exam grade. I felt at times as though I should buy stock in the Kleenex Company. This process is unpleasant and (can be) degrading both for you and for the student. Doing a careful job of grading in the first place, and posting carefully written solutions for students to see, can help to assuage much of student discomfort with grades.

Make the student read your (posted) solution before you agree to talk about the grading of a problem. Many times I, as an inexperienced instructor, have spent fifteen minutes haggling with a student over a problem only to realize that the student had not read the correct solution. Once he read it, his objections faded away.

No matter how fair and ethical and "right" your grading methods may be, the way you grade may run afoul of departmental policies. I know many a mathematician who has painstakingly prepared a grading curve for his calculus course and submitted his grades to the department, only to have the Chairman or the Director of Undergraduate Studies call him on the carpet for not following the "approved departmental grading curve". The usual party line is, "In this department, we recommend that you give 20% 'A's, 30% 'B's, 40% 'C's, and 10% 'D's. We discourage 'F's." Of course these specific figures are manufactured, but the scenario occurs all the time—probably more often at large state institutions (subject to various public pressures) than at small private ones. This is no laughing matter. Many math instructors will just go along with the chairman's wishes, simply because it is easier than bucking the tide. Others will stand on their right to autonomy in their own classroom.

This is a tough ethical call, and I cannot tell you what is right. You have to live in your department, with the colleagues and the policies that it has. Probably the best advice I can give is that you should find out what the departmental policies are *before* you begin to teach. If you think that you are going to have trouble living with the grading policies, then discuss them with departmental honchos and try to work out a position that everyone can live with.

It seems to be an increasingly common occurrence (see [WIE]) for a student to come to the instructor after the course is completely finished and say, "I got an 'F' in your course. Could we talk about how to raise my grade?" This is like buying a car, signing the papers, making the down payment, driving the car home, and then coming back a week later to see whether you can renegotiate the sale. It makes no sense.

A grade is supposed to be *an evaluation of the work that the student performed during the term.* When the grade is given, the work (or lack of it) is a done deal. The very notion that this is a point for haggling is a genuine travesty—it shows a true misunderstanding of the university's mission.

At a prominent university in St. Louis—not my own—there has developed a new process that is called "grieving". A student who receives a disappointing grade in a class will say "I'm going to grieve this grade." There is a dean who is in charge of such matters, the student makes an appointment with that dean, and then the young scholar puts on a dog and pony show to convince said dean that the grade is unfair. Then the dean changes the grade! Without consulting the instructor of the course!! Sadly, the practice of grieving amounts to an institutionalization of the wretched behavior that I described in the last two paragraphs. The institution in question is funded strictly according to its enrollment, and it takes great strides to see that its customers are happy. While I sympathize with the school's plight, I certainly do not endorse its practices.

2.12   The Syllabus (and the Course Diary)

Every mathematics course should have a syllabus. The teacher should give the course a little thought and planning before classes begin. What is the text? What will be covered? What are the prerequisites? How many exams will there be? How will the grade be determined? What is the instructor's name, office number, phone number, and office hour?

The syllabus should be in outline form—not paragraph form—and display essential information so that it is easy to find. A sample syllabus follows.

Syllabus for Math 411
Real Analysis

Course Description: This is a rigorous course on the foundations of mathematical analysis. Topics to be covered include set theory, logic, the real number system, sequences and series (both of scalars and of functions), compactness, topology of the reals, approximations, differentiation. We will cover at least the first five chapters of the text.

Course Prerequisites: Calculus and linear algebra.

Instructor: Steven G. Krantz

Office: 103, Cupples I

Phone: (314) 935-6712

FAX: (314) 935-6839

Office Hours: To be announced. Consult the course Web page.

Course Web Page: `http://www.math.wustl.edu/~sk/math411`

Math Department Office: 100, Cupples I

Text: *Principles of Mathematical Analysis*, 3rd edition, by Walter Rudin.

Class Meeting Times: MWF 1-2

Classroom: 115, Cupples I

Exams: There will be two midterm exams and a final. Exams will be scheduled by the university. Consult the class Web page or the department office for information.

Quizzes: There will be weekly quizzes. Schedule and format to be announced.

Homework: There will be regular homework assignments and these will be graded. Late homework is not permitted. I will drop your two lowest homework grades to allow for missed assignments.

Grading: The components of the course are weighted as follows:

Homework:	10%
Quizzes:	15%
Midterm I:	20%
Midterm II:	20%
Final:	35%

This is a "bare bones" example of a course syllabus. It contains the critical information that should be a subset of any syllabus. Your syllabus may contain more information about the course topics, or about the text, or about homework assignments, or about grading.

It is only courteous to provide students with the information that has been indicated. The syllabus is also a paper trail for your course. Try to stick to your syllabus as much as possible. If a student comes to you in the seventh week and says, "I didn't know that there would be two midterms." or "I didn't know that homework was such a big component of the course." then you can point out that these particulars were explained in the syllabus that you distributed on the first day of class and that has been posted all semester long. The syllabus serves as a sort of contract between you and the class. It keeps you honest and it keeps the students honest.

The syllabus should not be a *magnum opus*. In some large calculus classes, especially when there are several such classes being coordinated, instructors may find it useful to list the topic for every class period, relevant pages from the book, and the homework problems that are assigned for each day. That is fine in its place. For most courses, a syllabus of one or two pages is more than sufficient.

Ideally, the syllabus should be available in a stack outside your door, or perhaps in the undergraduate office, all semester long. If you set up a Web page for the course—and this is an excellent idea—then you should post the syllabus (together with all homework assignments, exam times, solution sets, etc.) there. This practice is just good business, like a restaurant posting its menu or a gas station posting its prices. There is admittedly a certain cachet to being completely disorganized and doing everything by the seat of your pants, but it doesn't pay. You end up wasting a lot of time covering your tracks, you create too many potential opportunities for aggravation, and it leaves students with a bad taste in their mouths.

For the same reason that you should have a course syllabus, I would also recommend that you keep a course diary. This device could be several sheets of paper that you tape inside the front of your copy of the text (and you should tape a copy of the syllabus in the same location), or it could be a separate little notebook that you keep together with your text and your grade book. Any time a student requests a makeup exam, or asks for an extension on an assignment, or whenever anything comes up in connection with the course—put it in the diary! When you and the class decide on a date for the midterm, put it in the diary. If you set up a review session, put it in the diary. If you are going to be out of town, and Professor Veeblefetzer has agreed to cover for you, put it in the diary. That is to say, make a dated entry with the item that you need to remember. It is even reasonable to put all appointments with your students in the course diary. That way you'll have a record of all transactions pertaining to the course, and you also will be less likely to forget them.

2.13   Choosing a Textbook

Choosing a textbook sounds like a simple motor skill. It is not. As a mathematician, you are trained to be rather self-indulgent. You may tend to choose a text that pleases you. If there is a cute new proof of Stokes's theorem, or if there are four color computer-generated graphics, or if the book sales representative takes you to a particularly nice place for lunch, then you are liable to be influenced.

Remember that, while teaching the course, you will not spend nearly as much time reading the book as will the students. Try to step outside yourself as you look at a text. Ask yourself whether the students will be able to understand it, and whether they will be well motivated to try to do so. Are the definitions well-formulated and the theorems clearly stated? Are important topics easy to find? Are the Table of Contents and the Index accurate and thorough? Are the examples well chosen, sufficient in number, and do they order from the simple to the complex (this device is known as "stepladdering")? Are the exercises suitable, and well-coordinated with the text, the lesson plans, and the examples?

There is a great temptation, especially in an advanced course, to choose a text that in your own mind is a "standard reference". But a *reference book* is not a *text*. Try to see things from the students' point of view. The text for your course should not be one of those books that is a pleasure to read *after* one has mastered the material. It should be a *text*, for someone who does not yet know the material.

If you choose a poor text, you will pay for it throughout the semester. Homework assignments will be difficult to design. They will be difficult to grade as well. Students will not perform well on exams if they do not have a good book to study. If the text is weak then the correspondence between it and your classes will be tenuous. This dilemma defeats class morale.

I will now give an example of a problem that exists in the current crop of calculus texts in this country. What does it mean to say that a function is discontinuous at a point $x = a$? Some calculus books state that a function is discontinuous at any point at which it is undefined. For example, the function $f(x) = 1/x$ is discontinuous at $x = 0$. But wait, it gets worse. Consider the function $g(x) = (x^2 - 1)/(x - 1)$. On the face of it, this function is undefined at $x = 1$ and hence discontinuous there. After division, however, the function becomes $x + 1$ which is both defined and continuous at $x = 1$. But are $(x^2 - 1)/(x - 1)$ and $x + 1$ the same function? Who knows? And beware: If you declare a convention different from that in the text, many students will become confused and some will be lost.

This issue of continuity is but one instance of the sort of mess that a textbook can get you into. Some people never use the first edition of a text because they reason that **(i)** all misprints will not be weeded out until the third edition and **(ii)** glitches like that described in the preceding paragraph will be mollified (in response to user protest) in later editions. In point of fact, the conundrum described in the last paragraph occurs in the third edition of a very successful calculus text. The (very famous and wealthy) author insists that this is the only way to handle continuity at a point where the function is undefined.

Let me expand on this last point. If a textbook uses notation or other conventions that you do not like, then don't use that book. You really are obliged to follow the notation and definitions and other paradigms in the text you have chosen. Otherwise all but the gifted students will be lost. If you repeatedly criticize the text as the course proceeds, then you will be sending a confusing message to the students: Why did you choose this book if it is obviously so full of flaws? Isn't it your job to select a text that you can teach from?

Now you can hardly be expected to read and digest every word of a text—especially one as cumbersome as a calculus book—before you use it in a class. The safest policy when selecting a text is to consult someone who has already used it. Don't be afraid to seek the advice of a more experienced faculty member when selecting a text—especially if it is for a class that you have never taught before.

Recently I used a text—in a multi-variable calculus class—that was a disaster. First, it was too formal. It stated results in a fashion that was far too abstruse. As an example, it defined multiple integrals on any spatial region whose boundary has "zero content". Do you know what that means? If you do then you must be a geometric measure theorist, for most people don't. Also, the exercises in the book had a great many small errors. Remember that young students read mathematics quite literally. They are unaccustomed, as are you and I, to making plausible adjustments to what they read. They do not readily generate error-correcting code. Perhaps the worst offense of the book is that I would encounter some strange piece of terminology on page 257, but could not find the page where it was originally defined. This irritation occurred repeatedly, because the book had a very poor index and because usually these obscure pieces of terminology were defined in some optional exercise. I was terribly embarrassed by the albatross that this book turned out to be for a very long fourteen week semester. I should have asked one or more colleagues for advice on which text to choose. Or I should have read a sample chapter before I adopted the text. Instead I chose the book because I know one of the authors and he is a nice guy.

When you are selecting the text for an undergraduate course, then *do just that*. Pick one book—not more—and set it before the students. If we want our students to develop the scholarly frame of mind and the discipline of which we so often speak, then we must give them a hand. We owe it to our undergraduates to put a single book before them and to teach them to read it. To teach an undergraduate course by first dipping into this book and then dipping into that book may be charming and lots of fun for you, but it just confuses the students. Please don't do it. (*Caveat:* Philosophy courses and literature courses and history courses are different. Of course a "history of western philosophy" course will have many sources. Mathematics is a much more compact and directed subject. Therefore a single, well-chosen text is usually just what the doctor ordered.)

It is a revelation for a young student to become completely involved with a book, to realize that—when he wrestles with page 200—he can go back to page 100 (which he has already mastered) for help. You assist this process by giving the students *a single book* and helping the students to plough through it. Choose that textbook carefully, for you will have to live with that book too—for an entire semester! You will have to fashion your lessons after the book, solve the problems, create the exams from the book, answer student questions, and so forth. In choosing a textbook you are defining your life for a non-trivial period of time. It is a bit like setting up housekeeping with the author—and you want to choose a partner who will not do you wrong. Choose wisely and well.

And be sure to select a text that was written for the students in your class, and for the sort of students that attend your university. You can teach calculus from the famous text of Courant and John [COJ], or from the amazing book of Spivak [SPI], but you had better have some pretty clever students if you are going to pull this off. Today there are calculus books at every imaginable level, ranging from *E. McSquared's Calculus Primer* [SWJ] (which is in fact a comic book about calculus) to very rigorous and advanced texts such as the two just mentioned. There is a similar, although perhaps less well-populated, range of texts for almost any undergraduate course. Give some careful forethought to what sort of students will be in your class, and whether they will be able to read the text that you choose.

Each textbook has its own *gestalt*, reflecting the author's view of the subject, and of the way that it ought to be learned. By the end of the semester you will have a pretty good idea what that *gestalt* is, but it is quite difficult to discern it during the brief perusal that will most likely inform your textbook choice. This is yet another reason why consulting an experienced colleague is a useful part of the textbook selection process. Let me discuss just one example to illustrate my meaning.

There is a noteworthy mathematical textbook series written by John Saxon. Saxon has actually appeared on the television show *Sixty Minutes* and wept openly because—if schools would only adopt his textbooks—the mathematics abilities of American students would be greatly enhanced. The hallmark of the Saxon books is that they have no chapters. The calculus book, for instance, has more than one hundred sections, each about three pages long. In any given section, the first page reviews what has come immediately before and the last page previews what is about to come. The lone middle page of the section contains a new little nugget of knowledge.

Can you see what might be wrong with a textbook of this sort? The answer— I think—is that the text is constantly holding the student's hand. The student has little need for retention, because the book is constantly reminding him of the old ideas, and only introducing new ones in minuscule quantities. The series, although well-thought-out and well-intentioned, has not been a success. As an experienced instructor, I can with little effort perform this sort of evaluation of a new textbook. If you are new at the game, then you will want to cultivate this sort of critical skill. While you develop it, consult experienced colleagues to help inform your textbook choices.

I once spent an evening fiddling with the CD-ROM version of a popular calculus text. It was a technological marvel. Using this disc the reader can, if he wishes, read each page (with black characters on a white background), much as though he were reading a hard copy text. Key words are hyperlinked—if the reader forgets what "local maximum" means, or what a "derivative" is, he can click on the highlighted term and a box will pop up reminding him of the main ideas. If that precis is not sufficient, then another click on an icon will move the user into a workbook. Another click will return the user to the place in the text at which the reading was interrupted. The reader can record marginal notes, and these are each marked (for future readings) with an icon that looks like a pencil on a notepad. Amazing. When doing exercises, the reader can click on any particular exercise and the solution will appear on the screen *instantly*.

I fell asleep that night with visions of sugar plums dancing in my head. I woke up in the morning convinced that I had seen the end of Western civilization as we know it. If a student can click on "derivative" any time he wishes—to remind him what the concept means—then I bet that he will *never* actually learn what a derivative is. He will never internalize it, and therefore never really be able to use it. The typical young student that I know, when "doing" exercises in this CD-ROM environment, will proceed in the following fashion: He will glance at an exercise, click to see the solution, say to himself, "Oh yeah, I could have done that," and then go on to the next one. In short, he won't do the exercise and won't learn anything. As things are at present, the student faces the small psychological block of having to turn 850 or so pages and find the answer in the back of the book. I would like to think that that barrier serves as an impetus for the student to try to do the problem himself.

I have twenty-five years of teaching experience that tell me that hard copy textbooks, more or less of the traditional form, work. New, untested teaching technologies make me nervous. Of course we have to try them, otherwise we will ossify and never learn anything new. But we should be aware of the pitfalls, and engage in careful self-evaluation, as we proceed.

And now a coda on cost. It is not impossible these days for a textbook to cost $100 or more. If the book that you choose is expensive, be prepared to defend your choice. Is there an equally good book that costs just half as much? (I'm not talking here about *your monograph* on *your subject*. Rather, I'm considering something like a linear algebra text, for which you should have less emotional involvement.) Checking the cost of the text for your class is just the sort of courtesy that you would have expected your instructor to show when you were a student.

2.14 Large Lectures

At many large state universities, and some private ones, it is common to teach some or all lower-division courses in large lectures. The large lecture situation offers special teaching problems. How can you, as the instructor, make yourself heard? How do you fill the room with yourself? How do you field questions? Do you really want the students to ask questions? If so, how can you encourage the students to do so? What about exams? How do you organize your teaching assistants?

Most large lecture halls come equipped with a microphone for the lecturer. Unless you have a voice like William Jennings Bryan, use it. If you don't use the mike, then either you will not be heard or you will be under such a visible strain that it will detract from what you are trying to accomplish. Like it or not, the instructor in a large lecture is putting on something of a performance.

Of course you are not dancing the macarena, nor are you singing the blues, but you *are* trying to get through to an audience, and to engage them in the learning process. Obviously there will be some technique involved. If your face is beet red, and perspiration is popping out on your forehead, and your armpits are soaked with sweat, and you look like a nervous wreck, then you will not be a success at this job. Learning to use a microphone will help you to avoid these obvious liabilities.

You *can* learn to become relaxed with a microphone clipped to your collar. When you have done so, you will be able to speak in a normal voice and to be heard clearly. You can then concentrate on getting your mathematics across to the students.

One way to develop student involvement in the classroom is to encourage questions (see Section 1.7 on the importance of student questions). The discourse that is built around student questions is a critical part of the learning process. But large lectures impose severe time constraints, and severe communications problems. You must learn to handle questions in your large lectures with care. Too many questions will bring the lecture to a grinding halt. But a good one can make everyone prick up their ears. You have seen Jay Leno and David Letterman walk out into the audience and engage in repartee with select members. Those in the audience who have been selected are usually happy to participate. Unlike our students, the participant in the Leno or Letterman show does not stare at the floor, nor hold his breath until his face turns blue. What is the trick?

It helps that Leno and Letterman are celebrities. Everyone wants to talk to a celebrity. If you think that's all there is to it then you are kidding yourself. Leno and Letterman know something that perhaps you do not. And that is how to handle all different types of people who behave in all sorts of ways. Imagine, for instance, that you have picked out a student and asked him a question and he plainly is embarrassed because he doesn't know the answer. The inept and counterproductive way to handle the situation is to stand there and persist and embarrass the student further. A more productive way to proceed is just as you would proceed in your home if one of your dinner party guests spilled his wine or dropped his potato on the floor:

- Make a joke of it, create a diversion, or pretend not to notice;

- Affect to get distracted and then ask someone else;

- Say, "I must not have formulated the question very clearly" and then try again on the other side of the room.

As any book of etiquette will tell you, the quintessence of politesse is to make everyone feel comfortable—under any circumstances. Never forget to use these same skills in your classroom! Most of the Hollywood movies that depict a college classroom invariably show the professor standing before his students denigrating their intelligence and abilities—and the students lap it up! This may be poetic license, but it is certainly not reality. Show your students the same courtesy that you would have wanted to be shown when you were an undergraduate.

The remarkable book *One L* by Scott Turow [TUR] describes the life of a first-year law student at Harvard. Central to the experience described in the

book is the notion that all first year law classes at Harvard are conducted by
the "Socratic method". Note that first-year law classes at Harvard typically
have an enrollment of 140. In such a class, the Socratic method consists of the
professor, in the first few moments of the class, getting in one student's face and
humiliating him for (as much as) the rest of the hour. And the students take
it! They hate it, but they take it. And the consensus seems to be that it makes
them better lawyers.

The way that Scott Turow describes the first-year law experience, he makes it
seem as though a class of 140 can be like a class of 10. The professor at Harvard
Law plays the class like a harp, and sets the students against one another in
draconian ways. At the end of the book, Turow vilifies the experience. But it is
clear that it has made him stronger.

I certainly do not advocate that you read *One L* and teach your classes
accordingly. Freshman calculus students have neither the proven abilities nor
the determination of first year law students at Harvard. But the book shows
that a large lecture can be taught both powerfully and incisively. There is a skill
to it, and it is one that we can all learn. At Harvard, if *One L* is to be believed,
the catalyst to success is fear. But my experience dictates that other catalysts
work as well. Communication, intellectual inquiry, and discourse are some of
these.

Of course you want to know that your class is alive—that it contains living,
breathing people. One way to do this is to make students comfortable with
asking questions. Once you have created an atmosphere in your class in which
people feel natural asking questions, then you have a foundation that you can
build on. If they are at ease asking questions, then they can move on to making
statements, formulating conjectures, suggesting lines of reasoning, or pointing
out errors in what you have written on the board. It is this atmosphere that I
strive to create in my own classroom, and one that I enjoy immensely.

But there is a trap. Left to their own devices, students will lapse into asking
questions of the rote form, "How do you do problem 6?" Such questions must be
discouraged in any class, but especially in a large class, as they are boring and
non-instructive. If you do get such a question (and if you want to consider and
then to answer such a question), then don't simply turn to the blackboard and
record the solution for all to see. Instead, engage the questioner in discourse.
"Have you tried the problem? How far did you get? Where did you get stuck?
Did any of the rest of you have any luck with this problem? Did you get stuck
in the same place? Does anyone want to suggest a way out of this mess?" It is
too easy for a question like "How do you do problem 6?" to degenerate into a
private discussion between a single student and the instructor. It is also difficult
for the other students to pay attention when you are addressing the needs of a
particular individual. Endeavor to turn an individual's question into more than
what it is. If you cannot use it as a catalyst for a useful classroom discussion,
then tell the student to see you after class.

What you really want are questions like, "Why don't we define 'continuity'
this way?" Or, "Why does the chain rule have this form rather than that form?"
It is up to you to *prompt* the students for the questions that you want. In order
that they not become rhetorical questions, you must put these issues to the class

and then *wait for an answer*.[8] It is not enough to say, "Why does the product rule have this form? Well, here's why." If you want a reaction from your class, you must draw it forth.

When a student asks a question, *repeat it* so that you can be sure that the entire class has heard it (and that you have *understood* it). Write the question on the board. This is sound policy even in a small class. If you do not repeat the question, then the interchange between you and the student becomes a private conversation. The rest of the class is excluded. If other students (those who are sufficiently interested) are interrupting with "Huh? I can't hear. What's the question?" then you are wasting valuable class time and also losing control. Other private conversations will start up. Your class will go badly.

But there is another important consideration to repeating the student's question. If the question is not optimally formulated (or just plain wrong) then the repetition gives you an opportunity to clean it up or rephrase it. Then treat the issue raised with respect and answer it directly.

It is especially important in the large lecture situation that students be sure that you will not belittle them or make them look foolish in front of the other students. Be prepared to make even a dumb question look smart. Writing the question on the board and repeating it out loud gives you a chance to turn the question into something that you can answer, and that will make both you and the questioner feel witty and wise.

If the first question that you field gives rise to a second, then say something like, "Let's go back to the lecture for a bit. I think it will clarify this point for you." Or you can say, "This question session is getting a bit too specialized. Why don't you see me after class?"

There are virtually no data to support the contention that small classes are "better" than large classes for mathematics teaching. Statistics do not indicate that students in small classes perform better or retain more. The statistics *do* indicate that students feel better about themselves and the class, and enjoy the situation much more, if the class is small. Another way to say this is that large lectures are not a good tool for *engaging students in the learning process* (see Sections 3.5, 2.10). Just as nineteenth century social theorists noted that factory workers on an assembly line easily can become alienated from the work process, so students who are educated in a large class (that smacks of an assembly line) can become alienated, and therefore will not learn as well. An awareness of this problem, on the part of the instructor, is an important first step in dealing with it.

The role of intuition in our lives should not be minimized. And our intuition is that small classes are better. Why is this the case, and why is our intuition inconsistent with the statistics cited in the last paragraph? The answer (as Len Gillman has patiently pointed out to me) is that the advantages of small classes are intangibles: The friendly give-and-take between instructor and students (eminently possible in a small class but quite difficult in a large one) is a form of

---

[8]This classroom technique has been studied in detail by experts in the psychology of learning (see [MOO]). One such study recommends that, after you formulate a question to the class, you wait 30 seconds—but not more. Less than 30 seconds does not give students enough time to formulate an answer; more than 30 seconds is wasting time.

encouragement, and a way to make the subject seem fun and exciting. In a small class it is likely that you can cover more material more efficiently, thus you will have time to treat ancillary topics that will whet the students' appetites, and may cause some of them to decide to be math majors. The main point is that a small class is *personal*, while a large one is not. You would not want to have your annual physical exam in a large auditorium with 200 other people. You probably do not want to learn calculus in that fashion either.

One device that works very well (when teaching a class either large or small), if you can manage it, is to make yourself available in the front of the classroom for fifteen minutes after class. Students feel much more comfortable talking to you when surrounded by their peers, and while their questions are fresh in their minds.

Never, ever get involved in a personal discussion of grades in front of a class of any size. If student T wants to know why problem n was given only p points, tell that student to see you privately.

It is imperative that the instructor for a large lecture course be extra well prepared. If you begin to get lost when doing an example or sidetracked with an incorrect explanation then you will quickly lose a large segment of the audience, a lot of talking will start to take place, and the room will soon be bedlam (see also Section 4.7 on discipline). Everyone has off days and makes mistakes, but you must take extra care in a large lecture to have the material down cold.

The teaching of a large lecture course offers complications of a special nature. Discipline and commanding attention are two of these that are treated elsewhere in this book (Sections 2.1, 2.4, 4.7). But there are others. You *must* have a syllabus for such a course. You *must* prepare your homework assignments and exams carefully—and well in advance. There are few things more unpleasant than facing down a hostile audience of 350 hungry freshmen right before lunch. Therefore do not give exams on which the problems don't work out; do not give homework assignments on which the problems don't work out; *do* plan ahead for *all* activities. Prepare and prepare and prepare some more. Have a fair and objective system of grading. If a student comes to you with questions about grades, then have a fair, consistent, and clear set of data to show the student.

If you are the professor in charge of a large lecture, then you will probably be in charge of a group of 2 to 10 or more Teaching Assistants (TAs)—refer to Section 2.19. You must exercise organizational skills with them as well. Meet with them once a week to be sure that they are covering the right material, are aware of upcoming assignments and exams, and to apprise them of any difficulties that have arisen. Make regular use of e-mail to keep in touch with your TAs.

If the TAs are helping you to grade an exam, then you must tell them how you want the papers graded (see also Section 2.11 on grading). Grade exams *horizontally* rather than *vertically*. This means that you should put TA #1 in charge of problem #1 (on all exams), TA #2 in charge of problem #2, and so forth. This is the only way to insure some consistency. If you let each TA grade a stack of exams from start to finish, then you will have wild inconsistencies and many student complaints.

Horizontal grading is also a useful device even when you are grading a stack of just twenty exams all by yourself. It will discipline you, it will make your work more accurate, and it will tend to make the job go more quickly.

At a large university—with 30,000 or more students—there may be several large lectures of the same calculus or linear algebra course running at the same time. If you are in such a situation, you will often find it convenient to work cooperatively with the other lecturers (in some departments you may have no choice). In fact students in this situation seem to value a sense of overall fairness and uniform treatment more than they value the flair and pizzazz of the individual instructor's personal style. Thus you may find it useful to meet regularly with the other lecturers and to hammer out some uniform policies, and even uniform treatment of the topics in the course. The article [MOO] contains a sensible discussion of how the precepts of the management concept TQM (Total Quality Management) can be used to guarantee that the students are treated fairly in this type of learning environment.[9]

Here is a trick that some professors teaching a large lecture have used with success. The professor recognizes that students in a large lecture can develop a feeling of alienation. Such students are afraid to approach the professor, afraid to go to office hours, afraid to ask questions in class, afraid to lodge complaints. Therefore, at the beginning of the term, the professor asks for two or three students to volunteer to be the class *ombudsmen*. As you know, an ombudsman acts as a go-between: he fields questions from the constituency and presents them (with no mention of the source) to the professor. It is not difficult to see that this little device can serve to open lines of communication. Students are much more prepared to present their concerns to a peer (thus preserving their anonymity with the instructor) than they are to confront the professor one-on-one. Good ombudsmen will be able to answer trivial concerns without even getting the professor involved (because the ombudsmen will, as a matter of course, be meeting regularly with the professor). The ombudsmen will also consolidate and clean up student complaints and present them to the professor in a manner which the professor will find agreeable.

Faculty who have used student ombudsmen in the manner just described report that (i) it is a decisive means for bridging the communication gap between professor and students when a class is large, (ii) students respond readily and well to peer ombudsmen, (iii) the ombudsman device is a good way for the professor to develop a close relationship with at least some of the students in the class, and (iv) students will volunteer to be ombudsmen if they are made to understand that a professor can write an especially good letter of recommendation for a student whom he has come to know in this capacity.

It is especially tricky in a large lecture to help the students get to know you as a person, to cut through the barrier that always exists between the person in front of the room and the large audience in the back of the room, and to keep the lines of communication open. You know that successful performers can do these things. You know that preachers can do it. And you know (see Sections 1.6, 3.3) that skilled teachers can do it. One of your career goals should be to develop this talent yourself.

---

[9]I note that TQM—a management technique borrowed from industry—is an emotion-laden topic in the educational setting. I shall not try to treat it here. But see [MOO] for a quick, if not impartial, introduction to the idea.

2.15　Small Classes vs. Large Classes

We all know, deep in our guts, that small classes are a much more stimulating venue for learning than large classes. After all, in a small class students can participate, they can feel much more a part of the process, they can get to know the instructor personally, they are more comfortable asking questions, and they experience a lot of one-on-one interaction.

All true, but (as noted in the last section) there are no studies that show significant improvement in learning, performance, or retention when students are taught in small classes rather than large classes. In fact the main difference that can be objectively verified is that students in small classes feel better about the class, feel more empowered, and have higher self-esteem than those in large classes.

This is certainly a situation in which we instructors must train ourselves to separate objective fact from self-evaluation and intuition. When we argue (with the dean, for example) for smaller calculus classes, we are motivated in part by a concern for student welfare, in part by a distaste for teaching large lectures, and in part by a desire to find objective reasons for hiring more faculty.

If the conclusions of the objective studies are valid, then perhaps we can learn something from them that will inform the way that we teach large lectures. I'm sure that there are few preachers who want smaller flocks, who would prefer to preach to an audience of 30 rather than 300. I wonder why that is so? Can we learn ways to make our students in a calculus class of a few hundred feel more involved, have a better opinion of themselves, feel good about the learning process?

The final word on these questions is certainly not in. But simply being aware of the information in this section and the last should give you food for thought the next time you face down a large class. What is it that makes teaching a large class seem to be difficult? Is it the size of the room, the population of students, the use of a microphone, the feeling that there is less room for error, or what? What can we as math teachers—not preachers or performers—do to make a large class still seem like a family? What can we do to make a large class still feel like a place where learning is taking place?

2.16　Problem Sessions, Review Sessions, and Help Sessions

At many big universities, the large thrice weekly lectures in a lower-division math course are supplemented by once- or twice- weekly "problem sessions" or "help sessions". Usually the lectures are delivered by a professor or instructor while the help sessions are staffed by graduate student teaching assistants (TAs).

Imagine that you are the graduate student in charge of a problem session. It is easy to fall into the trap of not taking the work very seriously. After all, student attendance at these sessions is poor in general and spotty at best. Students seem to be inattentive and their questions are often puerile. But the quality of any class or help session is largely influenced by the attitudes and efforts of the person in front of the room. If your attitude is to treat the help session casually or carelessly then you will get correspondingly disappointing results from the students. Consider giving weekly quizzes, sending students to the board, and other devices for livening up your problem session (see also Section 3.12). I wish to concentrate here on more mundane matters.

It is arguably more difficult to conduct a good problem session than to give a good lecture or class lesson. For the problem session presents all the difficulties of a class period, and more. At least in a class you are in complete control of the order of topics and can, if you wish, present them from prepared notes. In a problem session, if you really let the students ask what they wish, then you must be ready for anything. And you must be able to think quickly, on your feet, of the best way to present any given topic, give a hint on any problem, or handle any point of confusion. In a class or lecture you can always pull rank and say, "There is no time for questions now. See me in my office hour." (I don't recommend that you say this very often, but it is an option that is available). But help sessions are for questions.

If you are a novice, then it is probably safest to view the help session in the most naive way. Your role is to help students do their homework assignment for *that week*. Thus your preparation for a help session might consist of working all the homework problems for the week, or at least staring at them long enough to be sure that you know how to do them.

Be certain that the techniques that you present are consistent with those used in class and in the book. Some professors require their TAs to attend their classes, just to insure this consistency. Such a professor might even do a spot check of the grader's work, or drop in on help sessions to see how things are going.

I know of at least one professor who works closely with his grader and his TAs by attending, once per week, each problem session for his class *accompanied by the grader!* This requires some extra effort on everybody's part, but it shows real consideration for the student who has questions about the way that his homework was graded (or how the class, as a whole, is being conducted). It goes without saying that in order to use this device to good effect the professor will have to be well-coordinated with the grader on how he wants each homework assignment graded.

When you are helping with a homework problem that is to be handed in, don't give away the store. One reasonable answer to the dreary question, "How do you do number 14?" might be, "I'll do number 16 for you, which is similar." Another reasonable answer might be, "I'll get you started. You do the rest." A third is, "Here is an outline of the basic steps." The truly skillful instructor will turn this question-answer session into a team effort. Gently goading the students with his own prompts and questions, this instructor will resist simply doing the requested problem for the students. The trouble with just solving the problem— and nothing more—is that only the requestor and perhaps a few others will be paying attention. If instead the instructor can generate some repartee, and can get the students to want to pitch in, then there will be considerable student interest and a number of class members will learn from the experience.

There are subtle psychological forces at play in the scenario just described. If each student is worried about protecting his turf, and simply does not want to share what he knows, then you will have a hard time generating useful dialogue in your problem session. If instead the atmosphere is one of learning being a sharing activity, and of giving knowledge in expectation of receiving knowledge, then the problem session can be a worthwhile and nurturing experience for everyone. (We all know of mathematicians who collaborate easily and well, and of others who seem to be thoroughly incapable of collaboration. Perhaps these differences reflect attitudes similar to those being described here.) Of course you as the TA or instructor must set the example. If the signal you send is that *you* are not willing to help, that *you* are not willing to share, that instead you are like the oar master on an ancient galley, then you will get little in the way of cooperation and sharing from your students. If instead the example you set is one of patience and giving and caring, then you are likely to be the beneficiary of an enthusiastic response.

The advice to the TA (five paragraphs ago) to work all the homework problems the night before a problem session is one that I tender hesitantly. I never do this, but I've been teaching math for twenty-five years. I am rarely surprised by any question in a calculus class or help session and, even if I am, I can usually slug my way through whatever new features are present. If I am at a review session for an exam and a student presents a really difficult question then I always have the option of saying, "That's an interesting question, but one that could never be put on the test. Let's discuss it privately."

In your first few years of teaching you will have to strike a balance between being thoroughly prepared (by working all problems in advance) and spending too much time on preparation (see also Section 1.3). Just remember that a large part of your job is **(i)** to show the students how to do the problems and **(ii)** to persuade the students that the problems are doable (by ordinary mortals). If you fumble around and act baffled by the problems, then you are presenting a poor role model and, more to the point, doing your job badly. Students find appealing the fact that I can do all the problems and that, moreover, I invariably know where the difficult spots are and can help them to chart their way through them. This ability can only come with experience. It is the model that you should strive to attain.

2.17   Transparencies

I have already touched upon the topic of overhead transparencies (Section 2.3). With the use of transparencies, you can cover more material than you could by just using a blackboard. By using several overhead projectors, you can create an ambience similar to that achieved by several blackboards. By using color, overlays, photos from books, data printouts, computer-generated graphics, histograms, and the like, you can put on a dazzling display of information.

Many of the principles governing good blackboard technique also apply to overhead slides. But, in this somewhat different environment, they take on a new form. Of course you must be organized and write neatly. Be sure to write *large.* Since an overhead slide has the size of a piece of paper, you may tend to lapse into the habit of writing as though you are taking personal notes. Instead, write as though you were preparing a sign to be held up to a (small) crowd for viewing. Keep the top 1.5 inches and the bottom 1.5 inches of your slide free and clean. First, it is the top and bottom of a slide that tend to slip out of view, or off the projector screen. But, even when the slide is squarely on screen, those in the back of the room simply cannot see the material at the bottom.

It is a good idea to use abbreviations on a slide. A slide should have one or two short paragraphs on it, and each paragraph should concisely showcase a key idea. The notion of preparing a long manuscript in TEX—even in 12 point type—and then printing the manuscript on slides and showing it to a group is just crazy. It will not work. This is five times too much material per slide. If you really want to use TEX, then use SliTEX (a special version of TEX, with extra large fonts, that is designed for slides) and double- or triple-space everything. If you can print neatly, then you are much better off preparing your slides by hand. You can then use several colors easily, and can employ many useful displaying conceits as well.

Of course the sort of teaching environment that uses overhead slides and other technological marvels requires a lot of preparation. It may take several hours to amass the information needed to present a one hour multi-media event like that described in the first paragraph. It is not clear to me that the chain rule will thereby be any better conveyed than with a piece of chalk in the hand of a skilled instructor. In fact, I have some strong opinions about this matter that I would like to take this opportunity to share.

There is a danger, when you use overhead transparencies, that your students will pay less attention to what you have to say and more to trying to copy down everything on each transparency. This is a little bit like the computer trivializing what you are doing just because it does everything too quickly (Section 1.10). Part and parcel of the process of using overhead slides is that the material goes by too quickly—indeed it *glides* by. With chalk, the material is etched on the blackboard in discrete bites. To my mind, chalk suits the communication process more naturally. Transparencies tend to fight the process. You will have to judge for yourself which works best for you.

I want my students to take my class as an inspiration to go home, pick up a pencil, and do some math. If they get the impression, even unconsciously, that doing math requires a bunch of high tech equipment and software, then they may be disinclined to do it. If instead they just see a lone instructor with a piece of chalk doing math, then they may conclude that they can do it too.

On the other hand, there are some pictures (try sketching a graph of $f(x, y) = (x^3 + y^2) \sin[1/(x^2 + y^4)]$ without practicing) that are difficult to do on the fly. A prepared picture on a transparency can be a great help. With `Mathematica`, you can not only render a beautiful picture of a three dimensional graph, but you may also exhibit it from several different perspectives.

It is just about impossible to illustrate Newton's method adequately with freehand sketches using a piece of chalk. A computer printout on a transparency, or even an animation using a PC, can be a great help. Likewise Simpson's rule, the Runge-Kutta method, and other numerical techniques can best be illustrated with a computer and an attractive display.

BUT: If you are trying to teach your students to graph—to assimilate information given by the first derivative, the second derivative, intercepts, symmetry, and so forth—should you use `Mathematica` or similar software as a teaching tool? Should you teach them to use hardware to generate printouts and transparencies? I think not.

We do not want to teach our students to learn to push buttons. We want them to think critically and to reason analytically. It has been argued that `Mathematica` and similar software can be used to help students interact dynamically with the graphics: Vary the value of a in the equation

$$y = ax^2 + bx + c$$

and watch how the graph changes. That is not what I want my students to learn. I want them to understand that, for large values of x, the coefficient a is the most important of the three coefficients. And changing its value affects the first and second derivatives in a certain way. And, in turn, these changes affect the qualitative behavior of the graph in a predictable fashion. *After* these precepts are mastered, the student may have some fun verifying them with computer graphics. But not before.

Graphing is one of the basic techniques of analytical thinking. The picture is not an end in itself. It is the understanding that comes with the creation and analysis of the graph that is our goal as educators. Being able to push some buttons and render a beautiful picture or transparency of a graph in $\mathbb{R}^3$ is *not* the same as understanding the information contained in that graph. The way that you train to read a graph is by learning to create a graph—*by hand*.

Likewise, *your showing* the students a prepared graph of a surface does not teach them the analytical tools required to generate such a graph, nor does it teach the ideas needed to *read* such a graph. Thus my recommendation is to use prepared transparencies (and software) with restraint.

If you decide to present the ideas in your class using transparencies, then be careful to pace yourself. If the material is prepared in advance on overhead slides, then there is a temptation to shuffle them too quickly across the screen. And, even if you are writing the transparencies in real time during class, then you will be inclined to write too quickly. The physical scale of blackboard writing tends to slow you down a bit, and that is good. When you are using the overhead projector, there is nothing to slow you down. You must control the situation yourself.

2.18   Tutors

A commonly asked piece of advice, usually from a student having trouble in your course, is, "Should I get a tutor?" I have a very simple answer to this question: "No". It is almost unavoidable that the student will treat a tutor as a crutch. The student figures that, by paying $20 per hour (or whatever is the going rate), he is *buying* knowledge. And now looms the specter that to my mind should be the benchmark for all educational issues. *All learning of significant knowledge requires considerable effort on the part of the learner.* This fact has not changed since Euclid told Ptolemy (over 2000 years ago) that, "There is no royal road to geometry." Instead of just slugging his way through a new idea, the student finds himself thinking, "I don't get this. I'll have to ask the tutor."

I could go on about this point at length, but I will try to restrain myself. Go to any athletic facility and you will see young people spending hours perfecting their free throw or their skate board technique or their butterfly stroke. They don't hire tutors to achieve those goals. They also don't hire tutors for learning to build model airplanes or learning to modify their cars. The reason is very simple. There is plenty of peer support for these activities. Young people are highly motivated to be proficient at them. An eighteen year old understands clearly when an athletic coach says, "No pain, no gain." However the same concept makes little sense to him in the context of mathematics or another deep academic subject.

It is a sad fact that many students—and their parents too—view the university situation in the same way that they view buying a car: You pay your money and you take your choice. The professor is expected to deliver an education (in exchange for the big bucks) in the same way that your local Ford dealer is expected to deliver a car when you fork over your down payment. The college instructor who says, "I am a scholar, and I set a standard, and I expect my students to rise to it." may find himself in a very lonely place. I certainly do not think that such an instructor is misguided. Far from it, I agree with him wholeheartedly. But I also realize that if I expect my students to rise to a certain standard, then I must teach them that this is a worthy goal, and then I must show them how to do it. It's not so hard, once you realize that that is what you must do.

A good math student must be self-motivated. In most instances, the hiring of a tutor is an attempt by the student to buy his way out of some work. I've been a tutor. It's a great way to make money. But in most instances it is not a beneficial learning device. You might find it helpful to refer to Section 3.6 on math anxiety in connection with these issues.

Of course there are exceptions to what I am saying here. Some students are too timid, or too slow, or too far behind to catch up without help. Sometimes a student will have a legitimate and doctor-certified learning disability. If a student has been ill for several weeks, or has had a death in the family or some other personal crisis, then a tutor may be the only alternative.

It is a sobering thought to realize how different the students' point of view is from our own. There is at least one high quality large state university today where students routinely hire a tutor for each class that they take—*before they have even set foot in the classroom.* Clearly these students have convinced themselves that classroom instruction is inadequate, or that their own abilities are substandard, or that they do not know how to study and require a surfeit of hand holding. On days when you think that teaching is a straightforward process, stop and ponder this matter.

In any event, if you are a paid instructor at a college or university, then do not hire *yourself* out as a tutor for a student in a class that *you* are teaching. It is inappropriate, it is tawdry, it is a conflict of interest, and it might get you into trouble with your department. The safest policy is not to tutor students at your institution at all. The point is that you are already being paid a salary by your school to educate the students at that school. To further accept tutoring money from the students constitutes double dipping.

Even having to recommend tutors can put you in a position of conflict of interest. Most math departments maintain a list of qualified people who can tutor for math courses. This is done as a service for the students, but it is also done as a service for the faculty. When a student asks you about tutors, send that student to the departmental office and the official list. It really is the best policy.

2.19   On Being a TA

Being a Teaching Assistant (TA) provides some experience in being a teacher. But it does not provide much, and the background that it provides can be misleading.

When you are a graduate TA at a big state university, you are probably not your own boss. In most cases you work, alongside several other TAs, for some professor who is delivering lectures to a large audience. On alternate days, the class will be broken up into smaller "quiz sections" or "problem sessions", and you will be asked to teach one or more of these. You will also be asked to help with grading, with other assigned activities, and (primarily) you will be asked to do what you are told.

Being told what to do lifts a great deal of responsibility from your shoulders. But this also means that a TA has never really taught. You've had some experience standing in front of a group, organizing your thoughts, answering questions, developing blackboard technique, and so forth. But you will have never made up an exam, written a syllabus, designed a course, given a course grade, or any of the dozens of other activities that figure significantly in the teaching process.

However, if you have never been a TA (either because in graduate school you were on a fellowship that had no formal duties attached to it, or perhaps because you were educated in another country), do not despair. At least you are entering this profession with possibly fewer prejudices than are held by those who have stood as a TA before a hostile audience in this country. Perhaps reading this book will provide you with better information and a better outlook than having served as a TA under a professor who doesn't even care about good teaching.

Let me put an ameliatory note here. Some professors are well aware of the down side of being a TA and attempt to compensate for it. They give their TAs more responsibility. For instance, such a professor might write the first midterm exam for a class himself and then let the TAs write subsequent midterms (under close supervision). This is positive psychological reinforcement for the TAs, and good experience for them as well. Likewise, the TAs can be allowed to set the curve for grading (under supervision) and to perform the other ordinary functions of the instructor. The professor is not being lazy here. Rather, he probably has to expend more effort than if he were doing these tasks solo. But it provides awfully good experience for the graduate student TA.

At some schools, the TA is more autonomous. It is possible that the TA will be a free-standing teacher, creating his own exams and constructing his own grading system. If this description applies to you, then this section of the book does not. But the rest of the book does, and you may benefit from reading it.

For more information about the day-to-day duties of being a Teaching Assistant, see Section 2.14.

2.20 Advising, Letters of Recommendation, and Graduate School

Of course a substantial amount of your undergraduate teaching duties will consist of classroom contact hours and office hours. But that is not the whole enchilada. If you are a senior member of the department, then you may be asked to help with undergraduate advising. Apart from your official duties as an advisor, students may ask you for advice about their curriculum or about graduate school. And you will be asked to write letters of recommendation. (As you read on, refer to Section 4.9 about Advice and Consent.)

You are well qualified—indeed nobody is better qualified—to give your undergraduate students advice on what courses to take, or on how to prepare for various mathematical careers, or on how to select a graduate program. If a student wants to be an actuary, then one course of action is appropriate. If instead the student wants to be a software engineer then different advice would be the order of the day. Even if the conversation wanders beyond your area of expertise, you can surely direct the student to another faculty member, or to a guidance counselor, who can help. Too many undergraduates get the bulk of their advice from fellow students. Sadly, that advice is often based largely on rumor, innuendo, and misinformation. You really perform a great service when you take the time to provide an undergraduate student with expert advice.

Of course the advice you give students may be no better than what they can glean from their peers if you do not take the trouble to find out what you are talking about. Before you tell students to take this class rather than that, or this flavor of the math major rather than that, or to take an incomplete rather than a drop, *find out what the rules are.* Become acquainted with the requirements for the math major and minor. What are the rules for drops? For incompletes? What are the mathematics requirements for the physics, engineering, chemistry, and other majors? You can do a lot of damage if you offer advice without knowing whereof you speak.

Most undergraduate students don't have a clue about graduate school. They don't know how one gets in, how one pays for it, how long it takes, what it entails, what a Ph.D. is, how a Ph.D. differs from a Masters degree, what is involved in writing a thesis, and so forth. In general, their parents and their friends will know even less than they do. So, again, you perform a great service if you are willing to share your expertise. Once a student knows that he wants to get an advanced degree, he will need some real help in choosing a school and a program that suits his abilities, his interests, and his needs. You, the mathematics professor, are the best person to provide this information.

Part of this general circle of activities is that students will ask you for letters of recommendation. It is definitely part of your professional duties to field such requests. If a student is serious about getting into graduate school, or getting a job in a technical field, then he has few places to turn for recommendations beside his college instructors. You should make it clear to students that it is no burden for you to write a letter of recommendation, and you should agree to do so—UNLESS this is a student of whom you have a poor opinion, and for whom you would feel uncomfortable writing a letter. In that case you should say, "I'm sorry, but I do not feel that I could write a very supportive letter for you. Perhaps you should ask someone else." To agree to write a letter, knowing full well that you will never do so, is dishonest and unfair to the student. Please do not fall into this trap.

The companion volume [KRA] provides a lengthy disquisition on the chapter and verse of good letter writing. I shall not repeat the details here. But do endeavor to be supportive of your students—at least the good ones—and do endeavor to be honest. Your reputation as a letter writer is part of your professional *gestalt*. It will follow you around for the rest of your life.

Chapter 3

Spiritual Matters

Even paranoids have enemies.

Delmore Schwartz

Conservatism goes for comfort, reform for truth.

Ralph Waldo Emerson

If anything ail a man, so that he does not perform his functions, if he have a pain in his bowels even,—for that is the seat of sympathy,—he forthwith sets about reforming the world.

Henry David Thoreau

Reformers can be as bigoted and sectarian and as ready to malign each other, as the Church in its darkest periods has been to persecute its dissenters.

Elizabeth Cady Stanton

You can't make a Hamlet without breaking a few egos.

Anon.

If you think education is expensive, try ignorance.

Derek Bok

Bad planning on your part does not automatically constitute an emergency on my part.

Anon.

We, the unwilling, led by the unqualified, have been doing the unbelievable for so long with so little, we now attempt the impossible with nothing.

Anon.

Teaching has ruined more American novelists than drink.

Gore Vidal

3.0 Chapter Overview

This chapter addresses philosophical issues connected with teaching. How do students learn, how do they formulate questions, how should we answer those questions, what is the function of the mathematics teacher? An adequate instructor records the material accurately on the blackboard and then goes home. A truly dynamic instructor interacts with the students, excites their intellectual curiosity, and helps them to discover ideas for themselves. The material in this chapter should help you to pass from the first state to the second.

3.1   Personal Aspects

Like many activities in life, teaching is an intensely personal one. Some teachers have a lighthearted, informal, even jocular style. Others are more severe. Some give a rigid, structured lecture. Others conduct a Socratic interchange with the class. Some send students to the board to do problems (some, in the R. L. Moore style, do nothing but—see Section 1.12). Some instructors use overhead slides, computer simulations, symbolic manipulation software, and `Mathematica` graphics. Others do it all themselves, with just a piece of chalk. Some professors integrate a (computer-based) laboratory component into their courses. (In fact I would like to see mathematics become more of a "laboratory discipline". See Section 1.10.) All of these methods are correct. It is essential for you to be comfortable with your class. Therefore you should conduct the class in whatever fashion feels most natural to you.

However you should be willing to try new things. If you have never told a joke before, try telling a joke. If it works, you may be pleasantly surprised and may tell another. (But be forewarned: Eighteen year olds are insecure and are always worried that someone is making fun of them. Do not tell jokes that may be interpreted in that fashion. Do not tell jokes at anyone's expense. Do not tell sexist jokes. Do not use vulgar language or discuss offensive topics.) Try introducing the product rule with a story about how much trouble Leibniz had getting it right (see Section 1.7). Illustrate the subtleties of the constant of integration by integrating $\int 1/x\,dx$ by parts (without the constant) and deriving the assertion that $0 = 1$:

$$\int \frac{1}{x}\,dx = \frac{1}{x}\cdot x - \int x\cdot\left(-\frac{1}{x^2}\right)\,dx = 1 + \int \frac{1}{x}\,dx$$

hence $0 = 1$. Some of these endeavors will fall flat. Others will breathe new life into an otherwise old and (for you) dull topic.

It is important to me that my classroom have the atmosphere of an interchange of ideas among intelligent people. I would be most uncomfortable to stand for an hour reciting a litany of abstract nonsense to a sea of blank faces. Thus I am continually trying new approaches, new angles, new ideas. It is a way to keep my classes fresh, even in a course that I have taught ten times before.

I usually do not find it useful to send students to the blackboard to do problems. First, the time that it takes for the student to get to the front of the room, falter around, and sit down again, is too great for the benefit obtained. I do everything myself because I can teach a great deal even while I am doing the most mundane example. But others have been sending students to the board for years and swear by it. Do what works for you.

I have no use for overhead projectors. To me, part of pacing a class is to let the material evolve on the blackboard. Part of the dynamic of my classroom style is moving back and forth in front of the material. But others find that they can be more organized if they write out the material in advance on overhead slides. Still others write the material in real time on the overhead slide. Yet another group writes very little, but stands in one spot and delivers a strictly oral lecture.

Remember that you are delivering a product. Cadillac does this differently from Mercedes Benz. You must develop your own delivery. The over-riding consideration is that you be comfortable with your classroom style so that you can in turn make your class comfortable. The style, organization, and content of your class is a reflection of you and your attitude toward the class. If you stand in front of your calculus class facing the blackboard, mumbling to yourself, and writing "Theorem–Proof–Theorem–Proof", then what message are you sending to the students? If instead you do a stand-up comedy routine and get around to the mathematics in the last ten minutes of class, then what message are you sending to the students?

Not everyone agrees with the spirit of what I am saying in this section. For example David S. Moore, in his essay written in acceptance of a major teaching award (see [MOO]), declaims

> Some teachers may have charisma. Not I. My image of a sound teacher is that of a skilled craftworker, a master machinist, say, who knows exactly what she must do, brings the tools she needs, does the work with straightforward competence, and takes pleasure in a job well done. She does her work right every day, and every day's work fits the larger plan of her project. The craftworker's skill is quite separate from her enthusiasm on that particular day, which, as C. S. Lewis said in another context, depends more on the state of our digestion than on any more cosmic influence.

This point of view is worth considering. To be sure, a journeyman machinist must reach and maintain a certain level of excellence. He cannot claim—two days per week—that he slept badly, or fought with his spouse, or was not in the mood. He is turning out precision milled parts for demanding applications, and his work must always meet a rather exacting standard.

We, as teachers, have a bit more slack. We are more like play-actors. Our days of divine inspiration can, and usually are, balanced out by days of doleful plodding. I think that Moore's remarks can inspire us to set a basic standard to which we all can, and should, aspire. There really should be no lecture or class that we give that is less than competent, less than accessible, less than accurate. That is not too much to ask.

But the machinist is working with an inert substance—such as stainless steel. And his task is essentially a zero-one game. He either produces his precision part correctly or he doesn't. Students are *not* inert substances. Far from it, they are highly volatile. And our task—to impart knowledge, to excite, to set an example of learning and scholarship—is hardly a zero-one game. It is a multi-parameter process that requires skill and care to do well.

Teaching is an intensely human activity. Part of our job is to convey facts. But perhaps a more important part is to inspire and to ignite curiosity. If we want to influence students to be scientists, to be math majors, to be thinkers, to get caught up in mathematics, then we must be more than competent. We must be on fire ourselves. So let Moore's criteria be a lower bound. Let the heavens be an upper bound.

And now a coda on haberdashery. Speaking strictly logically, it should not make any difference whether you wear a suit, or jeans and a work shirt, or wear a loin cloth and carry a spear when you teach. But it does. Dressing nicely sends a subtle signal to the class that you are the person in charge. If you wear ripped, stone-washed jeans, a tie-died tee shirt, and have long, greasy hair tied in a bandana, then you may convey to the class that you just came from lube-ing your car, or that you have great empathy with sharecroppers, or that you are just plain "folks". But you may, inadvertently, also convey that you just don't give a damn.

Straightening your tie and combing your hair before going to class is like putting on your mortar board before going to graduation. You are pausing to say to yourself, "Now I am going off to do something important."

3.2   Attitude

I have long felt that those who cannot teach are those who do not care about teaching. If you actually care about transmitting knowledge and inspiring curiosity and a love for learning, then much of what I say in this book follows automatically. But some comments should be made.

Your students are a lot like you. When you enroll for a class, you have certain expectations. It is reasonable, therefore, that when you *teach* a class you should endeavor to live up to those same expectations. From this it follows that you should prepare, be organized, be fair, be receptive to questions, meet your office hour, and so forth.

On the other hand, your students are not like you. Especially in elementary courses, you cannot expect your students to be little mathematicians. Many of them are in the class *only* because it is a prerequisite for their majors. Try to remember how you felt when you took anthropology or Latin or biology. Not everyone has a gift for mathematics. Unfortunately, some people have an attitude problem to boot (this attitude problem is sometimes termed "math anxiety"—see Section 3.6).

So you must learn to be sympathetic and receptive, and you must learn to be patient. Teaching is part of your craft, and part of your job. Perhaps if you are Gauss, or if you have just proved the Riemann hypothesis, then you can justifiably say that you are above these pedestrian considerations about teaching. I'm betting that you are not either of these. If you call yourself a professor, and if you have the temerity to stand in front of an audience and profess, then you should show your audience some respect and consideration.

David S. Moore, an award-winning teacher already quoted in Section 3.1, noted the following when addressing an audience of teachers (see [MOO]) on the craft of good teaching:

> Our individual experience, both as students and later as teachers, is atypical. As students, we were the survivors, the fittest by some quite esoteric standards of fitness.

These are wise words. When we are teaching calculus, we must remember that we are not teaching present or future clones of ourselves. On the contrary, we are teaching people of modest mathematical ability, and perhaps equipped with only slight interest in the subject of mathematics. We must keep this insight in mind as we formulate our lessons and articulate our thoughts.

A recent article on education claimed that there are some people who will never be good teachers. Period. As proof, the writer offered the story of a mathematician from Japan who spent a semester at an American midwestern university. He lectured to a large calculus class by filling the blackboard with long row upon long row of tiny, but neatly written, script—uttering hardly a word. Half an hour went by, and there were a great many dense paragraphs recorded on the board. Suddenly the lecturer looked up with surprise and embarrassment on his face. He turned to the class, bowed, and said, "I'm sorry, I'm sorry" repeatedly and ran back and forth across the front of the room erasing little words and replacing them with other little words. This took some time. After about five minutes the lecturer turned to the class and offered his abject apologies yet again. "I am so sorry. I forgot that I was in the subjunctive mood."

I don't know whether I would go so far as to say that the man in the last paragraph would never be a good teacher. But he had a long way to go. His primary sin was that he was thoroughly and trenchantly unaware of his audience. I wonder what David Moore (Section 3.1) would say about a teacher like this?

I have certain favorite questions that I like to put to my lower division classes. I like to pose at least one of these per week. Some of these are, "Why do we study the real numbers? Why don't we do calculus on the integers?" Another is, "Are all numbers rational? Why or why not?" "Is there a number system that is bigger than the reals? Is there more than one such system?" "What is the interest of the complex numbers. Do they exist, or are they 'imaginary'?" Yet another is, "Why do we use radians to measure angles—is it just to confuse students and make them unhappy?" Well, you see the point. I want the students to realize that knowledge is open-ended. I also want them to know that everything that we do in mathematics has a reason, and one that has been well thought out. Learning is a lifelong process, and one that the individual can conduct for himself. Part of my job is to convey that spirit to my students.

To stand in front of a class, with the charge of holding forth for an hour or more, is a heady experience—especially for the novice instructor. It is an ego trip. If you are prone to showing off anyway, this is an opportunity to let your predilections get out of hand. There is a temptation to tell too many jokes, to give a monologue, to use off-color language, to emphasize points with pratfalls or physical humor, to wear grotesque or offensive T-shirts or funny hats, to dress up like Isaac Newton, or just to be silly. A good rule of thumb is, "Don't." There is a famous calculus teacher who used to wear a gorilla suit when he taught the chain rule. The idea was that the chain rule is so simple that even a monkey can do it. My view is that gimmicks such as this one distract from the task at hand, which is to convey knowledge. How can a student concentrate on the mathematics if the instructor is dressed like a gorilla and acting foolish to boot? If you want to wear a mask for the first couple of minutes of a class on Halloween, I guess that is all right. I know people who will occasionally sing a song to the class, or play something on the guitar. But do maintain your dignity. And do not introduce distractions into the classroom atmosphere.

A favorite story is of a math professor who taught in South Africa. He would let his hair grow for a full year, and then have his head shaved. He repeated this routine year after year. Imagine the reaction in his calculus class when he would show up on Monday with hair down his back and on Tuesday with a gleaming dome. This is a funny story, but also a good illustration of how *not* to behave with your class. It is fine to joke around, and even to make fun of yourself. But don't do things that undercut your dignity and your control of the class. Don't turn yourself into an object of ridicule. It only weakens your position as class leader.

I read recently of an assistant professor at a large university who, instead of lecturing, sings rock songs to his class—of course he professes to be a serious scholar, hence he changes the lyrics so that he is singing about the chain rule, or the method of substitution, or Green's formula. The journalist seemed to find this instructor's activities original and inspiring. I find them nauseating. Don't get me wrong. I have a song about the quotient rule that I sing to my class. They usually love it. But that's once per semester. The guy that I'm describing here sings all the time—that's his *lingua franca.* I'm against it.

Teaching is a serious business, and you should act accordingly when you do it. Of course you can have fun while you teach, but don't drag the subject, or the class, through the gutter.

For the length of a semester, you and your class are like a little family (if it is a large lecture, then read "large, loosely knit, family"). The class develops its own *gestalt* and set of attitudes. Things will go smoothly if the attitude in your class is that you and the students are working together to conquer the material. If instead it is you and the book pitted against the students, then you've got an attitude problem. If, on the other hand, you take repeated pains to criticize the text, then you are setting up another attitude problem. No text is perfect. Neither are the students and neither are you. But make it clear from the outset that you are on the students' side. You convey this attitude in thought, word, and deed: Prepare your classes, respect student questions, give fair exams, meet your office hours.

Implicit in this discussion is a simple point. A successful class is not a confrontation between the professor and the students. The professor and the students should be allies, with the former playing the role of mentor, in mastering the material at hand. You know the stuff and they don't. It is your job to mentor them and to help them learn.

Your classes should be friendly, but you do not want to be friends with your students. This sounds a trifle cold, especially to a graduate student or new instructor, but it is an important device in maintaining control of the class. A slight distance helps to preserve your authority. In particular, don't allow your students to call you "Bubbles" or "Squeak" or "Porky" or any other affectionate name. It is probably not even a good idea to let your students address you by your first name.

Of course you should not date your students. It is safest not to date any student at your college or university. Given the way that students like to gossip about faculty, if you date one student then you may as well be dating them all.

However, if you are inexorably smitten with one of your own students, then advise your love interest to transfer to another class or another section. Sexual harassment is an issue of great concern these days (Sections 4.9, 4.10). Professors are particularly vulnerable to charges of sexual misconduct. So beware.

It is natural to want to show students your human side. There is probably no intrinsic harm in having a cup of coffee with some students. A relaxed atmosphere can help to open lines of communication. There might be some harm in having coffee with just one student—examine your conscience before doing so. If you meet a student in a bar off campus for a drink, then—let me speak frankly—you've got more on your mind than teaching.

Now let's return to teaching proper. A recurring theme in this book is that you should *prepare* your classes. For a novice instructor, an hour or two or three of preparation may be necessary. For a seasoned trouper teaching calculus for the tenth time, as little as thirty minutes may be sufficient. The main thing is to be sure that you can do the calculations and and examples and that you have the definitions and theorems straight, and in the proper order. If you are the sort of person who freezes up in front of an audience, then be extra well prepared. I know experienced professors who get so locked up in front of a large group that they cannot remember their own phone numbers. If that describes you, then have the necessary information on a sheet of paper in front of you.

If it is evident to your students that you are winging your class, then they are receiving a counterproductive message: If it is OK for the instructor to fake it then it is OK for the students to fake it. To those math instructors who say, "I don't prepare because it is good for the students to see how a mathematician thinks." I say, "Nonsense". This is just laziness and/or self-serving rubbish. You must be a role model, both as an educated person and as to the way that mathematics is done.

The first few classes that you give in a semester-long course set the tone for the entire semester. You may be at a slight disadvantage because you are coming off summer vacation. Perhaps you are not quite yet in the mood to be teaching, and your lack of enthusiasm shows. We've all fallen into the trap of saying, *sotto voce*, "I'll just wing the first few classes and get things straightened out in a few weeks." This is a mistake. It is sending the wrong message to the class about your attitude toward discipline (both academic and personal), your attitude toward the students, and your attitude toward the subject matter. To repeat, you are a role model. In fact you are a model for the students of what an educated person is and should be. You've got to "walk that walk and talk that talk." If you come across as a bumbling clod then—when these students are productive, voting members of society—how will they remember their education? How will they cast their vote on educational issues?

What your students write on their homework and on their exams will be a derivative version of what you show them. If you do not work out maximum-minimum problems systematically then they will not either. I always set up six steps to follow in doing a max-min problem and follow them scrupulously. This step-by-step approach is an elementary device, but it is a useful one for keeping interest up. When doing an example, after I do step two I can say, "OK, so what do we do next?" This keeps the ball rolling.

Mathematicians fall unthinkingly into the use of jargon. Among ourselves, we frequently say "trivially", "clearly", "by inspection", and so forth. Do not do this—especially in a calculus or precalculus class. First, it sounds pretentious. Second, it is dangerous to assume that anything is either trivial or clear unless you make it so. Third, to say that something is trivial is a subtle put-down. In the popular psycho-babble this would be called "passive-aggressive behavior".

3.3 Caring

Your students need to perceive that you care. I do not mean this to be a sappy statement, of the sort that might be uttered at a "Power of Positive Thinking" symposium. Rather, it is a practical fact about teaching. If you trot into the room, throw some math on the blackboard, and trot out again then you have definitely sent the students a signal. And the signal is that you just don't care. You don't care who they are, or what their needs are, or whether they learn anything. You just don't care.

A colleague of mine once wandered into his class and began to lecture about Weyl chambers and universal enveloping algebras and irreducible unitary representations. After about twenty minutes he looked up, struck his forehead, and said, "Oh God! This is my calculus class." What does this say about how well he knew his class, or how much he cared about it?

Underestimating your students' abilities is a way of showing that you don't care. Maliciously competing with your students (or showing off to establish how much smarter you are than they) is a way of showing that you don't care. Not preparing a syllabus and sticking to it is a way of showing that you don't care. Not having a consistent grading policy is a way of showing that you don't care. Making a lot of mistakes in your class is a way of showing that you don't care. Wasting the students' time is a way of showing that you don't care. Not preparing for class is a way of showing that you don't care.

Even your weakest students—with no talent for mathematics—will pick up right away on the fact that you don't care. It will ruin the whole class, both for you and for them.

Not long ago, a chemistry professor at a prestigious private university was becoming increasingly alarmed over the nonsense that her child was bringing home from her high school chemistry class. The professor could only conclude that the high school chemistry teacher did not have a clue—she didn't understand the basic principles of chemistry. So one day the chemistry professor made an appointment with the high school chemistry teacher to express her disappointment and frustration. Imagine the professor's surprise when she instantly recognized the high school teacher as one of her former chemistry students!

One hopes that the experience just described caused the university chemistry professor to reassess how she taught, and how she evaluated her students' progress. I wonder whether she took time to think about how much she had formerly cared about her teaching, and how much she would care in the future.

It doesn't require much effort to show that you care about the class. If they do well on a test, celebrate with them. If they do poorly, commiserate with them. If they cannot do the homework, then help them with it. If they can't understand the concept of orientability of a surface, explain it to them. Show them a Möbius strip. They have probably all seen Möbius strips before, but never realized that they had a serious use. The point is that you should share their pain. You are there to help them learn. Act like it.

I once witnessed a conversation between two mathematicians—one American and one German. The first man was discussing John James *Audubon*, the American naturalist. The second was discussing the *Autobahn*, the famous high-speed German highway. Each scholar prattled away for ten minutes or more, discussing his own topic, blissfully unaware that his interlocutor was discussing something entirely different. How was this possible, you may ask? Well, how many mathematicians do you know?

Has it ever occurred to you that, sometimes, when you are talking to your calculus students, you are no different than the savants in the last paragraph? Here is the poor student trying to tell you that he can't even understand the *logic* in problem #6 on the homework, and you are busy trying to explain to him the circumstances under which the mean value theorem fails. Such a situation doesn't make you either a bad person or a bad teacher. But it certainly makes you less than optimal in helping your student to learn. Part of caring is to step out of your own shoes and, as much as possible, step into the student's. It may be a new experience for you, but it is one that you should seek.

3.4 Breaking the Ice

The first day of class is simultaneously a day of happy anticipation and a day of stress. It is the first of these (assuming that you like to teach) because you are, after a restful summer, jumping into something that you enjoy and that you do well. It is stressful because you don't know what this new group of students is going to be like, or whether they will play ball with you, or whether you can get through to them.

I am a teacher of long experience. On days of exceptional hubris, I convince myself that I am rather a good teacher. Yet most semesters, especially in the fall, I meet a new class with new students and I have to demonstrate to these people that I'm a good guy. We begin as total strangers, and my goal is to turn us into a working group. Usually this takes a while—often several weeks.

Since I so enjoy a class once we have all become friends, I find the period of tooling up to that happy steady state generally too long and too painful. What usually happens is that there is a period of two to five weeks during which the students look at me as though I am from Mars. They don't laugh at my jokes, they don't answer my questions, they don't seem to take me very seriously. If the class is to be a success, then some magical thing must happen to change everyone's attitude.

You should consider ways to make yourself seem like a human being to your students. Being playful, or impish, or making fun of yourself, is certainly one technique for accomplishing that goal. If that doesn't work for you, or makes you feel uncomfortable, then try something else. Read them some history. Tell them of Bishop Berkeley and his doubts about calculus. Tell them about fractals, or dynamical systems, or wavelets, or why[1] mathematicians don't get the Nobel Prize.[2]

Find some way to open up to your students so that they will open up to you. Some instructors hide their unease behind regimen. They take roll, or put together a seating chart, or ask each student to introduce himself. This routine is fine if you are comfortable with it. My view is that you should show students from day one that you are a person, and that you are going to spend the term doing your best to communicate with them. I don't think that taking roll is a good way to send that signal. It is better if you tell them what the course is about, or describe your grading policies, or give them some clues as to what *you* are like.

To repeat an important theme of this book: If your students are not talking to you it is probably because you are not talking to them. Set the tone on the first day. And never forget it.

3.5   Why Do We Need Mathematics Teachers?

When I wrote the first edition of this book, I had trouble formulating a cogent answer to this question. We all believe that teachers are necessary. Society must believe it too, for it deigns to pay (not very generously) a great many teachers. But *why* are teachers necessary? Two colleagues (Gary Jensen and Meyer Jerison)—both wiser than myself—have supplied a striking and memorable answer:

> The teacher
>
> **(1)** Sets a pace for the students;
>
> **(2)** Teaches students to read (mathematics);
>
> **(3)** Helps the students to become engaged in the learning process.

There is considerable wisdom in these simple observations. Let us consider them one by one.

(1) Watch a young person—or even an old person—attempt to learn to play the piano without benefit of a teacher. Such a person usually has neither the

---

[1]There are actually several versions of this story. The so-called French/American version is that a mathematician (Mittag-Leffler) ran off with Nobel's wife. The Swedish version is that Alfred Nobel was a practical man of the world who wasn't aware of mathematics as a discipline.

[2]If you are tired of the standard Nobel Prize story, then tell the lesser known story of the Mittag-Leffler Prize. Mittag-Leffler set it up, of course, to spite Nobel. He mandated that the medal would be twice as large, and the award twice as grand in several notable aspects. It was only awarded twice because Mittag-Leffler invested the funds in the Italian railroad system and German World War I bonds.

patience nor the discipline to first learn hand positions, then learn scales, then learn simple pieces, then learn chords, and gradually work up to sophisticated works. The neophyte wants to be playing Chopin's *Mazurka in D Major* or Jerry Lee Lewis's *Whole Lotta Shakin' Goin' On* right away. Without a well thought out program, the student will probably never succeed.

It is just the same with mathematics. Calculus has deep and stunning applications, but the student cannot begin to appreciate them until he has mastered a large number of preliminary steps. The instructor provides those steps, and forces the student to stick to them and to spend a proper amount of time on each. The instructor gives homework assignments, quizzes, and exams. He grades them, to be sure that the student is making progress. The instructor gives midterm exams and a comprehensive test at the end to insure that the student has mastered a body of knowledge.

(2) The instructor teaches the students to read. What does this mean? Take a simple example from linear algebra. The instructor writes on the blackboard

$$F : \mathbb{R}^2 \longrightarrow \mathbb{R}^3$$
$$(x_1, x_2) \longmapsto (f_1(x_1, x_2), f_2(x_1, x_2), f_3(x_1, x_2)).$$

No student—indeed no mathematician—is born with an ability to read this array of symbols. The student must be *taught* to discern the number of variables, and the number of components in an element of the range, and where they appear, and how to pick them out. In other words, the student must be taught to *read*. This need is probably more intense, and more essential, in mathematics than in many other disciplines. A student stares at a page of a calculus text and simply does not know how to extract information from it. He reads a chapter from a calculus book and does not come away with specific knowledge of what is important, what are the key ideas, what he needs to know. It is the teacher who explains and clarifies these matters.

Perhaps the most important thing that a student takes away from his college education is an ability to think critically: to evaluate arguments, decide what to believe in, formulate his own thoughts. It is these skills that separate the educated person from the dullard. Skill at reading is decisively linked to skill at critical thinking—see Section 3.14.

(3) The last of our three dicta is certainly not the least important. If you ever ask yourself, "Why won't my students talk to me?" or "Why is class attendance so poor?" or "Why won't they do the homework?" or "Why do they perform so poorly on the exams?" or "Why won't they attend my office hour?" then the answer is probably that they are sitting back and watching your class in the same way that they watch television. You know as well as I that many young people today grow up in households where the TV is a *constant* backdrop. It is on during breakfast, lunch, and dinner. People watch it while they get dressed, and while they bathe. The TV is a constant source of background noise. And many young people probably could not conceive of a household in which the TV is not forever droning away. They learn to make conversation with the TV going in the background, to make love with the TV going in the background, to have discussions and arguments with the TV going in the background. I've been to parties hosted by intelligent, mature people during which MTV was blasting away the whole time.

Thus it is not difficult to imagine that an immature, uninterested student might find himself treating your class as background noise. He knows that his parents are paying a whole lot of dough to send him to school. He knows that he is supposed to attend class. If he is an impressionable freshman, he probably *will* attend class. But he doesn't know what to do once he gets there. That is where your job kicks in.

What can you do to make the class interesting? What can you do to make the students *want* to learn this material? What can you do to arouse, indeed inflame, the student's inborn inquisitiveness? Evidence suggests that the American K–12 educational system has an insidious way of suppressing curiosity in students. What can we do to revitalize their interest, their intellectual vitality, their need to learn?

As I said in the Preface, I don't have simple answers for these questions; nobody does. Much of the teaching reform movement is predicated on the observations made in the last three paragraphs. Indeed, the battle cry of many is that "the lecture is dead." These are certainly words that one ought to ponder (but see Section 1.6). My reaction, when I first heard these words, was, "Your lectures may be dead, buddy. But mine are not." However, the notion probably deserves more respectful consideration.

I am not prepared to abandon lectures as a teaching device. They have worked for thousands of years, in many different societies, and in many different contexts. And they have worked well for me. But I am passionately interested in getting my students engaged in the material that I am trying to teach them. I am well aware of the observed fact that *lectures often do not succeed in engaging students in the learning process.*

In examining my concerns, and attempting to deal with them, I have found that I *lecture* less and less to my students and instead *talk* to them more and more. As much as possible, I stand among them and conduct a conversation. To be sure, I do most of the talking. But I treat the students as my family at the dinner table. I am the benign dictator, the philosopher/king. I run the show. But we are all participants. I encourage them to ask questions, to make conjectures, to throw out ideas. And I not only respond to what they say—I *develop* what they say. I make their contributions a part of the class.

The focus of many of the best teaching reform efforts has been to try to develop new and powerful ways to achieve goals **(2)** and **(3)** that were enunciated at the beginning of this section. If you are not a reformer, then you should consider what reform has to offer in light of these ideas.

Steven Zucker [ZUC] argues that an important rite of passage for any student is the transition from "high school style" education in mathematics to "university style" education in mathematics. He characterizes the former, not very charitably, as "programming" and the latter as the formation of a serious learner (see also the discussion in Section 1.2, as well as the Appendix that Zucker has written for this book). Many a high school student expects, after each class, instantly to be able to do the corresponding homework assignment. He expects the instructor to affix a bib about his neck and to spoon-feed the material to him. The university student should understand that the class is only a stepping stone to mastery of the material. He must study the text, go to the problem sessions, and *struggle* to get a grip on the material. In brief, in high school the primary responsibility for learning is with the teacher. In college it is with the student.

You may not agree completely with Zucker's characterization of the high school vs. university learning process, but there is certainly merit to his arguments. And an understanding of what Zucker is saying can certainly help to inform the college instructor as he teaches. If you don't know what it is that you are trying to accomplish, then how can the students know what it is, and how can you succeed? No computer, no self-teaching regimen, is going to turn unformed blocks of clay into real thinkers. But *you* can. You are a university teacher. You can take these minds and mold them. That is your goal. As I. M. Gelfand has expressed it (see [GS] and [GGK]), "The most important thing that a student can get from the study of mathematics is the attainment of a higher intellectual level." We should keep these words clearly in focus as we teach our freshmen and sophomores.

If you teach at a large state university, as I have done, and must teach calculus in a large lecture (with a population of 400 or more, as I have done), then you may be saying to yourself that Krantz is a dreamy-eyed fool. I humbly beg to disagree. Of course it is difficult to stand among 400 unruly eighteen year olds and treat them as your family at the dinner table. Of course you cannot field questions from every person among 400, nor can you integrate them into your lecture. But a lecture to 400 students can be lively, it can be engaging, it can make students want to learn. You know that Billy Graham can do it; you know that Norman Schwarzkopf can do it; Ronald Reagan could do it; Michael Atiyah and Christopher Zeeman and Ed Witten can do it. John Horton Conway can do it; Stephen Hawking can do it, and he fights unbelievable physical disabilities when he does so. Richard Feynman and Bertrand Russell could do it. So why can't you? You might be thinking that these are or were great historical figures, with the gift of gab. True enough, but the demands on you are much less. You are only trying to get through to a group of impressionable young freshmen or sophomores. You can do it too. Section 2.14 deals with special techniques for handling large lectures.

An education can be characterized as the product of the interaction of two first class minds. This idea is worth developing explicitly. A bright and eager student asks questions, processes the answers, adds those to his framework of knowledge, and then asks more questions. A good teacher anticipates questions, plants the seeds of new questions, and reaps the harvest. Of course you must train yourself to participate in this process. You must learn to adjust the idealized goals that I have presented in this section to the size of your class, to the sort of students who attend your school, and to the educational goals of your institution and your department.

You may wake up each morning and say to yourself, "If only I taught at Harvard, things would be different. I would have bright and eager students, who would hang on my every word. I would be inundated with brilliant questions and dazzling student work. My life would be bliss." Baloney. I know people who teach at Harvard, and I know people who teach at Tennessee Tech. Every school has its share of bright and eager students as well as its share of duds. Some schools may have a preponderance of students who cannot add fractions, while other schools may have a preponderance of students who are too arrogant to listen to anything that you have to say. A good teacher learns to grow where he is planted.

3.6   Math Anxiety

About twenty-five years ago the phenomenon of "Math Anxiety" was identified and described—by well-meaning people, educators endeavoring to explain why some people have more trouble learning math than others (see [TOB] and [KOW] for both history and concept). We don't hear much about math anxiety in math departments because such departments are full of people who don't have it. Math anxiety is an inability by an otherwise intelligent person to cope with quantification and, more generally, with mathematics. Frequently the outward symptoms of math anxiety are physiological rather than psychological. When confronted by a math problem, the sufferer has sweaty palms, is nauseous, has heart palpitations, and experiences paralysis of thought. Oft-cited examples of math anxiety are the successful business person who cannot calculate a tip, or the brilliant musician who cannot balance a checkbook. This quick description does not begin to describe the torment that those suffering from math anxiety actually experience.

What sets mathematics apart from many other activities in life is that it is unforgiving. Most people are not talented speakers or conversationalists, but comfort themselves with the notion that at least they can get their ideas across. Many people cannot spell, but rationalize that the reader can figure out what was meant (or else they rely on a spell-checker). But if you are doing a math problem and it is not right then it is wrong. Period.

Learning elementary mathematics is about as difficult as learning to play *Malagueña* on the guitar. But there is terrific peer support for learning to play the guitar well. There is precious little such support (especially among college students) for learning mathematics. If the student also has a mathematics teacher who is a dreary old poop and if the textbook is unreadable, then a comfortable cop-out is for the student to say that he has math anxiety. His friends won't challenge him on this assertion. In fact they may be empathetic. Thus the term "math anxiety" is sometimes misused. It can be applied carelessly to people who do not have it.

The literature—in psychology and education journals—on math anxiety is copious. The more scholarly articles are careful to separate math anxiety from general anxiety and from "math avoidance". Some people who claim to have math anxiety have been treated successfully with a combination of relaxation techniques and remedial mathematics review.

It would be heartless to say to a manic depressive, "Just cheer up," or to say to a drug addict, "Just say no." Likewise, it is heartless to tell a person who thinks he has math anxiety that in fact he is wrong—he is just a lazy bum. At the same time, mathematics instructors are not trained to treat math anxiety, any more than they are trained to treat nervous disorders or paranoia. If a student told you that he had dyslexia, then you certainly would not try to treat it yourself; nor would you tell the student that he just didn't have the right attitude, and should work harder. Likewise, if one of your students complains of math anxiety, you should take the matter seriously and realize that you are not qualified to handle it. Refer that student to a professional. Most every campus has one.

Never forget that you are a powerful figure in your students' lives. This fact carries with it a great deal of responsibility. If you were a follower of Dr. Kervorkian,[3] then you might take a troubled student in hand and say, "I know you are doing poorly in your math class. You must be in a great deal of pain, and suffering from shame. I have a solution for you—it's rather permanent, but it's painless." Chuckle if you will, but this is no joke. Problems of the psyche can be severe and dangerous. If a student comes to you with psychological problems then make sure that he gets help from someone who is qualified to administer that help.

Unfortunately, at some schools the "math anxiety" thing has gotten way out of hand. There are good universities where a student may be excused from a mathematics or statistics course (one that is *required* for his major) by simply declaring himself to be math phobic, or possessed of math anxiety. It is a sad state of affairs, but there is nothing that we math teachers can do about it. Because you and I are, by nature, good at mathematics, and because we do not suffer from math anxiety, it is difficult for us to empathize with people who suffer from this malady. It is really best to let those who know the literature, know the symptoms, and know the treatments to handle students who have this form of stress. Do not hesitate to refer your students to the appropriate counselor when the situation so dictates.

3.7 How Do Students Learn?

Psychologists, sociologists, education theorists, and anthropologists have debated the nature of the learning mechanism for decades. There is no general agreement on how students learn. See [ASI] for a modern discussion of some of the issues. This is a question on which serious scholars will spend their entire careers—and still not reach definitive conclusions. My only intent here is to share with you a few personal perceptions.

It is my opinion that the very best students tend to teach themselves. The instructor points out signposts for such a student, and then the student's native intellect takes over.

---

[3]Dr. Jack Kervorkian is the physician who has garnered considerable notoriety for assisting the suicides of severely ill people.

On the other hand, weak students are often quite dependent on the instructor and (perhaps) the text and the lectures. If you agree that these people are worth teaching at all, then you must be there for them. Provide good lectures and a reasonable text for them to work with. Set a good pace. Answer their questions. Be as helpful, and as encouraging, as possible to students who have the courage to come to your office for help.

Students of middling abilities are perhaps in the majority, and they share properties with both the best students and the worst. Know with certainty that you cannot please everyone. Students in the middle don't want to be clobbered with abstract theory, and they don't want to be bored with trivialities. It is not very difficult to follow your intuition and find a pitch that is appropriate for the middle level of a class. What is trickier is to do something for the lower end and for the upper end—and to handle the vast spectrum of student abilities all at the same time!

I like to pepper my classes with what I call "Culture Spots". I present one of these perhaps once or twice per week. These are tempting little bits that I throw out for the brightest, and the more curious, students to think about. If a culture spot sparks anyone's interest, then he can see me after class or in my office hour and we can discuss it further. An example of a culture spot is this: When I teach multivariable calculus, I of course must give the important example of the vector field

$$\mathbf{v}(x, y) = \left(\frac{-y}{x^2 + y^2}, \frac{x}{x^2 + y^2} \right),$$

which is defined on the annulus $\{(x, y) \in \mathbb{R}^2 : 1 < x^2 + y^2 < 4\}$, is closed, but is not conservative. Usually there is some student who asks me whether this is the only example. Great question! So I write "Culture Spot" on the blackboard, point out that this is an advanced topic that I'm only going to mention, and then tell them that one of the great theorems of twentieth-century mathematics is de Rham's theorem. I go on to say that de Rham's theorem says that the number of examples is equal (roughly speaking) to the number of holes in the domain of definition. Later on, when questions about different types of holes come up, I can refer to de Rham's theorem, and also make allusions to homotopy and homology theory.

Note that these advanced ideas are presented very briefly, and in such a fashion that students know that these are ancillary remarks, not part of the course, and are nothing that they will ever be examined on. But these are an entree for the gifted students to come talk to me, and to give them something to think about.

If you ever wonder why a calculus teacher should have a Ph.D. and be a research mathematician, then think about what I have said in the last two paragraphs. An instructor with minimal training from a junior college might be qualified to teach integration by parts to non-majors. But could he ever provide the sorts of insights that I have just described? I'll leave it to you to provide the answer.

Thus the lesson is to strive at least to provide some stimulation for students at all levels. The ability to do so, without an unreasonable amount of effort on your part, can come only with experience and determination. Even if you had a class with just three students, it is likely that their levels of ability would fall into two or more disparate categories. Thus there is no escaping the realities of having an uneven audience. Being forewarned, and being thoughtful, can help you to present your course so that you teach something of value to most of the students most of the time.

Let me cast these matters in a different light. An instructor who is demanding and difficult and who omits many details from his lectures will challenge the talented students and force them to go off and learn the material on their own. Certainly that was the nature of the graduate program that I attended, and it did me a world of good. On the other hand, an instructor who is lucid and who proceeds at a comfortable pace makes everything look too easy and can lull students into a false sense of security. That instructor will also bore the gifted students. How do you address both of these phenomena in your classroom?

Because at most universities we are training a widely diverse group of students, the issues raised in the last paragraphs are unavoidable. One solution, in a fundamental course like calculus for example, is for the math department to set up several tracks. There is a calculus course for scientists and engineers, one for pre-medical students, and one for business students (some colleges even have a "calculus for poets" course). This is a commonly used method to slice up student abilities so that the spectrum in any given classroom is not so broad. Another approach to the issue of widely varying student abilities might be to rethink the traditional classroom/lecture format and divide students into working groups in which they can seek their own level. The final word has not been said on how to deal with these problems. See Section 2.5 for more on the value of group work.

We should continue to explore new methods to teach mathematics. We are constantly hounded by certain observed facts. The failure rate in the large lecture system is unacceptably high. The retention rate is unacceptably low. America appears to be falling behind in the technology race because we are not training our young people thoroughly and well. Can we find a more effective way to teach—one that addresses the perceived problems?

In my view, it is simplistic to lay the blame for our teaching problems on the *formal method* by which we have been teaching mathematics. No methodology is perfect. But we all know something about how to lecture. We should consider how to make it a more useful tool before we discard it altogether and adopt new, unfamiliar tools.

Let us own up to the fact that many of us—especially those trained at high-powered research departments—are often not trained to care about teaching. Many of us do not. That is not the way the value-reward system is set up. Certainly none of us has had any real training in how to lecture, or how to get students engaged in the learning process, or how to teach discourse. The traditional methods of teaching still have much to offer, provided that they are being used by people who are properly trained and who care. But learning to teach is a lifelong process, and we should be open to learning new methods of teaching, and also to new facets of the old, familiar methods of teaching.

An instructor who stands in front of his class and drones on in an uninflected monotone, copying dry, discursive mathematical facts and theorems and proofs onto the blackboard is not teaching. He is not doing his job. In fact he is simply

hiding behind the mathematics. He does not have the courage to face his students
and to interact with his students.

I once had a colleague who so detested his students and his teaching assignment that he would handle his class as follows. Three minutes before the class
period, he would rip the relevant pages from his copy of the textbook. At the
appointed moment he would storm into the classroom, copy the material from
the eviscerated pages directly onto the blackboard, ignore all student questions,
and storm out of the room when he was finished. (I am happy to say that when
the chairman got wind of this behavior he yanked this tenured, full professor out
of his class—permanently—and docked his salary.) You may find this behavior
so bizarre as to be ludicrous. But if you examine your conscience carefully then
you will have to admit that we are all guilty of this sort of behavior from time
to time. On days when things are not going well, or you've been selected for an
IRS audit, or you cannot get interested in the topic of the day, or you cannot
figure out any good way to explain it, you just show up for class and blow the
whole thing off. You hide behind the mathematics, *and you do not teach.* Your
class is less like a learning experience and more like Kabuki theater.

So my message is a simple one. Stop hiding behind the mathematics. Don't
treat the mathematics as a screen or a foil. Step around and get out in front
of the mathematics, where the students are. Stand beside them and look at
the mathematics with them. Point out the interesting features. Help them to
appreciate what they see. Explain it. Talk to them about it.

Your class will be more involved with, and more engaged in, the learning
process if each class meeting contains activities in which the students can *participate.* If you copy a dry lecture onto the blackboard, discourage questions, and
then just walk out of the room at the end, then you have not offered the students such activities. If, instead, you engage the students in discourse (Section
3.14), then you have begun to break the ice. If you break the students up into
discussion groups, then you have gone even further to enable student activities.
If you send the students to the blackboard, either to talk to the class or to work
problems, then you have gone further still. The ideas in this paragraph are another illustration of the precept that in order to be a successful instructor you
must be consciously aware of what it is you are trying to accomplish.

Some universities these days are telling their faculty, "Don't be the 'sage
on the stage'. Instead be the 'guide on the side'." I am sorry to say that for
most university administrators these words do not spring from some deep well of
philosophical angst. Rather, they are seeking ways to teach more students with
fewer faculty. Even so, there may be something of value in this adage. You, the
instructor, should not be like the Wizard of Oz (see [BAU]), hiding behind a
screen and barking out profundities. Instead, you should be more like Dorothy,
standing with the students, helping them to ask questions and seek answers.
(See also George Andrews's Appendix on being a sage on the stage.)

3.8 Inductive vs. Deductive Method

It is of paramount importance, epistemologically speaking, for us as scholars to know that mathematics can be developed *deductively* from certain axioms. The axiomatic method of Euclid and Occam's Razor has been the blueprint for the foundations of our subject. Russell and Whitehead's *Principia Mathematica* is a milestone in human thought, although one that is perhaps best left unread. Hilbert and Bourbaki, among others, also helped to lay the foundations that assure us that what we do is (for the most part) logically consistent.

However mathematics, as well as most other subjects, is not learned deductively; it is learned *inductively*. We learn by beginning with simple examples and working from them to general principles. Even when you give a colloquium lecture to seasoned mathematicians, you should motivate your ideas with good examples. The principle applies even more assuredly to classes of freshmen and sophomores.

Beresford Parlett recently said

> Only wimps do the general case. Real teachers tackle examples.

This simple idea should be a guiding force whenever you are preparing to explain a new idea to your students.

Take the fact that the mixed partial derivatives of a C^2 function in the plane commute. To state this theorem cold and prove it—before an audience of freshmen—is showing a complete lack of sensitivity to your listeners. Instead, you should work a couple of examples and then say, "Notice that it does not seem to matter in which order we calculate the derivatives. In fact there is a general principle at work here." Then you state the theorem.

Whether you actually give a proof is a matter of personal taste. With freshmen I would not. I'd tell them that when they take a course in real analysis they can worry about niceties like this. Other math instructors may have differing views about the question of proof.

And by the way—you know and I know that C^2 is too strong a hypothesis for the commutation of derivatives. But, really, isn't that good enough for freshmen? If a bright student raises this issue, offer to explain it after class. But do not fall into the trap of always stating the sharpest form of any given result. Great simplifications can result from the introduction of slightly stronger hypotheses, and you will reach a much broader cross section of your audience by using this device.

Ralph Boas had these thoughts about the inductive method:[4]

> I once heard Wiener admit that, although he had used the ergodic theorem, he had never gone through a proof of it. Later, of course, he did prove (and improve) it.
>
> ......
>
> I do not think my story about Wiener is very surprising. One can't always be going back to first principles.

---

[4]Part of this quotation comes from a private communication between Boas and James Cargal, as quoted in [CAR].

I quite agree that—at least for *some* people (I am one of them) calculation precedes understanding. I have probably said before that I knew how to calculate with logarithms long before I knew how they worked. The idea that proofs come first is, I think, a modern fallacy. Certainly—even in this calculator age—a child learns that $2 \times 2 = 4$ before understanding *why*. The trouble with "new math" was (in part) the fallacy of thinking that understanding needs to come first.

Ralph Boas was a great teacher, and there is wisdom in what he says. Don't put the cart before the horse when you teach. A young student is ill-motivated to learn the inner workings of a mathematical idea before he has understood what it is and how to use it.

Now suppose that you are teaching real analysis (from [RUD], for example). One of the neat results in such a course is that a conditionally convergent series can be rearranged to sum to any (extended) real limit. When I present this result, I first consider the series $\sum (-1)^j/j$ and run through the proof specifically for this example. The point is that, by specializing down to an example, I don't have to worry about proving first that the sum of the positive terms diverges and the sum of the negative terms diverges. That is self-evident in the example. Thus, on the first pass, I can concentrate on the main point of the proof and finesse the details. After doing the first example, and thereby instilling the students with confidence and understanding, I easily can go back to the general result and describe quickly how it works.

Go from the simple to the complex—not the other way. It's an obvious point, but it works. An example of this philosophy comes from the calculus. Many calculus books, when they formulate Green's theorem, go to great pains to introduce the notions of "x- simple domain" and "y- simple domain" (i.e. domains with either connected horizontal or connected vertical cross sections). This is because the authors are looking ahead to the proof, and want to state the theorem in precisely the form in which it will be proved. The entire approach is silly.

Why not state Green's theorem in complete generality? Then it is simple, sweet, and students can see what the principal idea is:

Theorem: *Let $\Omega \subseteq \mathbb{R}^2$ be a smoothly bounded domain. Let $\mathbf{F}(x,y) = u(x,y)\mathbf{i} + v(x,y)\mathbf{j}$ be a smooth vector field defined on Ω together with its boundary. Then*

$$\oint_{\partial \Omega} \mathbf{F} \cdot d\mathbf{r} = \iint_{\Omega} \operatorname{curl} \mathbf{F}(x,y) \, dxdy.$$

When it is time for the proof, just say "to keep the proof simple, and to avoid technical details, we restrict attention to a special class of domains ..." This approach communicates exactly the points that you wish to convey, but cuts directly to the key ideas and will reach more of the students with less fuss. If a student asks you afterward what is lost by restricting attention to x-simple and y-simple domains, you can point out that more general (smoothly bounded) domains can be cut up into x-simple and y-simple domains. This is a *wonderful* answer, for the apt student will immediately begin to draw pictures and cut up domains, and will gain immense satisfaction from the process.

Here is a useful device—almost never seen in texts or discussed in teaching guides—that was suggested to me by Paul Halmos:

Suppose that you are teaching the fundamental theorem of algebra. It's a simple theorem. You could just state it cold and let the students think about it. But the point is that these are *students*, *not* mathematicians. It is your job to give them some help and motivation. First present to them the polynomial equation $x - 7 = 0$. Point out that it is easy to find all the roots and to say what they are. Next treat $2x - 7 = 0$. Follow this by $x^2 + 2x - 7$ (complete the square—imitating the proof of the quadratic formula). Give an argument that $x^3 + x^2 + 2x - 7$ has a root by using the intermediate value property. Work a little harder to prove that $x^4 + x^3 + x^2 + 2x - 7$ has a root. Then surprise them with the assertion that there is no formula, using only elementary algebraic operations, for solving polynomial equations of degree 5 or greater. (In the process, tell the students a little bit about Evariste Galois, and how he recorded his key ideas the night before the duel in which he was sure he would die—at the age of 21!) Finally, point out that the remarkable fundamental theorem of algebra, due to Gauss, guarantees in complete generality that any non-constant polynomial has a (complex) root.[5]

Notice how much depth and texture this simple discussion lends to the fundamental theorem. You have really given the students something to think about. Stating the theorem cold and then moving ahead, while *prima facie* logical and adequate, does not constitute teaching—that is, it does not contribute to understanding. As with many of the devices presented in the present book, this one becomes natural after some practice and experience. At the beginning it will require some effort. The easiest thing in the world for a mathematician to do is to state theorems and to prove them. It requires more effort to *teach*.

I once heard a splendid lecture by a distinguished mathematical physicist. He began by telling us that everyone thinks that someone else understands the Second Law of Thermodynamics. But in fact nobody really understands it. He went on to say that he had come up with a new formulation of the Second Law, and he could now say that he understood at least *this new formulation*. The rest of the talk concentrated on explaining his new idea. He began that portion of the discussion by saying, "Suppose you have a cup of coffee ..." Throughout the talk, he illustrated sophisticated ideas from mathematical physics by way of the cup of coffee.

Think a minute about what this speaker did for his audience. First, he exhibited incredible humility by saying that, up until this moment, he had never understood the Second Law of Thermodynamics. Then he went on to say that he found a new way to think about it that at least *he* understood.[6] To illustrate his new ideas, he spoke of a cup of coffee. Who in the audience would not be at ease after an introduction like this? Who would not be dying to hear more?

---

[5]I always take some delight at this point in telling my students that both of Gauss's sons settled in the St. Louis area.

[6]Bear in mind that this speaker is the holder of the Dirac Prize and many other high honors in the physics world.

Who would not feel that the speaker was welcoming him into his world? Isn't this what you want to do when you teach?

One could go on at length about the philosophy being promulgated here. But the point has been made. Saki once said that, "A small inaccuracy can save hours of explanation." Mathematicians cannot afford to be inaccurate. But, for the students' sake, we can simplify. We can reach out to our audience, and find a meeting ground. We can communicate.

3.9   Who Is My Audience?

I think that, collectively, we academic mathematicians have made some tactical errors in the past forty years. During that time, while our universities were undergoing explosive growth, and both student and faculty populations were increasing at a startling rate, we also saw a great deal of curricular development and change. And it happened, with some regularity, that other departments would approach the math department and say, "We think that our majors should know some mathematics. Can you help us?" Awkward discussions ensued. It came to pass that the "other departments" were unable to articulate their needs, and we mostly didn't care. So we would ultimately say, "Have them learn some calculus." The result is the huge, undifferentiated throng of students that populates our calculus classes these days. At some big state universities, as many as three thousand or more students enroll in calculus in the fall semester. But the hard truth is that we set up our service courses more to suit ourselves than to suit our customers.

Mathematicians of my generation were brought up to teach mathematics in general, and calculus in particular, in the French style. That is, we teach as though the classroom were full of little future mathematicians. We state theorems and we prove them. In the cold light of day, you know as well as I that we make little effort to speak to the business students in their own argot, or to the pre-med students in their own argot, or to the psychology students in their own argot. Schools with the requisite population and with adequate resources set up special sections of calculus for these different populations, and even use special textbooks. But the hard truth is that a calculus book for business students is little more than watered down calculus. Such a book is written just like a calculus book, except that most functions are polynomials, and the few exceptions are exponentials and logarithms. The business applications are usually just a tiresome appendage. A similar description applies to most calculus books written for the life sciences.

In fact the scenario that I've described above is a bit embarrassing for all of us. Probably a psychology major would get a lot more use out of a course that gave him some finite math and some statistics and maybe just a smattering of calculus. A business student needs to understand a statement like "the rate of increase of inflation is decreasing", but probably does not need to know the Weierstrassian $\tan\left(\frac{z}{2}\right)$ change of variable for integrals. A biology student needs to understand statements about average population density of mitochondria, but probably does not need to know how to analyze a hanging cable using hyperbolic trigonometric functions.

We have never given much thought to what our constituents really need, or really want. Recently at my own university the Dean of the Business School came to us with some uncertainties about our business calculus course. One upshot of the subsequent negotiations is that now when business students take our business calculus course they must concurrently enroll in a one-credit course—*given by the Business School*—in which a business instructor explains to the students what the calculus ideas mean *in their own language*. This is a wonderful idea, and one that I would like to see more widely adopted.

When we teach elementary mathematics courses, we must disabuse ourselves of the idea that our audience consists primarily of mathematics majors, or of future mathematicians. Even if you teach at MIT this notion is most likely incorrect. Like a good writer, you should be aware of your audience. And this awareness should inform your teaching methodology.

At a prominent public university on the West coast, the most popular calculus course (measured by the number of students enrolled) is the one taught by the Forestry Department. No kidding. A moment's consideration will explain why this is so. Forestry majors at this institution are required to take calculus, and they are not the most mathematically gifted people around. They cannot handle the calculus course offered by the mathematics department, so the Forestry Department created a course of its own. Meanwhile, students in other majors—chemistry, biology, nursing, and so on—are required to take calculus. But the regulations do not say *which* calculus course. So they vote with their feet—they gravitate to the easiest calculus course they can find.

If you are like me, your reaction after reading the fourth sentence of the last paragraph might have been "Well, we don't want to teach those students anyway." But you probably did not anticipate the way in which the paragraph would end. It is important for a mathematics department to look after its courses and, yes, keep control over them. At a large Midwestern University, the Engineering School teaches its own Japanese courses. This tells me that the Japanese Department is not minding the store. We should not be guilty of such negligence. Teaching students is what pays the bills, and what maintains our credibility with the administration. Teaching calculus to forestry majors may not be your favorite activity, but it is important for the welfare of your department. You should do it, and you should do it well.

Now let me return to the subject of being sensitive to one's audience. When I taught at Penn State, I once participated in a College of Science "Open House". The Open House was the university's way of reaching out to the community, and helping citizens become acquainted with the institution. Groups of young people were bussed in from all over the state of Pennsylvania to participate in the activities. Unfortunately, there was poor coordination between the administration and the individual departments. Therefore it was difficult to suit the planned programs to the different groups. In one particular hour, we planned to show a film about "turning a sphere inside out." The kids that showed up for this piece of entertainment were gang members from inner city Philadelphia. You can imagine that they were not fascinated by four Frenchmen in black suits, chain smoking Gauloises and discussing the niceties of low dimensional topology.

Being a quick thinker, I turned off the film and announced, "Let's talk about gambling. I'll show you some tricks with cards so that, when you gamble, you'll always win." One gang member fixed me with a grim look, pulled out a large knife, and said, "I don't need that %$!#&. I *already* always win."

As you can see, I thought I knew my audience, but I did not. The fact is that if I wanted to entice inner city gang members to further their education then I should have realized that mathematics—especially low-dimensional topology— may not be the medium that speaks to them. They might be more interested in law, or medicine, or engineering. As professors at universities, we take pleasure in being removed from the vicissitudes of daily life. But sometimes, to be good teachers, we must confront them.

3.10   Teaching Reform

The teaching reform movement had its formal beginning in 1986, when Ronald Douglas of SUNY, Stony Brook and Stephen Maurer of the Sloan Foundation organized a small (twenty-five participant) meeting, held at Tulane University, to lament the sorry state of lower-division mathematics education—particularly calculus teaching. Some of the early reports pursuant to Douglas's meeting, and the ideas that sprang therefrom, appear in [NCE], [DOU] and [STE1]. For a more recent assessment, see [STK], [LET], [TUC], [KLR], or [ROB]. Other points of view are offered in [WU1], [WU2].

Douglas is a great organizer, and he got people fired up with information that our attrition and dropout rate in calculus is embarrassingly high, that our failure rate is unacceptable, that our teaching is not optimal, and—yes—that the lecture is dead.

Douglas persuaded the federal government to be interested in the problem, and many different programs were set up at the national level to encourage mathematics faculty—at universities and colleges of all sorts—to rethink and re-invent their curricula. There are special programs sponsored by the National Science Foundation for work on

- calculus

- precalculus

- post-calculus

- high school curriculum

- instrumentation and laboratory improvement

- advanced technology education

- undergraduate faculty enhancement

- teacher preparation

and many others as well.

When Persi Diaconis was asked about his MacArthur Prize (see [MP1]), he said, "...if somebody gives you a prize and says that you're terrific, well, that's nice. But if someone gives you a prize and says here is $200,000 to show you that we mean it, maybe they really mean it." So it is with teaching reform. Government funding has made the reform idea into a movement, and has made it necessary for the rest of us to take it seriously. It has enabled the investigators to buy equipment, to run teacher-training workshops, to hold conferences, to write books, and to get the word out. It has really had an effect. Like it or not, the Harvard calculus book had the biggest first year sales of any calculus book in modern history—by a factor of about four. It has really had an impact, and has caused many of us to take a new look at calculus. That effect alone may justify saying that the money was well spent.

And the new methods have much to offer. Some of the hallmarks of the new methodology are these:

- Lectures, used as an exclusive teaching device, are not sufficient to adequately educate the college students of today.

- Traditional methods of teaching should be replaced by "cooperative learning".

- Students should discover mathematical facts for themselves.

- Students can help each other discover mathematical facts by working in small groups, with the instructor acting as a coach.

- Students can use computers to help them to visualize mathematical concepts.

- Students can use computers and calculators to take over the tedious aspects of mathematics (i.e., calculations and drill) so that they may concentrate on the conceptual aspects.

- Students can (and this is a significant new idea) use the computer to construct implementations of mathematical concepts in computer code, and thereby aid in the construction of those concepts in their own minds (see, for instance, [DUB], [DDLZ], [DUF], [DUL]).

Whether or not you agree with these precepts, they certainly merit your serious consideration.[7] I still use lectures to teach, but I do everything I can to let the students at least *think* that they are discovering the ideas. You certainly cannot stand in front of a group and *command* that they discover the fundamental theorem of calculus on their own (this is akin to a bald man commanding his hair to grow). But you can lead them through the woods, leaving a trail of corn that they can peck at and follow. You also cannot put students into a group and

---

[7]The Web site www.math.okstate.edu/archives/calcrefm.html offers information about many of the reform resources that are available.

say, "Discuss conservative vector fields and see what you come up with." You must provide considerable guidance and context.

Let me not mince words about one key notion. The British educational system—especially as it is practiced at Cambridge and Oxford Universities—does a terrific job of teaching students the art of discourse. Each student at these extraordinary universities is assigned a tutor, with whom the student meets weekly, who literally drills the student in the arts of written and oral communication. The student's weekly assignment is to write an essay on a topic chosen by the tutor. The student is to use all the resources of the university—lectures, other faculty, the library—in preparing his piece. Then he must meet with the tutor and defend it. It is a wonderful regimen, and serves to keep language and dialogue alive.

We in the United States have not traditionally done as well at this sort of training. Instead, we take eighteen year old freshmen, many of whom have never articulated a thought more sophisticated than an order for a burger and fries shouted at the plastic effigy outside of the local Jack-in-the-Box,® and we suddenly expect them to formulate and deliver cogent sentences about the chain rule and linear independence. Talking mathematics is a high level skill. We should not simply *expect* our freshmen and sophomores to be able to do it. Instead, we should *plan to train them to do it*. We should develop techniques for *teaching* our students to talk to us. Many of the teaching reform projects—notably the Harvard project (see [HAL])—have led the way in teaching students the art of communication, and particularly how to write (see also Section 3.14).

The Mathematics Association of America (MAA) has a project called CLUME (Cooperative Learning in Undergraduate Mathematics Education), the purpose of which is not simply to make faculty aware of new pedagogical techniques but also to help faculty to learn to use them. Information about this program may be obtained from the Web sites

>   http://www.math.upenn.edu/~gjporter/maa/cpd/clume.html

and

>   http://www.villing.com/clume/

I say this elsewhere in the book but it is worth repeating here. If you simply tack new teaching techniques on to your existing lectures and problem sessions then those new techniques probably will not succeed. If you want to use the new techniques then—to some degree—you will have to learn to teach anew.

The hidebound among us, those who are content with traditional teaching methods and who have little patience for the reform practices, are fond of suggesting that "money talks and baloney walks." If I were a real cynic, I would recall the story about Willie Sutton—the man who staged the great robbery of the Bank of England. When asked how he got into robbing banks, Willie said, "Well, that's where the money is." I know many of the principals who are involved in calculus reform. Surely they like receiving summer salaries and other pecuniary support no less than the rest of us. But these are people who are sincerely committed to bettering undergraduate education in mathematics. And they probably work a lot harder at it than the rest of us. Their conclusions are certainly worth our careful examination.

Some proponents of the reform movement cite the Meyer/Briggs Type Indicator, commonly known as MBTI (see [KRT1,KRT2]), to substantiate their views about the way that students learn. The argument goes as follows. The Meyer/Briggs shows, objectively and quantitatively, that there are sixteen different personality types. These types are based on four dichotomies: extroverted/introverted, sensing/intuitive, thinking/feeling, and judging/perceiving. These various types have different values and aspirations and needs. Yet mathematicians have always taught basic mathematics to suit one personality type (i.e., themselves—people with a penchant for mathematics). So, argue these reformists, it is up to mathematicians to acknowledge that they need to develop and learn different ways to teach mathematics to suit the different types of people that there are.

This is an interesting line of reasoning, and one whose acquaintance I only made recently. Certainly any successful teacher knows that there are students with mathematical ability and there are those without; there are students who are quick and students who are slow; there are students who think scientifically and there are those who do not. However, these are not the personality differences that Meyer/Briggs picks out. Rather, Meyer/Briggs focuses (as indicated in the preceding paragraph) on rather "personal" aspects of the personality. Should we adapt the teaching of calculus (for instance) to these different types of people?

My view is that one goes to college to get an education, and a substantial component of education is to learn different methods of discourse (refer to Section 3.14). The methodology and language and style of discourse of mathematics is tried and true. It conceptualizes certain types of problems in a precisely formulated language and and it provides tools for solving them. We would certainly be doing our students—and ourselves—a disservice to abrogate that hard-won value system and methodology in an attempt to sociologize our subject. Our job is to teach logic and problem solving and the mathematical method. It is what we know and what we are good at. And it is what we can teach and teach well.

My thoughts in the last three paragraphs are not meant to be a slam at the overall value system of teaching reform. Instead, it is a discussion of one particular set of values promulgated by a special set of people who profess to subscribe to reform. Since I am a great believer in critical thinking and the art of discourse, I offer this discussion as grist for that particular mill. Frankly, I think that it is far beyond the ken of the typical mathematician to become involved in assessing various personality types and adapting teaching to those types. Most of us are not very good teachers to start. To make the job vastly more difficult—by endeavoring to adapt our teaching to a spectrum of personality types—strikes me as self-destructive.

In reading articles about reform, and listening to discussions about reform, I find that there is a confusion between *content* and *motivation*. To my mind, there is nothing essentially wrong with the content of any standard lower-division math course. However, the ways that we motivate the material, or endeavor to make it real for the students, or attempt to make the material speak to the students' interests, is hit-and-miss at best. Most of us have given very little thought to the matter, and do not have the necessary knowledge from other disciplines that would be necessary to really make mathematics come alive for a business student, or a biology student. I think that the reform movement can teach us a great deal about how to help the students to become *engaged* in the learning process (see also Section 3.5 on the role of the mathematics teacher). But we should not confuse all the good features that reform has to offer with the (I believe misguided) fear that parts of the existing curriculum may be lost in the process of reform. Traditionalists will be more willing to accept reform if they are not beset by fears that the curriculum that they know and believe in is going to be lost in the shuffle. Reformers may be more willing to talk to traditionalists if they can believe that traditionalists are willing to listen.

Debate about reform tends to be quite emotional. I believe that the vituperation is due partly to the fact that people are worried about defending their turf. Also, middle aged math professors probably don't want to discard the teaching techniques that they have been using half their lives and then retool. But I have also found (and this phenomenon often holds when people disagree vehemently and hopelessly) that different participants in the reform discussion actually have different concerns and goals.

- Some are concerned with teaching a well-defined body of mathematics to a given audience.

- Others are interested in maintaining and promoting the students' self-esteem, and in empowering certain groups.

- Still others wish to cut the attrition rate, so that more students can advance to upper-division mathematics courses—no matter what the loss in content and curriculum.

I am not judging the goals described in the last paragraph. But if one reviews the somewhat alarming debates about mathematics teaching that have been going on in California (see [AND], [JAC1], [JAC2], and [ROSG]), one sees in the end that frequently the warring factions are discussing entirely different matters. It is essential that we communicate, and find grounds for cooperation rather than for hostility (see particularly the article by Judith Roitman in [GKM]). In the end, we all want to learn how to teach our students so that they learn what they need to learn and go on to success in their later activities.

Let us return now to the mainstream of the reform movement—the Harvard calculus project. I certainly do not agree with all of the tenets of the Harvard project [HAL], and I would probably have trouble using the Harvard book as a text. But at least the Harvard people have *done something*. They have formulated an approach to lower-division teaching, and they have written some books implementing that approach. They have run workshops to train people to use the materials that they have developed. Surely what they have done is much more constructive than the collective whining of which many of the rest of us are guilty.

Of course the Harvard book is not the final word on teaching calculus—nowhere close. As noted previously, one of the new calculus books on the market is Stewart's "Reform" edition (see [STEW]), which professes to be an artful marriage of reform and traditional methodologies. In fact I believe that the publication of Stewart's new book marks the beginning of the "second generation" of reform. Thus, as noted in the Preface to this new edition of *How to Teach Mathematics*, we will now see concerted dialogue between reform values and traditional values. I would expect and hope that other texts professing to be a miscegenation of tradition and reform will appear, and that we will thereby converge, as a collection of professional teachers, to a new platform of insights about teaching.

As I have said, the Harvard group has dealt with teaching reform, and calculus reform in particular, by *doing*. There are others, notably Ed Dubinsky and his collaborators, who have studied the theory as well as the practice of mathematical teaching. Worth particular note are the references [ASI], [BDDT], [DDLZ], [DUB], [CDNS], [ACDS]. The first of these posits a theoretical framework for mathematical curriculum development. The second, and the third as well, concentrates on cognitive aspects of student learning of binary operations, especially in the context of elementary group theory. The fourth describes a computer programming language—ISETL—which Dubinsky has found particularly useful in helping students to build ideas in their own minds. The fifth considers how to help students understand what a limit is. All of these are questions that ought to be of interest to the mathematics instructor. Reading these articles, you may agree or disagree with their findings. But the articles are bound to be useful in helping you develop your own ideas about teaching.

Franklin Roosevelt said that "Politics is the art of the possible." There is wisdom in these words. Nobody is prepared to discard wholesale the traditional teaching techniques that we have used for so long, and likewise no thinking person can ignore the new needs of our students, the new technologies that are available, and the new ideas about teaching that are being developed. Teaching is a lifelong passion, and we spend our entire lives developing and growing as educators. We continue to learn both from the reform movement and from rethinking traditional methods of learning and teaching. Our ideas will continue to evolve.

3.11 Mentors and Neophytes

I've had many mathematical mentors, but never a teaching mentor. I've learned how to teach by making virtually every mistake that can possibly be made and then trying to learn from those mistakes.

Having a mentor—someone experienced in the craft of teaching—can save you a lot of travail. I encourage you to seek such a guide. Your mentor will not always give you the answers that you want, nor possibly any answer at all. But you can benefit from the mentor's experience.

I have frequently served as a teaching mentor, for faculty both young and old. This has come about partly because of my willingness to play the role of mentor, and partly because I have written this book. I have derived some satisfaction from the mentor role, and have often felt that I've done a little good. But not always. For example, when I visited Australia, the chairman of one mathematics department took me aside and told me of their special problem. The freshmen in their large lectures—especially the males—act like escapees from the Asylum at Charenton. They run through the aisles. They throw paper airplanes, and fruit, and books—both at the instructor and at each other. And they shout and scream and carry on at great length. I expressed some confusion and dismay at this description, and allowed that I knew of no American university with this problem. After some discussion, the chairman told me that their freshmen all live at home with their parents and are therefore still *extremely immature.* This information certainly helped to explain the situation, but gave no clue as to how to deal with it (my only thought was massive and regular doses of thorazine). In the end, we agreed that what they were already doing was probably the best that they could do. Instructors had to learn to be tough and harsh from day one, and to deal with miscreants both directly and severely. New instructors were given a crash course in this technique. Those who lasted the first year were survivors in the strong sense of the word.

Fortunately, most of the teaching problems that you will face in your first year or two at the job will be more prosaic. A mentor can show you the ropes when you are writing a syllabus or exam, help you to choose the right text, explain when you can skip a topic in a course, suggest how to handle disruptive students, or tell you what to do if you are running out of time in a course. A mentor can help you curve the results of a test. To ask for help is nothing to be ashamed of.

Some mathematics departments now assign a senior teaching mentor to each new young faculty member. If your department is one of these, then take advantage of your mentor's experience and perspective. You don't have to follow all of the advice that is tendered, but at least you will have food for thought. If you are not assigned a mentor, then take the initiative and approach a senior faculty member. He will probably be quite pleased and flattered to be asked, and you will have initiated a useful relationship.

In 1998, in American business schools, it is fashionable to speak of the *sempai-kohai* relationship. Borrowed from the Japanese, this terminology refers to the notion that one should revere older people, and heed their wisdom, just because they are old. If you had asked me to respond to this notion when I was twenty, then I would have quoted the Yardbirds:[8]

> The shapes of things before my eyes
> Just teach me to despise.
> Will time makes man more wise?

Now that I am sliding into my golden years, I have a slightly different view of the world. Nobody is *a priori* correct just because he is old. But experience is a valuable commodity, and anyone who is willing to share the benefits of his

---

[8]The Yardbirds were a popular rock group in the 1960s—famous in part because guitarist Eric Clapton was in the group.

experience is someone worth listening to. Find a mentor, ask him questions, and listen to the answers. And then use this input to craft your own style of teaching.

3.12   How to Ask, How to Answer

If a pollster asked the average American voter, "What do you think of the up-coming election?" then the resulting answer would probably not be very en-lightening. If you turn to your calculus class one day and say, "OK, now we've covered Chapters 3 and 4—any questions?" then you will get a bunch of blank looks. By the same token, if a textbook salesman hands a new calculus book to a math professor and says, "What do you think?", the professor will probably say, "I dunno; they all look the same to me." By the same token, students come to professors with questions such as, "Like, you know; I don't think I understand any of this stuff we're doing."

It is a strange facet of the human condition that most of us don't know con-sciously what we think about most things most of the time. A skilled questioner learns to ask *specific questions* in order to obtain meaningful answers. Rather than asking your class if there are questions about Chapters 3 and 4, ask them instead if they are comfortable with the chain rule, or if they can do related rates problems, or falling body problems. The material in a person's memory is hung on hooks. You must reach for those hooks to get useful answers to your questions.

The same principle applies when you are holding a review session—for a midterm exam, let's say. If you are serious, if you *really* want to help the students, then it is simply not good enough for you to stand before the students and say, "Any questions?" They *do not know* what they want to ask. And, even if they think they know, they are timid about doing so. You must prompt them: "Do you understand integration by parts? Can you do partial fractions? What about the u-substitution? Is Section 7.5 confusing? Was the second homework assignment particularly difficult?" You, the instructor, must understand that your having said these things will **(i)** jog their memories, and **(ii)** make it OK for them to ask about these topics. I find that it breaks the ice for me to write a list of topics on the board. This is just one way to get the "review session ball" rolling. Remember: You must poke the students and prod them and, if necessary, embarrass them a little. Never forget the psychological aspects of teaching.

We implement these dicta naturally when writing an exam. You would never set an exam question for freshmen that said, "Tell everything you know about differential calculus." Instead you ask very specific questions. You want to train yourself to do the same when talking or lecturing to students. More, you want to train yourself to do the same in reverse when you are trying to elicit questions from students.

There is a gentle art of getting your students to pose questions. And I don't mean questions like, "Will this be on the test?" I mean the kind of meaty, well-thought-out questions that we all live for. Perhaps the most common complaint that I hear from disillusioned mathematics instructors is that they cannot develop any participation from their lower-division classes (see Section 4.5 on Frustration). The matter of garnering good questions is a non-trivial issue, and one to which an entire separate book could be devoted. You are going to have to find methods that suit your personality, and your teaching style, and that work for you. (See the Appendix to this section for some specific suggestions on how to increase student participation and inject some life into your class.)

The devices that you use can be quite simple. For example, giving a good quiz once or twice per week is a device for focusing student attention on some *particular issues*. The quiz is a little bit like a traffic officer pulling you over and threatening you with a citation. We are all aware—in a general sort of way—of the traffic safety laws. But if a cop gets in your face and starts telling you things that you are doing wrong then suddenly the penny drops.

The devices that you use can also be complex. You could have each student develop an ongoing, long-term project. Such a project might have the property that it must include material based on the ideas from each week of the course. And each student must be prepared to report to you at any time on the status of his project.

You may very well think that quizzes are too trite and semester-long projects are too massive for you to consider. Fine. I use quizzes frequently in my own course, and I'm frankly too lazy to do semester-long projects. Finding a way to get students to participate is something that you must do for yourself. Consider wheedling, threatening, cajoling, joking, challenging, priming. You can get through to your students by making them like you, or by scaring the hell out of them, or by conning them, or by being gruff with them. I am not necessarily recommending any of these. But if you want to be an effective teacher then you must find something that works for you.

As you experiment with ways to liven up your class, bear in mind the nature of the enemy. One enemy is that young adults, for the most part, are quite unsure of themselves. Unlike an experienced mathematician, who in effect makes a career out of asking (often stupid) questions, the student is deathly afraid of looking silly in front of his peers. He is not intellectually mature, and not experienced. He is not expert in the art of discourse (see also Section 3.14).

This last point is worth developing. If you have survived in the academic game, then you have learned to ask questions. You would never go up to a member of the National Academy of Sciences and say "Duh. I was trying to prove an interior regularity theorem for the Laplacian, and I just cannot seem to do it. I tried integrating by parts, but I couldn't decide what to do with the boundary term." Your friend the National Academy member would—justifiably—probably conclude that you were an idiot. A safer way to pose the question would be: "I've been thinking about interior regularity for the Laplacian. I know the classical ideas, but what is the modern approach? What would be a general context in which to fit this type of question?"

If you know something about elliptic partial differential equations, then you are probably not sent into paroxysms of ecstasy by the second question either. But it certainly sounds more intelligent than the first. And it gives the questioner some room to maneuver. Students simply don't have this skill at discourse, so they resort to the obvious subterfuge—they clam up. Part of your job as teacher is to help your students learn to engage in scholarly discourse. Help them to ask questions. If a student asks a weak question, help him to turn it into a better one. Try to create an atmosphere in which you and the students are co-explorers. Convey that you will sometimes make false starts, and so can they. It's a knack, but you can learn it.

Another enemy, besides the observed fact that students are uncertain and don't want to talk, is that mathematics *can be* (it is not by nature) a dry, forbidding subject. Part of your job as teacher is to make the subject come alive and to motivate the students to want to learn the material. This book supplies a variety of techniques for achieving that goal (Sections 1.7, 1.12, 3.1, 3.3, 3.5, 3.7, 3.12, 3.14).

APPENDIX: SOME SUGGESTIONS FOR ENCOURAGING CLASS PARTICIPATION

This appendix contains several techniques, drawn from the literature or from my own experience, for bringing your class to life. Take them for what they are worth. Some may appeal to you, and some may not. But reading about them may give you ideas of your own. Note that the activities discussed here are designed for classes of manageable size. They do not lend themselves well to a large lecture of 350; see Section 2.14 for a consideration of techniques suitable for that environment.

In lower-division political science courses, it is common for the instructor to begin a class by saying, "Today we are going to be a medieval village. Who wants to be the mayor? Who wants to be the executioner?" And so forth. It is quite natural for a mathematician to react to that type of classroom activity with derision, to observe that it appears to be childish and non-productive. Perhaps, but such devices are a wonderful way to get students involved with the subject matter. What can we do in our math classes that will **(i)** teach the students something of value and **(ii)** get them involved with the subject matter? Here are some possible answers.

1) Get students to go to the blackboard. I have noted in Section 3.1 that this is not necessarily the most efficient use of time. But it *is* a way to get the students to participate. If you wish, and if it is feasible in your learning environment, you could record problems on the board before students come into the classroom. Those who wish can go to the board—even before class begins—and work problems. To avoid having the same old students monopolize this activity, you could institute a rule that no student may work a problem at the board twice in one week. Of course the *entire class* should discuss the various solutions that are so recorded.

2) Have students prepare oral reports or mini-lectures. This activity is usually best reserved for the last part of the semester, when everyone is tired and students are receptive to a change of pace. Since most of the students will be inexperienced in activities of this nature, I recommend that you assign students to each give a fifteen minute lecture on a very specific topic. Time considerations show that this activity is only feasible in a rather small class.

3) Have students take turns writing and grading quizzes. It might be appropriate to assign a team of three students to each quiz. Not only will this activity cause the students to think critically about the material that they are studying, but it will also imbue them with an appreciation for the sorts of things that you, the instructor, must do.

4) If a student *cannot* do a problem, and brings this fact up in class, then have him go to the blackboard and explain what he tried and where he got stuck. It is certainly true that some students will be too shy to pull this off, but most students will be secretly thrilled to be treated like fellow scholars. You can orchestrate a similar activity for a student who *does* know how to do a problem.

5) Use "Minute Notes". These work in the following manner. Once every week or so, ask students to jot down on a slip of paper anything that is bothering them—problems that they cannot do or concepts that they cannot understand or anything else that pertains to the class. You give them just one minute for this task (hence the name). Do it at the beginning of the class hour, and collect the notes right away. Read them on the spot. You will suddenly have a much clearer picture of what is going on in the class, what concerns the students have, where you should go from there.

Perhaps more importantly, you will have given the students a feeling of empowerment. You will have helped them to understand that their input is a constructive part of the class. After a few weeks of Minute Notes, you will generally find that students are much more willing to raise their hands in class and make meaningful contributions to the learning experience.

6) If you are truly daring, then you can design your course so that it is more like a literature course. That is, you give the students regular reading assignments and homework assignments, but you do not lecture directly on a linearly ordered sequence of topics. Instead you come to class each time with an air of, "Well, what shall we talk about today? Who would like to begin?" The idea is that your classroom is a marketplace of ideas. You need to really know your stuff, and have an engaging manner, to pull this off. But it is bound to be great fun.

7) Have guest instructors. To use this tool well, you must work closely with the guests to be sure that they will talk about material that is salient to the class, and will present it at an appropriate level. If you think of the fourteen weeks (give or take) of your course in the same way that I have discussed single lectures or classes (see Section 3.7), then having guest instructors is a way to prevent your course from being an "uninflected monotone". You can also consider roles that graduate students, teaching assistants, and "teacher's aids" (i.e., teachers in training doing their practicum in your class) might play in livening up the atmosphere.

8) If you have the resources, and the breadth of acquaintance, or if your department has the contacts, you could bring in guest speakers from industry or government or business. Imagine a calculus class in which you bring in someone working on the NASA space station project to talk about how calculus is used to design the work platform for the engineers in space (I'm not making this up; there really is such a project). Students would really wake up and smell the coffee when confronted with such a class experience.

9) This technique was devised by Jean Pedersen. She asserts that it works extraordinarily well for her. It is called the method of "mathematical POST-IT$^{®}$ notes".

We all know that POST-IT$^{®}$ notes are those little squares of colored paper that easily can be affixed or un-affixed to a document for the purpose of making remarks or memos. The idea for the application of these devices in a math class is that the professor comes to class with a tablet or two of these notes, each having the professor's name (or some other identifiable epithet) stamped on it. Whenever a student asks a good question (not "Will this be on the test?" or "What is this stuff good for?" or some pseudo-question that the student just cooked up), then he is rewarded with a POST-IT$^{®}$ note. "So what?" you ask.

When the next exam comes around, the students are instructed to bring their POST-IT$^{®}$ notes along. They are to affix them to the front of the exam that they hand in. The student then receives two extra points (or some number to be pre-determined) for each POST-IT$^{®}$ note.

Reports are that, when this policy is announced in class, it is as though a jolt of electricity has run through the room. Suddenly hands are waving in the air, and previously uninterested students become the life of the party.

Now let me be the first to admit that this teaching device, like any other, is not perfect. Some students who are already alienated will become more alienated if they are unable to garner any POST-IT$^{®}$ notes. Other students may object that they are being treated like children. Think carefully before you try this, or any, new technique.

10) I have saved the most frivolous suggestion for last. Although you probably will not choose to use it yourself, it may suggest analogous techniques that more naturally suit you and your classroom. And, although the technique is a bit silly, it is currently in use by at least one successful math teacher.

On the first day of class the instructor announces that he is very embarrassed to report that he simply cannot spell. Students should feel free to correct his dreadful spelling. Then he begins to lecture, spelling "line" as "lien" and "book" as "buk". Students are so delighted confidently to be able to correct the professor's spelling that participating constructively in the mathematics portion of the course becomes very natural.

I find this last technique of deliberate misspelling to be a bit dishonest, but it's hard to argue with success. In my own classes, I endeavor to create the feeling that we are all creating the lesson together. I do this with a constant line of patter, much like that used by a magician or an illusionist. With this technique, I have the students talking all the time as well. If a mistake is made in class, then it is *our mistake*, and we fix it together. If a problem is solved correctly, then that is our shared triumph.

The key to bringing your class to life is to become involved with the students and to make learning a shared activity. Perhaps this is one of the great lessons of the reform movement. It is not an ideal learning environment to have the teacher as stick-man preaching before an audience of sponges. Learning should be done symbiotically, and it is up to the instructor to structure his class accordingly.

3.13   Teaching with the Internet

The Internet is a marvelous tool for making information available to a large body of people quickly. For example, if a mathematics department subscribes to an electronic journal, then as many people in the department who wish to do so can read the journal at the same time. Those who want to study a paper carefully can download it, compile it, and print it out. When you prove a new theorem, you can post your abstract (or your entire paper) on an electronic bulletin board. Your results are then instantly known around the world.

The Internet is also a useful teaching device. Create a Web page for your class. Put the class syllabus on the Web page. You could have a page about prerequisites for the course, or ancillary reading, or ways to prepare for exams. Post homework assignments and due dates on your class Web page. Put information about upcoming exams there. If you need to write up a correction to something from class, or disseminate a list of errata to the text, or post homework solutions or exam solutions, then the Internet is just the ticket.

I once read a proposal for an "Internet Mathematics Curriculum". The premise was that, at certain universities with a great many part-time and commuter students, absenteeism is a problem. Students have families and jobs and cannot always make it to class. In the electronic age, modes of communication are changing—so why not take advantage? The proposal was that the professor would still give his lecture, and those who could attend would do so. But there would be assigned note-takers who would post official notes on the Internet. The Internet could also be used to cut through the problem that students will not—or are too shy to—participate in class. The math class would have its own electronic bulletin board(s), and students could post their queries there—anonymously or not. Other students, or the professor or the TA, could answer the queries as they saw fit. Since many students have the same questions, this use of a bulletin board would allow the professor to use his time more efficiently.

The proposal that I just described was not funded. In fact it didn't even make the first cut. I think there is real merit to some of the ideas just described. But

I also think that the concept of an Internet University abrogates much of what the learning process is all about. Classes are held for a good *reason*, and it is this: Many things that we do in life have a ceremonial aspect. We hold funerals to come to grips with someone's passing, and to create a sense of closure; we have graduation ceremonies to pause to think about an important moment in a young person's life; we select people for prizes (the Nobel Prize, the Cannes Film Festival Award, etc.) in part to recognize talented individuals and in part to ponder the human condition and what we are trying to achieve. Just so, we hold classes so that the students will take an hour, go to a special place, sit in a controlled environment, and think in a focused manner about a particular subject under the guidance of an expert. If this were not as important as you and I know it to be, we would not do it.

My point is that Internet classes, while they may have their place, eliminate what is powerful about attending a class. Glancing at prepared lecture notes for your calculus class on your computer screen is (for the student) a bit too casual, and too much like turning on the radio. The student attempting to learn in this manner could be interrupted by the telephone, the doorbell, a pot boiling over, a baby crying, or any number of other exigencies (again, this is why traditional classes are a good thing). A mature and disciplined person with suitable scholarly training might be able to learn successfully from an Internet class. I'm not so sure about inexperienced eighteen-year-old students.

You can use the Internet as the nerve center of your class, to keep everyone informed of up-to-the-minute information and last-minute changes, to post new homework assignments, to post grades, to change your office hours, to give last minute room or seating assignments for the upcoming exam, and so forth. The concept of fielding questions over the Internet, or with *e*-mail, is a fascinating one. The one obvious impediment is that most students don't know how to enter mathematics using the keyboard.[1] This certainly is more efficient than trying to remember to photocopy the information and bring it to class, it avoids the class time wasted when you distribute handouts, and it is more permanent (that is, the material can always be found right there on the Web for the duration of the term).

I believe that the full picture of the value of the Internet as a teaching device is yet to be determined. But I caution you against thinking that it can be a substitute for classroom learning.

3.14 The Art of Discourse

Ask yourself this question: If a student has a successful and fulfilling college education, then what does he take away with him? Twenty years after graduation,

---

[1]The software product *NetTutor* by Link-Systems is designed to cut through this problem. It presents the student with a white board on which to write his query by hand. Or else the student can click on icons to pull down mathematical symbols. The student can submit a question anonymously or not. The professor can answer questions in real time or at his convenience—and he can do so publicly or privately. The professor also can, with little effort, create a database of frequently asked questions that he can allow the students to access.

what does that student still retain? What intellectual framework does he have to build on?

Comic Don Novello, in his role as Father Guido Sarducci (on the television show *Saturday Night Live*), gave the following answer. "If you majored in Economics, all you remember is 'Supply and Demand'. If you majored in French, all you remember is 'Parlez vous Français?' If you majored in Physics, all you remember is 'Every action has an equal and opposite reaction.' " (He might have added, "If you majored in math, all you remember is 'Take the exponent and put it in front.' ") So Father Sarducci proposed that people not spend four years and $100,000 on a traditional university education. If this is all you are going to retain, argued the good cleric, Father Sarducci will teach it to you in five minutes—and charge you much less. He called his solution "Father Guido Sarducci's Five Minute University."

We who devote our lives to university teaching hope fervently that there is considerably more to higher education than Father Sarducci's droll diatribe would suggest. In this section I am going to endeavor to say what that "more" is.

The naive answer to the question "What does a student get from his education?" is that the student receives career training. Certainly career training has significant value, and should not be dismissed lightly. But if we take the long view then we can see a larger picture. We can see depth and texture. What a student ought to take away from college is **(i)** critical thinking skills and **(ii)** knowledge of and experience with discourse (see also the discussion in Section 3.10). These two aspects of education are essential, and they are not disjoint.

In college, a student declares a major. And that is the area in which the student obtains advanced training. But most of the student's courses are *not* in the major. In those other courses, the student is learning philosophical discourse, humanistic discourse, the discourse of social thought, and scientific discourse. The student is learning *different modes* of critical thought.

For example, Renaissance philosophers considered the questions, "What is the world we see and what is the world we experience and what is the world that is *actually out there*? Are they one and the same world? If not, then how do they differ? And how can we tell?" Renaissance mathematicians studied algebraic equations. Renaissance musicians studied the lute. All of these are valuable avenues of inquiry, and they are all quite different. An important part of gaining an education is learning about these different modes of thought.

When we teach undergraduates—especially lower-division students—we are primarily teaching non-majors. It can certainly inform our teaching, and remind us of what we are about, to be cognizant of our goals when we teach. When you teach calculus to a pre-medical student, or finite math to a business student, you are endeavoring to acquaint him with modes of mathematical thinking, with our special method of reasoning and analysis.

In fact, when I teach my undergraduates, I have in mind a much larger and more ambitious goal. I want to teach my students that the world need not be a place in which they are passive observers. They need not spend their lives "letting things happen." Put in other words, we do not—at least should not— live in a world in which some nebulous *other people* generate ideas, and hold office, and make decisions. In fact it is *we* who are to become educated, to assume the positions of leadership, and to make the important decisions. To my mind, this is the role in society of an educated person. Perhaps we instructors have, in our own lives, realized this truth. But we should determine to pass it on to our students.

Surely it is more constructive, and more fulfilling for everyone involved, to bear these thoughts in mind as we lecture these unformed lumps of clay. Do not view the teaching process as a sorry labor—akin to shoveling out the Aegean stables. You are not trying to turn these eighteen and nineteen year olds into little mathematicians. Instead, you are trying to *educate* them, to stretch their minds, to teach them to analyze and to think critically.

3.15   What about Research?

Mathematicians of my generation were taught, both explicitly and subliminally, that there is only one thing worth doing—and that is mathematical research. Teaching and curriculum and departmental service and university service are all fine and well, but not among serious mathematicians. I have no reason to believe that more recent generations have been trained any differently.

OK. I believe it, you believe it, we all believe it. But the observed fact remains that there are precious few of us who can sustain a vigorous career of mathematical research from cradle to grave. On a festive occasion when he turned 50, A. Besicovitch (1891–1970) commented that, "two thirds of my mathematical life is over." Twenty years later, someone observed to the contrary that Besicovitch had written more than half of his papers—many of them among his best—since turning 50. He hastened to remind Besicovitch of the foolish remark that the great man had made at age 50. Besicovitch's reply was, "Well, numerator was correct." A great and inspiring story, but few mathematicians can claim that their best work was done between the ages of 50 and 70. We get tired, we run out of steam, we want for ideas.

As you read the words of this book on how to teach, you may be thinking, "This is all just dandy. But I am a research mathematician; that is what I do. I have neither the time nor the inclination to worry about teaching. And I'm frankly surprised that Krantz has the time to do so." I don't know whether I truly have the time. My research program is going rather well right now, thank you very much. But I *made* the time to write this book. I think that teaching is important.

I also think that our profession would be healthier if there were room for us to pursue other activities besides research. To put it in different words: If you can do research, if you want to do research, if you've got the ideas and the drive and the stamina, then that is what you should do before all else. But, instead of spending the waning years of your mathematical life staring forlornly at the wall, why not think about the curriculum, or think about creating a new course, or think about writing a textbook, or think about how to improve your teaching, or how to help others in the department improve their teaching? We all know that we get some of our best mathematical ideas while mowing the lawn or taking a shower. I have also gotten them while preparing a calculus lecture or while walking to class. You are no less a mathematician if you are interested in teaching and curriculum and writing as well as in research. Likely as not, such interests will make you a better scholar, and will give you more choices as your interests and abilities change.

Thinking about things other than pure mathematics is NOT a waste of time, nor is it a "substandard pursuit". These activities are still part of the world of mathematics, they keep your mind humming, and they can be productive and useful. Just as we are slowly learning that it is perfectly reasonable for our Ph.D. students to get jobs in industry or with the government (because the number of academic jobs is growing ever fewer), so we are *very slowly* learning that there are many different and worthwhile mathematical activities. If your research is going great guns, then stick with it. But if it is not, then perhaps you could spend a little time doing *other mathematical things*. And perhaps the rest of us could show respect and support for those who do such things. Who's it going to hurt? And how many can it help?

One of the striking weaknesses of the American mathematical infrastructure is that the best people tend not to write our textbooks or to get involved in curricular issues.[9] Go have a look at the current crop of calculus books, or linear algebra books, or ordinary differential equations books, and you will have no trouble convincing yourself of this assertion. Wouldn't the world be a better place if our mathematical leaders in research also were our mathematical leaders in didactics? In curriculum? I wonder what shape teaching reform would take if the members of the National Academy were behind it?

3.16   Do I Have to Teach Calculus Again?

Yes, you do. Especially if you teach at a large state school, chances are that you are going to spend a non-trivial amount of your time teaching calculus and statistics and linear algebra and sophomore ODEs (ordinary differential equations). It's all in the numbers. Just count the feet and noses and divide by three. Most students who take a math course—if they are not taking *pre*-calculus—take calculus or statistics or linear algebra or ODE's.

Of course you can, if you wish, treat this task as a dreary duty. And hardly anyone would blame you. But the harsh reality, at the close of the twentieth century, is that your department's reputation around campus hinges on how well you and your colleagues teach these three magic courses. And the dean will know what that reputation is and will act accordingly. There was a time when we could thumb our collective noses at the students and the administration and say, "We prove good theorems and we all have research grants. Go away and leave us alone." I'm afraid that, at most schools, this attitude will no longer fly. Administrators and parents and students—*and society*—have new expectations for the denizens of the universities. Like it or not, we live in the real world.

---

[9]It is also the case that the best people often are not attracted to go into teaching, either at the college or at the K–12 level, but that is a subject best saved for another venue.

Doing a creditable job with lower-division teaching is not a lot of trouble. You may neither desire nor seek the reputation of "teaching hero", but everyone—including your colleagues but especially including your chairman and your dean—will appreciate it if you pull your weight in the teaching game. And if you don't create trouble or cause extra work for others.

Teaching is really a team effort. While it is certainly true that each instructor is completely in charge of his one or two or three (or more!) classes per semester, it is also true that the aggregate of professors is responsible for educating all the students. You are part of a faculty that is offering a curriculum of mathematics. It is a great help, and a boost to morale, to think of your teaching assignment as part of this group effort. If you do a good job, it will make your life easier. But it will also make your chairman's and your colleagues' lives easier. I hope that reading this book will help you in that task.

Chapter 4

Difficult Matters

Nothing in education is so astonishing as the amount of ignorance it accumulates in the form of inert facts.

Henry Adams

Education makes a greater difference between man and man than nature has made between man and brute.

John Adams

It takes most men five years to recover from a college education, and to learn that poetry is as vital to thinking as knowledge.

Brooks Atkinson

Lilies that fester smell worse than weeds.

William Blake

I saw the best minds of my generation destroyed by madness, starving hysterical naked . . .

Allen Ginsburg, *Howl*

This is not the end. It is not even the beginning of the end. But it is, perhaps, the end of the beginning.

Sir Winston S. Churchill

It takes two to tell the truth: one to say it and one to hear it.

Henry Wadsworth Longfellow

4.0 Chapter Overview

Any activity that involves people interacting with people is going to generate occasional friction. In teaching, this friction could result from student cheating, or from late work, or from excessive mistakes in the lecture, or from ill behavior (by the student *or* the teacher). An experienced instructor knows how to dispatch these situations both decisively and efficiently. The inexperienced instructor is liable to bungle the matter and make the situation worse. Such errors can annoy everyone, and make the instructor's life unnecessarily miserable.

This chapter offers advice on how to handle many of the most common "problem situations" that arise in teaching.

4.1   Non-Native English Speakers

Many new instructors in this country are not native speakers of the English language. Besides worrying about all the usual stumbling blocks that dog a new teacher, such a person needs to worry about **(i)** lack of fluency with the language and **(ii)** lack of familiarity with classroom techniques in the United States.

If you are not a native speaker of English then fluency is entirely your responsibility. Every university offers a short crash course on English as a second language. Take the course. Watch television and movies. Read books in English. Converse with American colleagues. One friend of mine, from Vietnam, perfected his English by listening every morning to the wartime speeches of Winston Churchill—and regaling us with quotations every day at lunch!

If you are a non-native trying to learn English, then don't seek out and spend all your free time exclusively with people of your own ethnic and national background. It is of course natural for us to seek the company of people like ourselves. We all should do it from time to time, if only to relax. But if you want to learn English then you must first learn some discipline, and you must force yourself to talk to native speakers. It is not enough to know the words and the syntax of the language. You must learn both to speak and to understand it in its natural speed and rhythm. If you are going to succeed as a teacher in this country, then you must speak the language well enough to **(a)** be understood and **(b)** be able to field questions. It is point **(b)** that causes more trouble than **(a)**. Most students can get used to a teacher who has less than perfect proficiency with the mother tongue. But if you as instructor cannot understand their questions then you will be a complete failure in and out of the classroom.

Some students are prone to complain, and will use any excuse to justify a complaint. I speak and articulate accentless English. Yet, when I taught at Penn State, some students complained that I did not speak with the local accent. Some students complain about teachers with British accents and teachers with Australian accents. There is nothing that you can do about such complaints, so you should not worry about them.

If you are organized, if you speak up, if you treat students with the respect that *you* would desire from an instructor, and if you show some enthusiasm for what you are doing, then students will forgive a lot. Your foreign accent will fade into the background. Nobody will hear it any longer. And you will be a successful teacher.

Remember that there is a big difference between **(i)** speaking good English with an accent and **(ii)** not knowing how to speak English. Many people who fall into the second category rationalize their lack of ability by saying to themselves that they actually fall into the first category. This is a trap, and you should avoid it.

Many professors who received their training in other countries are (justifiably) impatient with our students. American students do not specialize as early in their education as do, say, European students. Even in a sophomore differential equations class in America, there is a broad cross section of students that includes pre-medical students and others from outside the mainstream of mathematical science.

I once taught a junior/senior level real analysis course. One student's primary interest was chemistry, but he was studying for an advanced degree in statistics, and this in turn required that he take real analysis. Fine. He was a bright and hard-working student, and I couldn't help but like him. One day I gave a rigorous definition of "continuous function" and he raised his hand and said, "That's not what I think of as a continuous function." A part of me wanted to beat him over the head. But he was coming from a different world, and he had posed a serious comment that demanded a serious answer. I was really on the spot. I had to *defend* my definition. I certainly learned from this dialogue. And I have, as a result of this experience, become more open to such questions. I would encourage you to do the same.

My message is this: Learn to be patient. Students will ask you to repeat terms. Students will ask you "non-mathematical" questions. Students might seem less able, or less well prepared, than those in your country. But they are bright and they are willing. You must learn to work with them. *After* you have learned how the American education system works, and what the students are like, you will find that your colleagues are receptive to your thoughts about its shortcomings. Before you have made this acquaintance, you are working in a vacuum and you should keep your own counsel.

In some countries it is the style of the university professor to stand at a lectern in the front of the room and to read the textbook to the class. Questions are considered to be a rude Americanism. An extreme example of a teaching style that is virtually orthogonal to what we Americans know is one that has been attributed to the celebrated Hungarian analyst F. Riesz. He would come to class accompanied by an Assistant Professor and an Associate Professor. The Associate Professor would read Riesz's famous text aloud to the class. The Assistant Professor would write the words on the blackboard. Riesz would stand front and center with his hands clasped behind his back and nod sagely.

My point is that in the United States, for better or for worse, we have our own way of doing things. The style here is to indulge in discourse with the class. Some professors make the discourse largely unilateral. That is, they lecture. Other professors encourage more interchange between the students and the teacher. Reading this book will help you to become acquainted with the traditional methods, and some of the newer methods, of teaching in this country.

4.2   Late Work

Late work is a nagging problem. The easiest solution to the "Can I hand just this one assignment in late?" dilemma is to "Just say 'no'." But what of the student who has a *really good excuse*? What if there has been a death in the family or some other crisis that the student cannot avoid?

The trouble with making one exception is that it tends to snowball at an exponential rate to N exceptions. In a large class this can be catastrophic. One possible solution is to tell the students that, when you calculate their cumulative homework grade, you will drop their two worst grades. That means that any student can miss one or two homework assignments with essentially no penalty.

It's a remarkably simple solution to an otherwise difficult problem.

There are a number of other possible answers to the late homework problem. You can downgrade late assignments, or you can assign extra work. You can just forget the missing assignment and base the student's course grade on the remaining course work. The point is that you should think about this matter in advance, and formulate a policy that you will use consistently. A choice of incorrect policy toward late work could lead to a lot of extra effort and/or aggravation for you. Don't be afraid to ask a more experienced colleague for help in this matter.

4.3   Cheating

Cheating is a big, and probably unsolvable, problem. Academic dishonesty is demoralizing for the teacher and for the non-cheating students. Honest students react to cheating with emotions that range from outrage to pity to melancholy. What is the point of studying so hard if cheaters can get good grades through skulduggery? And the cheaters' inflated grades affect the grading curve, which in turn affects everyone. On the whole, cheating is a moral outrage—for both instructor and student alike.

You will find it difficult to deal with the sort of students who cheat, for they may be dishonest with themselves and with others in a number of aspects of their lives. You want to be firm and fair and just all at the same time. But you *must* deal with them, and you must do so directly and firmly. As with late work and other difficulties, you must have a clear and consistent policy to apply to cheaters. Fortunately (see below), the university may have already formulated such a policy for you.

You may wish to set a moral tone against cheating by making an announcement on the first day of class. For large lectures, this may be especially important. Declare that you consider cheating to be an egregious offense—against yourself, against the other members of the class, and against the university. While you admit to the class that you may not be able to catch all cheaters, you assure the students that anyone caught cheating will be punished to the full extent of the law—*including expulsion from the university when appropriate.*

Be forewarned: Most American universities have set policies about handling cheaters. You are not free to act as you please when you catch a miscreant. In particular, there are due process procedures set up (to protect the rights of the accused cheater) that you must follow if you wish to punish a cheater. You do not necessarily have the right to tear up the student's exam, to give the student an "F", or to mete out other retribution. Check with the director of undergraduate studies in your department to determine the proper course of action when handling a suspected cheater.

One rule of thumb is that you should not be lenient with cheaters. Cheating cuts at the very fiber of what university education is about. When you catch a cheater, you must send a strong message that this behavior is intolerable.

At one Ivy League university, entering students are required to sign an oath that they will adhere to the university's Honor Code. Part of the honor code is that, at the start of any exam, the professor will record on the blackboard the statement "I pledge my honor that I have neither given nor received information during this exam." Each student is to copy the pledge verbatim onto his exam sheet and then sign it. The instructor is then required to leave the room for the entire duration of the exam. He may, if he wishes, return briefly in the middle of the exam to answer questions. The critical part of the honor code that the student signs at the outset of his education is that he pledges not to cheat and he also pledges to turn in any other student whom he observes cheating. Since the instructor must leave the room at exam time, we see that the entire onus of catching cheaters is placed on the students themselves!

An interesting policy, and one that would not work at every institution. A notable feature of making each student copy and sign the pledge on his exam is the following. If he is planning to cheat, then the university is forcing him to lie as well. Having served on university committees that adjudicate cheating, my experience with students is that they are disinclined to rat out their peers. Most people want to be told what to do most of the time, and the students whom I have known prefer there to be an authority figure who will identify and deal with cheaters. This means you, so you had better figure out how to do it.

The best defense against cheaters is offense. Give your exams in a large room. Space the students far apart. Check picture ID's to make sure that students have not sent in ringers (substitutes) to take the exam for them. Patrol the room. Avoid turning the exam into a power trip situation. Just maintain control.

Another aspect of cheating is plagiarism. Plagiarism is not as likely to arise in a mathematics class as in, say, a history class. But you should be aware of what it is and how to deal with it. Plagiarism is the appropriation of another person's words or ideas. It is too large to treat in any detail here, but see [MLA] and the Web site

http://www.cas.ilstu.edu/English/145web/DprtInfo/Plag.html

One advantage from your point of view is that you do not have to handle plagiarism in real time. You have the plagiarist's work, together with the putative source material, in front of you. You may consider it carefully, show it to colleagues, ask your undergraduate director how to proceed. The best policy is not to attempt to act alone.

One could easily write another book about techniques to catch cheaters. In some departments, exams are photocopied (or at least a sample of them is photocopied) before they are returned to students. This is to dissuade a student from altering a graded exam and then coming back to the instructor to request more points. Some departments (such as my own) use elaborate statistical procedures to detect unnatural correlations among students' answers on multiple choice exams. (A student caught by means of such a mathematical technique finds it quite difficult to defend himself!) Many other devices are available.

The point is that it is worth spending a few moments thinking about how you will handle cheaters. There are many pitfalls to be avoided—in particular, you must respect the accused cheater's rights as specified in your university's code of conduct. There is nothing very pretty about a situation involving cheating.

Just remember that you are not free to act on your own. Become acquainted
with your university's procedures.

4.4   Incompletes

The profession of teaching, while certainly a stimulating and rewarding one, is
littered with nasty little details. One of these is the "incomplete". The theoretical purpose of an incomplete is to provide a vehicle for handling certain problem
situations. Perhaps a student *has* completed a substantial amount of material in
the course, but has been ill or has suffered a death in the family or some other
setback. He needs to defer completion of the course work until the next term.
The professor fills out an "Incomplete Form", and records the student's term
grade as an "I" or "Incomplete", to formalize the understanding that the student will complete the work at some pre-specified future time. Many universities
find it convenient to let professors administer incompletes as they see fit. As a
result, there is much inconsistency and abuse.

Frankly, I've given a lot of incompletes in my life and very few of them were
ever completed. Students get busy with the next semester's work, and never get
around to things past. In fact I did not complete the only incomplete that I ever
took as a student. It is also unfortunately the case that certain students will
simply blow off a course and then ask for an incomplete at the end of the term.
Often it is easier for you as the instructor to just grant the incomplete, given
that an otherwise undisciplined student is not likely to complete it (the grade
then usually, but not always, reverts to an "F"). You may very well wonder what
is the point of engaging in a long interview with such a student to determine
whether the incomplete is merited.

All this having been said, it is probably best, as with all matters in teaching
that impinge on fairness, to have a uniform policy for handling incompletes. But
think this through. Are you going to require that the student provide *proof* of
his excuse? This sounds reasonable, but what if the student says, "My mother
is dying of cancer." or "My grandmother just died and I cannot concentrate on
my work." I know professors who will demand a letter from the physician or
the undertaker, but this strikes me as a bit extreme. It could also prove to be
uncomfortable or embarrassing for all concerned.

One convenient way to handle the request for an incomplete is to instruct the
student to approach a professor teaching the same course the following term. The
student should ask whether he can audit the course, having his work graded. The
new professor of course will *not* submit an official grade for this student (after all,
the student is not registered in his class). Instead, he will transmit the resulting
grade to you. You then fill out a form to remove the student's Incomplete grade
and replace it with that letter grade. This is clean and simple, and it works.
You certainly don't want to have to re-teach some or all of the course for the
benefit of just one student.

You are the academic analogue of a middle management executive in the
business world. Executives exist, presumably, because they are smart enough
to handle exceptional circumstances. Teaching is loaded with all of the sorts

of exceptions that are connected with dealing with *people*. I have used the "incomplete" here as but one example of the problems and potential enigmas that can arise. Your department probably has set policies, or at least guidelines, for handling incompletes. Become acquainted with the routine procedures before you give your first "I".

4.5 Frustration

One of the most commonly heard complaints of college mathematics instructors, especially experienced instructors, is this: "Math 297 is a prerequisite for the course that I am teaching yet the students don't seem to know anything from Math 297." A variant of this is "My calculus students cannot[1] add fractions" or "My calculus students don't know how to expand $(a + b)^2$.".

Indeed, these are valid complaints. It is also valid to complain about the high cost of living, or about death and taxes. The peccadillos described in the first paragraph are facts of life and we as math instructors must deal with them. The truth is that we instructors think about math all day long, every day. We see the entire curriculum as a piece. For the experienced math instructor, there are no seams and creases between linear algebra, calculus, differential equations, and so forth. We swim effortlessly through the ideas, using whatever tools are needed. (By the way, if this doesn't exactly describe you then don't panic—I'm using a bit of poetic license here.) Students are different. They think about math when they are in the math classroom and (one hopes) for a few designated hours outside the classroom, but they are not inured in the subject.

So what is the point? It is simple. If you are in the second week of freshman calculus and you need to add two algebraic expressions that are fractions, then gently remind the students how to do it. If you need to expand the expression $(a + b)^2$, then say, "You remember how this works—right?" After a few gentle reminders, most of the students fall into the flow and they *will* remember how it goes.

Take a break and watch the "Tonight Show" or "Late Night". Listen to the monologue. If the host is going to crack a joke about someone slightly less famous than Bill Clinton, then he gently reminds the audience who it is that he is talking about. It's just good sense. These television hosts can be even less sure of how well informed their audiences are than we can be in our math classes. They guarantee that their viewers will *understand* by providing a bridge.

Contrast these recommendations with the following rather common alternative. The professor needs to add two fractional algebraic expressions, so he just barrels through it (rapidly and without comment, as one would do for a colleague). After a few moments some hands are raised, some hesitant questions are asked, and it soon becomes clear that many students are lost. The professor says, "What is the matter with you people? This is high school stuff. Am I a

---

[1]The Bank of America in Westwood Village, Los Angeles used to regularly place an advertisement in the UCLA student paper each spring. The purpose of the ad was to encourage members of the latest graduating class to consider a career with B of A. The ad read in part "applicants must be able to add fractions." So it is not just math teachers who are plagued by this problem.

baby sitter or what?" (I'm not making this up; I have colleagues who do just this.)

In my view the behavior of the mathematics instructor in the last paragraph is the mathematical analogue of shooting one's self in the foot. The professor (perhaps unconsciously) *sets up* a situation for failure. And there is no good or useful point to it. It not only sets a bad tone for that day, but also for the remainder of the course. The instructor needs only to expend just a little extra effort to anticipate these pitfalls, and to devote a few seconds to allaying them. And it does a world of good.

In fact, for me, helping the students recall how to add fractions or to expand quadratic expressions (or any other analogous elementary operation) is a form of protective coloration. It is too easy for me to make errors when doing these elementary operations. If I whip through them, leaving most of the students in the dust, then I in fact increase the likelihood that I will make an error, and also abrogate any sympathy I might have garnered when my error was detected. If instead I slow down and walk the students through the calculation, then it becomes "our calculation". They help me to check it, and the chance of an error is reduced to virtually nil.

As indicated at the beginning of this section, these frustrations also present themselves at more advanced levels—even with math majors. As an instance, linear algebra is often a prerequisite for multi-variable calculus. And well it should be, for matrix language is a natural vehicle for expressing the derivative, the chain rule, and so forth. But it is an artifact of the American mathematics curriculum that linear algebra is often taught in a vacuum. The students have no hooks to hang the ideas on, and they do not remember them very well. There is no alternative, if you want to keep your multi-variable calculus course on an even keel, to giving a whirlwind review of the salient linear algebra ideas as you use them. Here, by "whirlwind", I mean a five or ten minute snapshot, on the fly, of the relevant idea right before it is used.

Here is another way to look at the matter of frustration. You and I have become accustomed, when we visit our physician or our attorney or our psychotherapist, to a certain amount of professional decorum. Often the doctor or lawyer or counselor meets us in a well-appointed office, dressed in a suit or other formal attire, and he exudes courtesy, detachment, and professionalism. Sadly, academics do not seem to have bought into this game. If you had a fight this morning with your spouse, or got a speeding ticket when driving to work, or got a rejection of your latest paper in the mail, then you are liable to take it out on your students. Your office may look like a pigsty and your haberdashery like something from Zola's *La Terre*. But you have set a standard for your students and if they don't meet it then you may lose your objectivity or your patience and you may react.

If you have been dutifully teaching your calculus students for several weeks running, and if their latest midterm shows that they've absorbed very little of your wisdom, then you are liable to vent your spleen at them. You would never expect your family doctor to start hollering at you about losing weight, nor your lawyer to scream at you about paying your taxes on time, nor your psychiatrist to excoriate your for being too neurotic. But you and I sometimes find ourselves altogether losing it and—more is the pity—giving hell to our students.

Of course there are notable ways in which doctors and lawyers differ from academics. What makes us special is that we endeavor to impart knowledge to our students *and we expect them to radiate it back at us*. When they fail to do so, then we are disappointed, sometimes angry, and certainly frustrated. What I am suggesting here is that it can serve to your advantage to set yourself apart from your students. Maintain some objectivity. Try not to become emotionally involved. As award-winning teacher David S. Moore observes (see [MOO] as well as Section 3.1), teaching is a job: You prepare your class and you go do it. If there are problems, you deal with them. If the students aren't learning then you teach harder. It is part of the academic milieu, and part of our training, to think of ourselves like operatic divas: If things don't go as they should then perhaps a tantrum is in order. Not so. Be strong.

If the students are not working hard enough, nor absorbing the ideas at a pace and depth to suit your ideals, then too bad. But it's too bad for them; it's really no big thing for you. Teaching freshman is like mowing your lawn. No matter how good a job you do this Saturday, you are going to have to do it again next Saturday. Yelling at the lawn doesn't help.

Of course I am disappointed when my students—despite my best efforts— can't do three-dimensional graphing, or can't understand Stokes's theorem, or can't apply the ϵ-δ definition of continuity. But my job is to teach and I just get in there and do it. If I have to cover a tricky topic twice, or even three times, then that's the breaks. Part of being a successful teacher is gaining your students' trust. Go watch the movie *Stand and Deliver* about the legendary calculus teacher Jaime Escalante. He was tough on his students. He told them when he was disappointed and he worked them hard. But he never belittled them, and he never lashed out at them. He showed genuine pride and enthusiasm when they did well. The most important thing he did for his students is that he made them believe in him. They worked hard for *him* because they trusted *him*.

The frustration problem described here is one of the few in this book that plague the experienced instructor somewhat more than the novice. Novices are usually drunk with youthful enthusiasm for teaching. Middle aged folks like myself are often just tired. We tend to lose our patience, and to forget the struggles of the uninitiated. An instructor who has been dealing with, and teaching, the ideas for twenty years cannot understand why students don't remember what they have already seen once. *Once!* The key to success here is to try to develop (or remember) a little sensitivity to the point of view of the students.

As a closing thought, consider the following. If your students are not speaking to you then it is probably because you are not speaking to them. You may be lecturing *at* them, you may be exhorting them, you may be talking down to them, or you may be venting your spleen and verbally abusing them. But you are not relating to them as people. You are not teaching. Try it—you'll like it.

4.6   Annoying Questions

At several junctures in this book I have mentioned some spine-tingling, bone-chilling, conversation-stopping questions that students can and will ask. One of these is, "What is all this stuff good for?" Another is, "Will this be on the test?" Another is, "Why don't you prepare your lectures more carefully? You are wasting our time."

If you are asked the third question then the fault probably lies with you. You should have done a better job preparing your class. If things are really going dreadfully, you might say to the class, "I apologize. This class is going very badly. Let's quit for today." Nobody will take this amiss, and it is probably the most diplomatic way out of an uncomfortable situation—but do not use this device more often than about once per semester. The best policy is to use forethought to prevent such an encounter.

Dealing with the first two questions, and others like them, is something that you will learn to do through bitter experience. In America in the 1990s, we endeavor to educate a broad cross section of the population. We cannot assume, as perhaps a don at Oxford could one hundred years ago, that our students are at the university primarily to learn to become refined citizens—and that they are happy to consider whatever we set before them. In particular, today's students are prone to *challenge* what we are doing. It is a part of your job to be prepared to answer their challenges. The challenges are not generally hostile, but having respect for your audience requires that you be prepared to provide a thoughtful response. If you accept my premise—repeated throughout this book, but particularly in Section 3.14—that getting an education is learning the art of discourse, then you should set an example for your students. If one of them poses an intelligent, well-thought-out question (even though it may be one that you don't particularly want to hear), then you should endeavor to answer it in a manner that is both correct and intellectually stimulating. Now let me say a few words about the particular queries pinpointed at the outset of this section.

Let us consider the question, "Will this be on the test?" One option that you have for an answer is the obvious one. Tell them that it *will* be on the test and then, indeed, put it on the test.

Now let us look at the opposite situation. If you are going to present something to your students and have no intention of testing them on it, then you have two choices. You can tell them up front that they will not be tested on it—they should just sit back and listen. Or you can tell them that they will be tested on it and then *don't* test them on it. Know consciously which choice you have made before you proceed.

If you choose the first route indicated in the last paragraph, then you might put the exercise in context like this: Explain that some ideas are difficult and deep. It requires several exposures to such an idea before it begins to make sense. This is an opportunity for the student to begin to ponder something important. Students are pleased to be treated like fellow scholars, and will usually act accordingly.

If you choose to tell the students that X will be on the test but in fact you have no intention of putting X on the test ... That is OK, but don't over-use this privilege. It will irritate the students and could damage your credibility.

The question "What is all this stuff good for?" is treated in Sections 1.7, 1.11. You, the mathematician, can get so wrapped up in your mathematics that such a question can catch you entirely off guard. Spend a few moments arming yourself against it.

The main point is this. You are not lecturing fellow mathematicians, who are inured to your point of view. You are lecturing students. Students will challenge you and ask questions. Some of these questions are difficult. If you want to retain the students' respect, then you must be prepared to deal with their queries and to understand their point of view.

4.7 Discipline

One hopes that, in a college environment, discipline will not be a big problem. But there are difficulties that can arise (see Section 3.11 for an extreme example).

In a big class, with two hundred or more people, student talking can get out of hand. Many students read the newspaper, or knit, or eat their lunches, or write letters to their friends. Some students, when the lights are turned low, engage in romantic activities. Students come in late and leave early. Students sleep.

I see no point in making a spectacle over a student who is not causing a disturbance. If a student is quietly eating lunch, then that is no problem for me. If a student comes in late or needs to leave early and does so in an orderly and non-disruptive fashion, then I let that student alone. (A truly courteous student will tell me in advance that he needs to leave early or come in late, and I am suitably appreciative. If you find this custom attractive, then tell your class that you want to be notified in advance of such temporal irregularities.) To make a scene will only alienate the whole class.

How should you handle a student who causes a disruption? First try something gentle like, "OK, let's quiet down." Another technique is to simply stop talking until you have everyone's attention. If one or two applications of "nice" is ineffectual, then come down on the offender quickly and sternly. Example: "Mr. Trump, if you want to talk then please leave the room." or "Ms. Lewinsky, everyone else is here to learn. Please keep quiet." If you deliver these injunctions firmly and with confidence, then they will have a chilling effect.

Often you can arrange for the other (non-offending) students to be the bad guys. If you simply sit down and wait for silence and cooperation, then the other students will "shush" the offenders. You can just provide a modicum of stern looks.

William James was both a father of twentieth-century psychology and also a renowned teacher. When he felt that his class was not cooperating—either inattentive, or talkative, or simply dull—he would fold his arms and lie down on the floor. After a while, the class would fall into puzzled silence. Then he would announce, "Education goes both ways. You have to participate too!" And he would try again. As you can imagine, a clever teacher could make something like this device into a productive game with his class. I have a special place on the wall of my classroom where I ceremonially bang my head when my students are being slow. You should create your own tricks—ones that suit your personality and your style.

On those absolutely rare occasions when a class is beyond control, you might throw down your chalk and say something like, "This class is hopeless. I'm through for today. We'll try again on Friday." I have never used this last device, and I fervently hope that I shall never have to resort to it. But somehow it gives me strength to know that it is at my disposal. If you do take this extreme measure, you had better let your chairman know what you have done.

I have seen large mathematics classes (of about 400 or more students) which looked like a cross between a rock concert and a Hieronymous Bosch painting (see also Section 2.14). Private conversations and mini-dramas were taking place all over the room while groups of students roamed the aisles. The professor stood at the front of the room, bellowing away on his microphone, while a small percentage of the students attempted to learn something. Such a situation is plainly unacceptable. Nobody can ask a question, nor can there be any interchange of ideas, in such an atmosphere. Certainly the question of learning the art of discourse is all but absurd in this context. Generally speaking, a situation like this comes about *gradually* over the course of the semester. It happens because someone (most likely the professor) lets it happen.

Some professors prefer to deal directly, and in advance, with the discipline problems connected with large lectures. On the first day of class in a large lecture, these instructors tell the class that large classes present special organizational problems. In order to make the experience as beneficial as possible for everyone, the instructor goes on to prescribe certain rules of behavior in the classroom. These include no eating, no talking, no reading of the newspaper, no coming in late, no leaving early, and so forth. Remember that the first few lectures of your class are your chance to set the tone. You may wish to take the opportunity (gently) to "lay down the law".

Of course if you are going to use the stern system described in the last paragraph—and it is a perfectly reasonable one—then you must follow through on it. If a student breaks one of the rules you have laid down, then you must call him on it: "Mr. Goerring, I said no eating in class." or "Ms. Flowers, if you need to leave class early then you shouldn't come at all."

Because you are an authority in your field, because you give out the grades, and because you hold sway over students' lives, you have both moral and *de facto* authority in the classroom. As a result, if you comport yourself like a concerned, dedicated professional, then you should have relatively few disciplinary problems. If you nip behavioral problems in the bud, and handle them with dispatch, then they will not get out of hand and your classes will go smoothly. But your antennae should be out for trouble. You will figure out quickly who the wise guys and troublemakers are. When they start to rattle, you start to roll.

Some instructors, in extreme problem situations where there are continuous interruptions and talking during a large lecture class, enlist a confederate. The confederate can sit in the back of class (unidentified to the students) and take notes on student behavior when the (official) instructor's back is turned. That way, after consultation with the confederate, the instructor can be absolutely sure who is doing the talking, throwing the spit wads, and causing the trouble. And he can act accordingly. I was always impressed by my grade school teachers who had eyes in the backs of their heads. They knew when I was going to misbehave even before I started thinking about it. Most of us have not developed such a skill, and may find it (on rare occasions) useful to invoke the device described here.

Not long ago I had a student come to me and tell me that he'd skipped the previous two weeks of class. But he was now returning and would do his best to catch up. I said "fine"—if he needed some guidance, then he should let me know. Indeed, he showed up in class that day. I began the lecture by saying, "OK, let's have another look at Stokes's theorem." My prodigal friend, who had just been in to see me, said, "Could you give a quick review of this concept?" (I had been discussing Stokes's theorem for most of his two week absence.) I am ashamed to say that I lost it. I said, "No. Not for someone who hasn't been to class for two weeks." The other students supported me in this. I could tell by looking in their eyes. But I felt like a rat. When I conducted my personal debriefing after class, I wondered whether I had done the right thing. It all came out well, because a few minutes later he came to my office and apologized to me, I apologized to him, and everything was hunky dory. In retrospect, I think I should have said, "We've been studying Stokes's theorem for two weeks, and it doesn't lend itself to a quick summary. See me after class if you want more help."

Always remember that you have the power to command respect, but you cannot *demand* it. If you present the image of an organized, knowledgeable scholar who is trying to do a good job of teaching, then most students will play ball with you. If instead you are a bumbling, unprepared clod who clearly doesn't care a damn for the class, then you can expect like treatment from the students.

I used to have a colleague who handled late arrivals to a large lecture in the following fashion. He would cease to lecture and make a show of timing how long it took the student to get seated. Then he would say, "There are 150 students in this class. It took you 90 seconds to get seated. Thus you wasted 3.75 hours of their time. Now each student here has paid N dollars to take this course. Let us next calculate how many of their dollars you have wasted." And so forth.

Let us consider the effect of this practice. Certainly any student who planned to come in late in the future would wear a bag over his head. But I cannot help but think that this sort of arrogant behavior on the part of the instructor suggests a serious attitude problem. Students will lose respect for an instructor who behaves in this fashion.

As the instructor in a classroom, you are in charge. You have every right to demand a certain type of behavior from students, and to enforce discipline. But you must not, in the course of disciplining a student, diminish his self-respect. Students are young adults, and should be treated as such.

We have all seen parents who cannot control their children. We have also seen 95 pound fathers who hold tremendous sway over their 250 pound linebacker sons. Parents of the latter sort understand the difference between *demanding* respect and *commanding* respect. The first is easy, is a convenient way to vent your spleen, and often doesn't work. The second is more gentle. It is an art that you need to cultivate. The techniques suggested in this book should help you in that task.

Remember this: You should be conscious of maintaining discipline in class from day one. This is not to say that you should be an unbending authoritarian— far from it. But if you let a class slide out of control for six weeks and then try to use the techniques suggested here (or other techniques) to take back the reins of power, then you will have an extremely difficult and unpleasant time.

The famous mathematics teacher R. L. Moore (see Section 1.12) is said to have once brought a Colt 45 to an unruly math class, set it conspicuously on the table, and then proceeded into his lesson in a room so quiet that one could have heard the sun rise. This technique may have been suitable in Texas fifty years ago. These days, however, I would recommend the use of more civilized techniques for keeping order.

4.8   Mistakes in Class

The most important rule to follow before giving a class is to prepare (Section 1.3). How much you prepare will depend on you—on your experience, your confidence, your training, and so forth. Being fully prepared gives you the flexibility to deal creatively with the unexpected.

But nobody is perfect. No matter how well prepared you are, or how careful, you will occasionally slip up. In the middle of a calculation, a plus sign can become a minus sign. An x may become a y. You will say one thing, think a second, write a third, and mean a fourth. It is best if you can handle these slips with a flair, and particularly without sending the class into a tailspin.

I endeavor in my classes to create an atmosphere in which students are comfortable to shout out, "Hey, Krantz, you forgot a minus sign." Or, "Is that a capital F or a lower case f?" This is a form of participation, and it can be a very constructive one. If you handle these situations badly, then students will be less inclined to ask questions or to approach you on other, more important, matters.

If mistakes are small, and occur in isolation, then they will not damage the learning process. But if they are frequent or, worse, if they snowball, then you will lose almost everyone, give a strong impression of carelessness, set a bad example, and (to oversimplify) turn off the class.

You may endeavor to bail out of an example that you are lousing up by saying, "Well, this isn't working out. Let's start another example." It won't work. This is in the vein of two 'wrongs' not making a 'right'. The only solution here is not to make mistakes and to handle those that you make anyway with a certain amount of finesse.

However: If you can see that the example you are working on is getting out of control, if you *know* that it is going from bad to worse, that you are so bolloxed up that you will be unable to bail out of it, then what do you do? Do not spend the rest of the hour trying to slug it out. Doing so is uncomfortable, counterproductive, and will not teach anyone anything. Instead apologize, say that you will write up the solution and hand it out next time (or put it on the class Web page!), and move on. My advice here may seem to fly in the face of Section 2.8, and to contradict the last paragraph, but it is only meant for extreme situations. Making mistakes is one of the surest ways to lose control of a class. It is the mathematical analogue of an equestrian letting go of the reins. Strive not to do it.

Besides preparing well, there are technical devices for minimizing the number of errors that you make. When I am working an example in a lower-division class, I pause *frequently* to say, "Let's make sure this is right" or, "Let's double check this step." I often pick out a student (who I know will respond well) and ask him whether that last step was done correctly. This procedure provides a good paradigm for the students. It also allows note takers to catch up and allows the bright students to strut their stuff in a harmless manner.

One of the most common ways that students make mistakes in their work is by trying to do too much in their heads. Therefore you should set a good example. Write out all calculations. Point out *explicitly* that you have had many years of experience with this material yet you still use lots of parentheses and write out every step.

4.9 Advice and Consent

If your students take a shine to you, and many of them will, then they will view you as looming larger than just "the math teacher". They will come to you for advice on all sorts of things, from the purchase of a computer or computer algebra software, to advice on the purchase of a car, to advice on how to handle their parents, or advice on very private matters.

A good rule of thumb for you as teacher is to stick to things that you know. You are probably well qualified to give guidance about math books, which section of calculus to sign up for next semester, which computer to buy for which purposes, or whether MACSYMA is preferable to Mathematica. If you are an auto buff you could give advice about wheels. But being in a position of authority and being asked for advice by a semi-worshipful student is heady stuff, and you had better be careful.

When you are advising students as to which math class to take, it is easy to fall into the trap of unintentionally (or, more is the pity, intentionally) criticizing your colleagues. The practice of this indiscretion is unfortunately rather common. Please do not fall into it.

The students are your clients, but in some sense you work for their parents (since they probably pay the freight). You are almost certainly out of place to advise your students on how to behave toward their mother and father. Do so at your own risk.

When a student starts asking you about private matters then you are in dangerous territory. It is often difficult to discern the difference between (i) a student asking how to deal with a significant other and (ii) a student making a come-on. Unfortunately, sexual harassment and political correctness, are a part of life these days. Defending yourself against an allegation of either is one of the loneliest and most miserable battles that you may ever have to fight. It can threaten your self-respect, your career, and your marriage. A word to the wise should be sufficient in this matter. Section 4.10 treats sexism, sexual harassment, and related topics in more detail. In any event, if a student has personal problems then he should be sent to a counselor who is qualified to deal with these matters.

Many undergraduates enjoy having a faculty member as a friend. If you are open to it, you can have two or three students hanging about your office at just about any time of the day (or night). It's a boost to the ego to have these attentions, young people are often quite refreshing, and this device provides a convenient way to rationalize wasting a heck of a lot of time. You will have to decide for yourself how you want to handle this trap.

Consider the matter a bit differently. If you make yourself available all day long to help your students with their math (never mind getting involved in their personal lives), you will indeed attract customers. Make yourself available to other professors' students and you will have even more customers. You can provide tutorials, make up extra homework assignments for students, and find innumerable other ways to while away the day. But the set of activities that contribute to a successful academic career and the set of activities that I have just described have a rather small intersection. That sounds rather Machiavellian, so let me be more gentle. Almost all learning is ultimately accomplished by the individual. I've engaged in nearly all the activities described here. They have produced few lasting results and, in the end, truly have helped very few students.

Most of those who end up surviving in the academic game are people who decide that, no matter how much they love their students, they love themselves a bit more. Your students won't like you any the less for saying, "I have to do some work now. Let's talk at another time." You will figure this out for yourself eventually, but you heard it here first.

Finally, all of the advice in the last three paragraphs must be filtered through the value system of the institution at which you teach. There are certainly colleges—whose mission is primarily teaching—at which professors are *expected* to have an open door all day long. Such a policy is consistent with what such a school is trying to accomplish, and how it expects its faculty to spend its time, and what sorts of academic activities its faculty might pursue. Most research universities have a different policy toward faculty office hours. A common policy is that you should have 1.5 office hours each week for every course that you are teaching. Apart from those designated hours, your time is yours (although the presumption is that you are doing something scholarly during the other work hours).

Swarthmore and M.I.T. take different approaches to education—and both are excellent. Do be sensitive to what is expected of faculty at your institution.

4.10 Sexism, Racism, Misogyny, and Related Problems

Nobody, certainly no educated person, thinks of himself as a sexist, or a racist, or a misogynist. That is what is so insidious about these sins when they come up in a university environment. People often are not aware that they are being offensive.

I don't want to preach about any of these topics. Rather I would like to be pragmatic and to mention some pitfalls. Common complaints from students are these: (1) The professor calls on male students more than on female students, (2) The professor will not answer questions from female students, (3) The professor shows favoritism toward female students in his grading policies, (4) The professor talks down to minority students, (5) The professor seems to believe that women are less able than men, (6) The professor seems to believe that Caucasians are more able than minorities, (7) The professor demands more from Asian students, (8) The professor makes suggestive cracks in class, (9) The professor uses vulgar language.

None of these complaints is cooked up. They have all been tendered, very seriously, by genuinely outraged students. It goes without saying that, in all instances, "male" may be switched with "female", "Caucasian" may be switched with "minority", and so on. The issue here, and I cannot emphasize this too strongly, is not that any particular class of people is persecuting any other particular class of people. Rather, the point is that every instructor has his foibles and shortcomings and biases and these will often be perceived by students through the filter of whatever issues are currently in the air.

Matters regarding the issues being discussed here are rarely clear. Sometimes a student with one of the complaints described above is doing poorly in the course and is looking for an excuse or a scapegoat. Sometimes the professor just doesn't realize that he is behaving in a manner that some students find offensive. In short, there is plenty of room for misunderstanding.

You should never physically touch your students. An arm around the shoulder or even a prolonged and enthusiastic handshake easily can be misinterpreted.[2] But many people, especially those new in the teaching profession, are not aware of the subtleties involved in the legal definitions of sexism, racism, and so forth. Let me surprise you with some other aspects of harassment that you may not know.

- If you are in the habit of saying, "I don't want to blow smoke up your dress." or "This is a pregnant idea." or "This problem is a bitch." or "Bulls—!" during your classes then, by a strict interpretation of the statute, you may be guilty of sexual harassment. This is true regardless of your sex or the sex of the members of your audience. In fact, if you even tolerate this language from others who are in the classroom, then you also may be guilty! The spirit of the law is that if someone *feels* offended then they are offended.

- If you have displayed in your office a *Playboy* centerfold, or a (closed, unread) copy of *Playboy* sitting around where others can see it, or a *Frederick's of Hollywood* catalog, or a Chippendale's poster, or in fact a great many of

---

[2]Several years ago the Bank of America instituted new policy strictly forbidding all back-patting, arms around the shoulder, and prolonged handshakes. The new rule came about just because there had been a large number of sexual harassment suits.

the other posters that can be seen in popular stores (even in the campus bookstore), then you may be guilty of sexual harassment.

- Many items that you may think of as art—paintings or sketches or sculptures or frescos—may be seen by others as suggestive or offensive. Look at the effect that the work of Robert Mapplethorpe has had on support for the National Endowment for the Arts.

- Signs or posters that have religious, political, emotional, or sexual content may be deemed offensive.

You may feel that being forced to monitor your language or other behavior this closely is an abridgment of your First Amendment rights, and you may be correct in this feeling. My view, much as I hate censorship, is that there is no percentage in going through life wearing a "kick me" sign. You can use this language, or display these artifacts, all you like when alone or in private circumstances. As a teacher you are something of a public figure and must suffer certain restrictions.

A number of universities in this country have distributed detailed guidelines to their faculties about the issues being discussed here. Some have gone further, and indicated specific words or phrases that ought not to be used. For instance, you should not say "snowman" but should use a suitably laundered asexual alternative. Various standard English phrases are suggested to be off limits and substitutes are recommended. The point is that we are dealing with very delicate and emotion-charged issues of legality and morality here and one has no choice but to take them seriously.

It recently came about that an offer that had been tendered by a big state university to a job candidate was rescinded because the new employer learned that the candidate had been found guilty of sexual harassment at his previous place of employment. Note that the offer had been tendered *in writing*, and note that it had been accepted *in writing*. Furthermore, the job candidate was never found guilty in any court of law. Rather, he was sanctioned in a private university proceeding. Yet his new putative employer felt that it was too great a risk to put this man in a position of *in loco parentis* before eighteen year old young women. Of course the university risked a lawsuit from the spurned job candidate, but it felt that that was a lower risk than what might transpire if he were allowed to teach in its math department.

The book *The Lecherous Professor* [DZW], a serious but inflammatory study of the sexual peccadillos of male faculty at American universities, played a key role in the sanctioning of the instructor described in the last paragraph. A glance at Section 4.9 and the present section of this book will show you that emotions run high over questions of sexual misconduct. I am not suggesting that sexual misconduct is not a serious crime—it most certainly is. But we must all be aware that the gun could be pointed at any of us—even if we are innocent. Most accusations of sexual harassment or misconduct are cases of one person's word against another's. There are rarely any witnesses. It is just as easy for an allegation to be made against you as it is difficult for you to thereafter clear your name. Hence forewarned is forearmed.

It is no fun having to deal with issues of racism or sexism or sexual harassment or misogyny. A complaint lodged against you is, in effect, an attack on your character and your integrity. So be aware that these areas are a potential problem for all of us. Behave accordingly. If you are called on the carpet for any of these matters, then do not become defensive. Show respect for the complainant. Get help from your department head or from your dean of human resources. This is serious business.

4.11 Begging and Pleading

Some students will come to you with unreasonable requests. They will tell you, after doing poorly on an exam, that if they do not pass this test—or this course— with a certain grade then they cannot continue in the pre-medical program, or the microbiology program. Of course you, as professor, can verify rather quickly whether this claim is true. But it does not matter. If the test was so important to the student then the student should have studied harder. The student should have come to your office hours for help before the test. If the student's homework was weak then the student should have seen the writing on the wall. Be cautious of these pleas. While you do not want to be heartless and unsympathetic, you also do not want to find yourself gradually being drawn into an ever more complicated morass of tricky moral dilemmas (see [WIE], and also Section 2.11, for a rather hard-nosed view of this sort of student negotiation).

Students are sometimes *unable* to see the writing on the wall. A student will come to me and ask why he got a grade of "D" in a certain course. I will look at his record and say, "Well, you got a 'D' on the first midterm, a 'D' on the second midterm, a 'D' on the homework, and a 'D' on the final. So a grade of 'D' seemed to be in order." Amazingly, this line of reasoning never occurred to the student (no, I am not making this up). So I have to be patient and explain how the world works. It is a little like explaining the birds and the bees to a slow child; but somebody has to do it.

A desperate student will offer you all sorts of inducements to change grades. Discretion prevents me from enumerating what some of these may be, but they range from the pecuniary to the personal. *You must brush off these attempted bribes with the disdain that they deserve.* If you act as though you are considering and then rejecting them, then you are looking for trouble.

It really is true that if you look and/or act like a student then students will find you more approachable. They will more readily come to you with propositions that they wouldn't consider broaching with a more detached faculty member. In short, younger faculty are more vulnerable. This is one reason for dressing differently from students and maintaining a slight distance. Again, this may sound cold. But I speak here from hard personal experience.

As has been mentioned in Section 4.10 and elsewhere, you must be sensitive to sexual harassment issues. Sexual harassment is not a pretty subject, but acting receptive—even mildly so—to any proffered inducements is only looking for trouble.

Sometimes a student—who has done poorly on an exam—will ask to be allowed to take the test again. You simply cannot allow this type of favoritism. For one thing, it is unfair to the other students. Second, if the others find out then they will become angry—and justifiably so. You *can* give the student a second try off the record and go over the test afterwards with the student. This artifice can be a device for giving the student some encouragement. You might tell the student, "I can see from this unofficial exam that you know the material better than your official exam suggests. If you do well on the final then you can still probably get a grade of 'B'." However you must engage in this charity sparingly, if for no other reason than it can use up large chunks of your time. Also, it is too easily misinterpreted.

A favorite student response to a poor test grade is, "I did very well on the homework but my test grade does not reflect what I know." Of course some students may "choke", or panic, on a test. As a student, I have done so myself. But unfortunately many students do their homework by copying examples from the text, or the lecture—merely changing the appropriate numbers. This leads to a minimum of understanding. You must stress to your students that, when they study for a test, they should reach proficiency *without* recourse to the book or notes. If the book and notes are necessary, then the technique has not been mastered. For interest's sake, refer to Tom Banchoff's technique, described in Section 2.10, for handling students who choke on a test.

One of the most common student remarks is, "I really understand the material but I cannot do the problems." A variant is, "I can do the problems on the homework but I cannot do the problems on the test." Consider if you will these analogous statements: "I really understand how to swim but every time I get in the water I drown." and "Playing the piano sure looks easy when Arthur Rubinstein does it. I wonder why I cannot do it?"

If you are a good teacher, then you will make the material look easy, or at least straightforward. Thus you can lull students into a false sense of security. You must continually warn them of the importance of mastering the material *themselves*—and of *practicing*. And this point leads to the second ludicrous statement recorded above. The fact that a student can do the homework problems is meaningless *unless* the student can do them cold, and with the book closed.

This last observation seems so obvious that it hardly bears mention. But recall that this is a book about the obvious, and this point bears not only mention but repeated mention. Many students view the learning process as a passive one—something like getting a massage. You must constantly remind them that this attitude will not do. You can remind them by just telling them. Or you can remind them by giving pop quizzes. Or you can remind them by giving an exam and watching them flunk. But, one way or the other, you must attempt to break through this psychological impasse.

In a related vein, many students think that

(**i**) studying

and

(**ii**) just sitting in front of the book

are one and the same thing. We all know that studying requires discipline, tenacity, and hard work. It is not something that just happens to you. It is instead something that you *make happen.* Helping your students to understand this point begins with being conscious of the problem yourself.

It is an observed fact that most students—but especially freshmen and sophomores—have no idea how to study. David Bressoud (see his Appendix to the book) recently conducted a survey in which he asked students in large calculus lectures how they studied. What he learned is that the student conception of "studying" is to attempt to do the assigned homework problems by emulating a cognate example from either the text or the class hour. This is certainly a meaningful activity, but it is only a minuscule portion of what true studying is. It is worthwhile to consider how you, as a mathematics instructor, can force the students to read the text and to turn the ideas over in their own minds. Some instructors have required their students to keep journals recording the development of ideas in the course. Lab activities—really tightly constructed ones that force the students to reinvent the ideas for themselves—are another way to accomplish this goal. Group discussion is a third method.

In my opinion, most students want to be told what to do. Your job as teacher is to tell them. Don't make them guess what are the important topics in your class. Tell them. Don't make them guess what they will be tested on. Tell them. Don't make them guess how to study for an exam. Tell them. Don't make them guess what are the pitfalls in studying. Tell them. Is there any reason not to do this? Would you rather deal with the begging and the pleading?

Chapter 5

A New Beginning

Education is not a product: mark, diploma, job, money—in that order; it is a process, a never-ending one.

Bel Kaufman

Education, like neurosis, begins at home.

Milton R. Sapirstein

I think the world is run by "C" students.

Al McGuire

He will wonder whether he should have told these young, handsome and clever people the few truths that sing in his bones. These are

1. *Nobody can ever get too much approval.*

2. *No matter how much you want or need, they, whoever they are, don't want to let you get away with it, whatever it is.*

3. *Sometimes you get away with it.*

John Leonard

Teach the students you have, not the students you wish you had.

Clarence Stephens

What would you pay for all the secrets of the universe? But wait! You also get a free twelve quart spaghetti cooker ...

An infomercial announcer

The truth is lies, and we jail all our prophets.

Charles Manson

5.0 Chapter Overview

At this point in time, American academe is at a crossroads. Both the university and society as a whole are making new demands on the professoriate. One of these is that we be directly accountable for our teaching. The purpose of this book is to point out that such demands are not antithetical to our scholarly pursuits. In fact your teaching activities can complement your research activities rather nicely.

I have discussed both philosophical issues and pragmatic issues in this text. Certainly good teaching is the sum of many particular skills, but it is also the product of attitude and purpose. I hope that reading this book has sharpened your focus on teaching.

151

5.1   The Role of the University Professor

A distinguished mathematician—well-known to us all—joined the University of Chicago Mathematics Department, as an Assistant Professor, in the early 1960s. As he was settling into his office, the chairman came by and chatted him up for a few minutes. When the chair departed, he waggled his finger at the new faculty member and said, "Remember: Our job is proving theorems."

In retrospect, one wonders why the chairman felt moved to make such a statement. Chicago is and was one of the pre-eminent mathematics departments in the country. In the early 1960s, the teaching reform movement was still a twinkle in somebody's eye. Teaching evaluations had not yet been invented. Everyone agreed with Paul Halmos that proving theorems was not just the main thing—it was the only thing.[1]

If we were to make a sequel to this movie, filmed in 1999, then the scene (at least at many universities, and especially public institutions) would be a bit different. The chairman would still drop by to chat up the new faculty member. He would remind the newcomer that he was hired for his ability with mathematics, and for his achievements in research. (Proving theorems, and learning new mathematics, is the highest and finest thing that we do. This fact has not changed, and I hope it never will.) But as the chairman departs, he will now waggle his finger and say, "But don't forget: It's teaching that pays the bills around here. Undergraduates come here expecting to be taught. And parents pay tuition because they want their children to be educated. I expect you to do a creditable job with your teaching. And I don't want to hear any complaints from students or parents. If I do, you will be making my job more difficult, and I in turn will make your *life* more difficult. A word to the wise should be sufficient."

Again, I am not trying to sound sappy. And I am also not endeavoring to sound draconian. If you are new to the mathematics profession, then you may as well know what sort of world we now inhabit. You have a choice: You can prove the Riemann hypothesis or you can learn how to teach.

If you have been in the profession for a while, and have never given any thought to teaching, then perhaps it is time you had better do so. You no doubt have your own ideas about the subject, but perhaps viewing the ideas presented here will give you food for thought.

My own experience is that my teaching meshes rather nicely with my research. I've had good ideas (for a research problem) while preparing a calculus class, and I've gotten inspiration for my calculus class from serious mathematics that I was working on. I suppose that this is the way it is supposed to be, and I believe that a part of the reason that these different facets of my professional life interact so well is that I am open to such interaction. I encourage you to foster this symbiosis in your own life.

---

[1]Personal communication.

5.2 Closing Thoughts

Sometimes the easiest way out, when we are faced with some difficult or distasteful task to perform, is to resort to cowardice. We are all guilty of this sort of avoidance. At one time or another we have all lied or engaged in subterfuge to avoid unpleasantries.

Our students, of course, suffer from their own shortcomings. One of my colleagues had a student knock on his door and ask for some help with calculus. The professor said, "I haven't seen you in class for three weeks. Why do you come to me now?" The student replied that he didn't need to go to class—he had the book. "Well, do you read the book?" intoned the impatient professor. The student replied, "Well, I could."

What are you going to do? I would tell the student that when he wanted to have a serious conversation he should phone me up for an appointment. Until then, he should not darken my doorstep.

I do not wish to dwell here on human frailties. But I think that the method of teaching that many of us use—and I have been guilty of this to a degree with certain classes that I really did not want to be teaching—is a form of cowardice. We just skulk into the room, write the words on the board, and convey with body language and voice and attitude that we are not interested in questions or in much of anything else connected with this class. Then we turn tail and skulk out of the room. I was once told (tongue-in-cheek, I think) that the secret to success in undergraduate teaching is, "Never let a student get between you and the door." Not an admirable attitude, but one that many of us have held from time to time.

How to Teach Mathematics has been an effort to fight this form of cowardice, both in myself and in others. Teaching can be rewarding, useful, and fun. To make it so does not require an enormous investment of time or effort. But it does require that you have a proper attitude and that you be conscious of the pitfalls. It does require being sufficiently well prepared in class so that you can concentrate on the *act* of teaching, rather than on the epsilons. And it requires a commitment.

We must believe that being a good teacher is something worth achieving. We must provide some peer support to each other to bring about this necessary positive attitude toward teaching. The last thing I want is for mathematicians to spend all day in the coffee room debating the latest pedagogical techniques being promulgated by some well-meaning educational theorist. I want to see mathematicians learning and creating mathematics and sharing it with others. But those others should include undergraduates. That is what teaching is about.

Appendices

Twelve scholars, who have dedicated a significant portion of their academic lives to teaching and the study of teaching techniques, have agreed to contribute Appendices to this new edition of *How to Teach Mathematics*. It is safe to say that most of these good people disagree with parts of, or in some cases much of, what I have to say. Others agree with me, but think that I am not sufficiently conservative or sufficiently liberal (depending on the point in question). Of course there is no "correct" position on any teaching issue.

I hope that these Appendices illustrate the precept that there are many ways, and many styles, of thinking about good teaching. Certainly the variety of ideas offered here enriches what this book has to offer.

The Irrelevance of Calculus Reform:
Ruminations of a Sage-on-the-Stage[1]

George E. Andrews

Penn State University

It is not an appealing task to rain on the enthusiastic Calculus Reform Parade. Indeed most of the efforts described on these pages have been undertaken by competent mathematics teachers filled with good will and good intentions. Most projects have succeeded in some sense or other. Who could possibly object?

Let me briefly state the issues I wish to address. Then in the remainder of the article I shall amplify my concerns.

First, the real problem in undergraduate education (not just mathematics) is that students are not studying enough; only grade inflation (i.e. lowered standards) allows them to pass. Second, by not addressing directly the failure to study, calculus reform is, in the long run, irrelevant. Third, attention to the amount of time devoted to homework should be a major concern of each teacher. Fourth, institutions of higher learning undermine high standards by: **(1)** supporting large lectures, **(2)** requiring computerized student evaluations, and **(3)** never considering homework policies in any personnel decisions. Fifth, there are major institutional projects impeding high standards; these often come masked with names suggesting support for high standards: Total Quality Management, Outcome Based Education, the proposed NCTM assessment Standards, the mathematics teaching standards proposed by the National Board for Professional Teaching Standards, etc.

First, STUDENTS ARE NOT STUDYING. Most of us suspect this subconsciously if not consciously. A major survey of undergraduate study habits at my own university (Penn State) suggests that 2/3 of our students study less than 15 hours a week. The Pace Report by the Center for the Study of Evaluation documents a similar phenomenon at the national level. Now if a standard load is 15 semester hours, then the old rule of thumb (2 hours outside class for each hour inside) suggests an average study time of 30 hours. Very few students are studying close to 30 hours a week.

Assuming this first point, I would argue that it is symptomatic of low academic standards and of moral abdication. We have *failed* to teach our students the virtue of hard work. I grant that a hypothetical student could study for 30 hours a week and learn little; so getting students to study is not our ultimate object. Our object is educated graduates. However we can be fairly certain that the non-studying majority is not getting educated adequately.

Given this major problem, how does calculus reform respond? At best, calculus reform addresses this question indirectly. It extols the joys of group learning, sings the praises of thrilling technological innovations such as computerized laboratories, and preaches the virtues of open-ended problems and term projects. Now each of these items has merit. Each is especially likely to interest a teacher who has become bored with standard methods. Each may affect the amount of

---

[1]This article appeared, under the same title, in the January, 1995 issue of *UME Trends*. It is reproduced here with the permission both of the publisher and the author.

time students study; however I have never seen any of these innovations touted primarily for its substantial positive effect on high standards and study habits. My impression is that calculus reformers believe they are making calculus so interesting that everyone will, in the natural course of things, just work a lot harder and do better.

Increased Achievement

I would welcome evidence that increased achievement is a substantial effect of many of these reforms. I have yet to be reassured. Mostly I see remarks like those of Tom Tucker in *Priming the Calculus Pump*: "...in every case we know where students from experimental sections took a common final exam with the rest of the calculus students, the experimental students did just as well or better." Of group learning, Davidson and Kroll report: "Less than half of the studies comparing small-group and traditional methods of mathematics instruction have shown a significant difference in student achievement; but when significant differences have been found they have almost always favored the small-group procedure." Surely we could hope for much more than this from expensive pilot projects. I suggest that part of the reason for the barely discernible improvement is inadequate attention to standards and study habits.

Furthermore, the technology aspect of calculus reform is especially disturbing. Many of our students have pitiful skills in arithmetic, algebra and trigonometry. While our brightest may gain much from MATHEMATICA projects, we must be vigilant lest the $B-$ or $C+$ students replace basic math skills with button pushing. (This issue has significant philosophical roots and is one aspect of the debate between the Artificial Intelligence enthusiasts and those of us who regard computers as pencils with power steering.)

If you suggest that the computer has obviated the need for facility in arithmetic, algebra and trig, then please provide firm evidence for your views. Part of the reason the New Math floundered was its unyielding contempt for anything remotely resembling "mere rote learning." If "mere rote learning" were renamed "essential drill," we might give it the respect it deserves.

For brevity, I shall combine my third and fourth points. If calculus reform is irrelevant, then what should be done individually and institutionally to enhance the education of our students?

Each instructor should think a lot about how to increase the amount of work students do outside of class. Clearly both traditionalists and reformers can address this problem directly. Institutionally we assess teaching in ways that either ignore the homework question totally or work against it. How many times have you participated in a promotion or tenure decision where homework policies were discussed as a major component of teaching effectiveness? The correct answer is zero! Indeed, what you probably discussed were computerized scores from student evaluations. ("This candidate only scores 3.82 on the Overall Instructor Rating against the college norm of 4.91.") I contend that *computerized* student evaluations are a menace to high standards and demanding teaching. Is it not human nature for students to prefer teachers who make life easy for them?

The temptation, almost always unspoken, lurks in the background; I'll you're a good student if you pretend I'm a good teacher.

I also believe that small classes are important. The strong personal relationships between teacher and students (which are impossible between a teacher and 400 students) can be used to support active homework policies.

External Pressures

Finally I wish to draw attention to some of the external pressures against high standards posed by movements and institutions with only the most benign intentions. ("If I knew for a certainty that a man was coming to my house wither the conscious design of doing me good, I should run for my life ..." Henry David Thoreau.)

To begin with we have the newly proposed NCTM Assessment Standards for School Mathematics (to date, I have only seen a draft copy). This 243 page document is *horrifying*. It extols almost any variant of politically correct, socially aware mathematics no matter how insipid while heaping contempt on traditional forms of teaching. It bashes current forms of standardized testing not for being inadequately objective measures but rather because they are objective measures at all. If the NCTM Assessment Standards are eventually published in more or less the form I have seen, they will produce entering college students even more poorly prepared than the ones you have now. If you don't believe me, take a look.

The same consciousness-raising mentality that animates the draft of NCTM Assessment Standards also afflicts the Total Quality Management partisans and the Outcome Based Educators. Again and again, programs with names suggesting high standards show little concern for high standards and much concern for a visionary social agenda. One cannot have high standards and egalitarianism as co-equal priorities: They work at such cross-purposes that one must dominate.

In conclusion, I would like to suggest that we Sages-on-the-Stage are not opposed to progress, despite all rumors to the contrary. We are anxious to encourage and support programs that enhance achievement. Indeed most of us hope that all you Guides-on-the-Side will really be able to respond convincingly to the issues raised in this article.

Mathematical Content

Richard Askey

University of Wisconsin

1. Introduction. Steven Krantz has given us a lot to think about in both his first and second editions. Fortunately, he has left a little to be said. Two of my main concerns are the mathematics which is taught, and the standards which need to be set. In a time when rapid changes are being made, both in the curriculum and in teaching methods, it is important to make sure that the essential parts of a mathematics program continue to upheld. That is the reason for my concern about content and standards. As will become clear, there are good reasons to be concerned about the mathematics which is being taught. Let me start with an example where I was not as careful about this as I should have been.

About 25 years ago, my wife was taking a course in statistics in the Sociology Department. Pocket calculators were very expensive, and the students needed to compute some squares roots, so the professor announced a review to teach how to do this. I called and asked him how he was going to teach this. This method was based on

$$(10a + b)^2 = 100a^2 + (20a + b)b,$$

although this is not how he described it to me. I explained an alternate method, of guessing an approximate root, dividing and averaging to find a better approximation. The next day he cancelled the review, thanked the unknown student who had a friend in the Mathematics Department, and passed out notes on an easy way to compute square roots.

There was a weekly laboratory session. That week one student complained to the teaching assistant about having to learn a new way to take square roots. She knew the other method, and did not want to have to learn a new one. The teaching assistant agreed with her, and said he had learned the other method as a freshman student of engineering at Washington University. He never put together the name of one student, Elizabeth Askey, with the name of the young instructor who had taught him this. It was reassuring to know that he remember this, but I should have thought a bit about different ways to teach how to compute square roots.

2. Learning trigonometry. One of my concerns is the mathematics which prospective teachers learn. One subject which high school teachers need to know very well is trigonometry. Here is what I think high school teachers should know about trigonometry.

The first idea is similarity of triangles. For right triangles, similarity allows the trigonometric functions to be defined as real valued functions.

The Pythagorean theorem gives

$$\sin^2 \theta + \cos^2 \theta = 1$$

161

and the other related identities. These, along with definitions and elementary algebra, show that any two of the six trigonometric functions are related.

The next idea is that an arbitrary triangle can be decomposed into two right triangles. This decomposition is usually used to prove the law of sines and the law of cosines. It can also be used to prove the addition formulas for $\sin(\theta + \varphi)$ and $\cos(\theta + \varphi)$.

Consider the following picture.

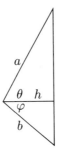

The area of the large triangle is

$$\frac{1}{2}ab\sin(\theta + \varphi).$$

It is also the sum of the areas of the two smaller triangles, so is

$$\frac{1}{2}ah\sin\theta + \frac{1}{2}bh\sin\theta$$

$$= \frac{1}{2}ab[\sin\theta\cos\varphi + \cos\theta\sin\varphi]$$

I learned this argument from a forthcoming book on trigonometry written by I. M. Gelfand and M. Saul, [9]. It appears elsewhere, but is not well-known.

A similar argument using the law of cosines leads directly to the addition formula for $\cos(\theta + \varphi)$. I would assign this as a problem. The addition formulas are so important that a different type of proof should be given. Another one which uses an important idea is to use the invariance of the unit circle under rotation. Both of these proofs should be given in what used to be a course in trigonometry. Now, in our high schools, this is usually done in precalculus.

High school teachers should know other proofs. A simple one comes from

$$e^{i(\theta + \varphi)} = e^{i\theta} \cdot e^{i\varphi}$$

and

$$e^{i\theta} = \cos\theta + i\sin\theta.$$

This can be done in detail in advanced calculus, and can be done in calculus if multiplication of power series and convergence of series of complex numbers are done. I think the first should be done, and the second sketched, since both are very useful topics.

There are also derivations of the addition formulas for $\sin(\theta+\varphi)$ and $\cos(\theta+\varphi)$ from the differential equation

$$y'' + y = 0,$$

and from the multiplication of two rotation matrices in the plane. I do these derivations in the courses where linear differential equations and rotation matrices are treated.

There should be many applications of trigonometry in addition to the standard problems of solving for missing parts of triangles. One use of the addition formula for $\sin(\theta + \varphi)$ is to show that

$$a \sin\theta + b \cos\theta = (a^2 + b^2)^{1/2} \sin(\theta + \varphi).$$

The addition formula for $\tan(\theta - \varphi)$ gives the angle between two lines in terms of their slope. I would ask students to find the angle between two lines whose slopes are known. It is not too hard to see that $\tan(\theta - \varphi)$ is useful, when $m_1 = \tan\theta$, $m_2 = \tan\varphi$. The formula for $\tan(\theta - \varphi)$ would not have been derived, so something would have to be done. By this time in a course, students should have learned that when trying to simplifying a trigonometric expression which contains $\tan u$, it is natural to replace $\tan u$ by $\sin u / \cos u$. From there on, it is a matter of using the addition formulas for $\sin(\theta - \varphi)$ and $\cos(\theta - \varphi)$, and then trying to find a way to simplify the resulting expression.

Tony Gardiner has a very nice set of notes titled "Recurring Themes in School Mathematics" [8]. He makes the point that meaningful simplification is one of the goals in life one should start to learn in school. The necessary simplification needed above is a good instance of meaningful simplification.

Here is a use of a consequence of the addition formula to an interesting problem in algebra. The problem is to find the polynomial $p_n(x)$ which satisfies

(2.1) $$|p_n(x)| \leq 1, \quad -1 \leq x \leq 1,$$

and takes on the largest value for each fixed $x > 1$. This is a result of Chebyshev. The related problem of finding the polynomial of degree n satisfying (2.1) which has the largest derivative at $x = 1$ was solved by Mendeleev when $n = 2$, and by A. Markoff for general n. See Boas [6] for a nice treatment.

This problem can be started as follows. First, use the addition formula to show that

$$\cos 2\theta = 2\cos^2\theta - 1.$$

Graph both sides as a function of θ, and graph the right hand side in x, with $x = \cos\theta$. Then, have the students derive the similar formulas:

$$\cos 3\theta = 4x^3 - 3x$$

and

$$\cos 4\theta = 8x^4 - 8x^2 + 1,$$

and graph these as functions of x for $-1 \le x \le 1$, and for $0 \le x \le 2$. When $-1 \le x \le 1$, these polynomials satisfy

$$|p(x)| \le 1, \quad -1 \le x \le 1.$$

Students should be asked to explain why this is true. A graph on a graphing calculator is not sufficient to explain this without a lot of extra work. There is a more fundamental reason, which also allows polynomials of higher degree to be generated. These reasons are:

$$\cos n\theta = T_n(\cos \theta)$$

is a polynomial of degree n in $\cos \theta = x$, and

$$|\cos n\theta| \le 1,$$

so

$$|T_n(x)| \le 1.$$

It is not too hard to see that if $p_2(x)$ is a polynomial of degree 2, and $|p_2(x)| \le 1$, $-1 \le x \le 1$, then

$$|p_2(x)| \le T_2(x) \quad \text{for each fixed} \quad x > 1.$$

A slightly more complicated argument, which is mainly counting zeros of $p_n(x) - T_n(x)$, can be given to show that

$$|p_n(x)| \le T_n(x) \quad \text{for any fixed} \quad x > 1$$

when $|p_n(x)| \le 1$ for $-1 \le x \le 1$.

To show that $T_n(x)$ is a polynomial of degree n in x, one can use

$$\cos(n+1)\theta = 2 \cos \theta \cos n\theta - \cos(n-1)\theta$$

or

$$\cos n\theta + i \sin n\theta = (\cos \theta + i \sin \theta)^n.$$

Either of these identities can be proven from the addition formula for $\cos(\theta + \varphi)$.

One problem about teaching this material to prospective high school teachers is that most of them will have learned trigonometry in high school and not take it in college. One course in which this can be done is a course on proofs. J. Rotman has written a book [14] for such a course which contains many of the necessary parts of trigonometry. He also included the rational parametrization of the circle by

$$\cos \theta = \frac{1 - t^2}{1 + t^2}, \quad \sin \theta = \frac{2t}{1 + t^2},$$

and use of these formulas to obtain Pythagorean triples. He pointed out that this parametrization can be used to verify many stated trigonometric identities. For example, he used these formulas to prove that

$$\frac{1 + \cos \theta + \sin \theta}{1 + \cos \theta - \sin \theta} = \sec \theta + \tan \theta,$$

He then quotes Silverman and Tate [15]:

> "If they had told you this in high school, the whole business of trigonometric identities would have been a trivial exercise in algebra!"

This is not true, for what should be taught when students are learning how to prove an identity is to teach how to do meaningful simplification. See [4] for comments on this, and an example of how not to teach trigonometric identities.

If high school teachers have not learned what the essence of trigonometry is, one could hope they would see this in the textbooks they use, or in articles in "Mathematics Teacher". There were books which had very nice treatments of trigonometry. One such book is [9]. However, many books have a treatment not too dissimilar to another one mentioned in [4], so one can not count on textbooks.

Here is a brief excerpt from an article in Mathematics Teacher, [12].

> The trigonometry teacher can use the graphing calculator in teaching identities. These equations can be used:

$$y_1 = \sin(2x)$$

$$y_2 = 2\sin x$$

$$y_3 = 2\sin x \cos x$$

> Have students graph y_1 and y_2. Two appear! Next have students graph y_1 and y_3. The students are now learning identities not by the rote method of pencil and paper but by experiencing and *seeing* an identity.

This appeared in a section titled "Sharing Teaching Ideas". The last sentence incensed me, so I wrote to two officers of NCTM to suggest that they write an article pointing out the improper use of the word "learning". Nothing appeared, either as an article or in the "Readers Reflections" section.

Hung-Hsi Wu wrote a paper about mathematics education [16], and commented on the quotation above. He noted that nothing was said about proving this identity. Then he continued with: "Now if the authors had said that 'in addition to proving the identity $\sin 2x = 2\sin x \cos x$, using the graphing capability of a calculator can reinforce students' confidence in the abstract argument,' we could have applauded them for making skillful use of technology in the service of mathematics."

Jeremy Kilpatrick [11] wrote a rejoinder to Wu's article. Here is one paragraph.

"[Wu] also hits some inappropriate targets. For example, he castigates two high school teachers [12] writing in the *Mathematics Teacher* for their attempt to help students see the functions involved in a trigonometric identity before establishing its validity. Quite apart from whether every article in an official journal of an organization promoting reform must reflect reform views, one can reasonably ask whether seeing the graphs of these functions alone might not help students understand the identity. And how can Wu be so certain that teachers who are having students use graphing calculators are neglecting proof just because it is not mentioned in the article?"

Unfortunately, I feel Kilpatrick has misread this article, and the authors meant exactly what they wrote when they wrote: "The students are now learning identities not by the rote method of pencil and paper but by experience and *seeing* an identity."

The lack of a proof of the double angle formula is not the only drawback in [12]. The graphical picture gives no idea why the identity is true, so no clue about how to extend it. Also, it gives the false picture of mathematics that formulas come out of the air, or are given by the teacher, with the student not expected to understand how anyone could dream up such a formula. Important formulas are almost never found in such a way.

After waiting almost a year and a half, I wrote a short note about [12] and submitted it to "The Mathematics Teacher". This note included the decomposition proof above in the special case when $a = b$, and suggested that to see if students had learned something they should be able to derive the addition formula for $\sin(\theta + \varphi)$ by a similar argument. This note was turned down with the following report from a panel member.

"This adds nothing to the Sharing Teaching Ideas article
– it is just a list of complaints."

3. Recurring themes in school mathematics. Tony Gardiner's booklet "Recurring Themes in School Mathematics" was mentioned earlier. Here is an example of a recurring topic which he does not treat.

In late elementary school, students should learn about the connection between decimals and rational numbers. They need to do some calculations from rational numbers to decimals, to see why the resulting decimal either terminates or repeats. The number 5/17 is a good example to give, but not as a first example. In a course for prospective elementary school teachers, I gave this as homework, with the students to say whether the decimal expansion repeated or not, and to explain their answer. They were to do the calculations by hand. A substantial minority said it did not repeat, and others said it did but could not explain why other than to say that it did because of their calculation. A few were able to explain why it had it, since the remainders would eventually repeat.

In addition to this direct problem, the inverse problem of going from a repeating decimal to a rational number should be considered. In [8], Gardiner made the point that inverse operations are frequently more important than the associated direct operations. That is true here when one considers repeating decimals rather than decimal approximations to rational numbers. The reason for the importance is that when this operation is done correctly, it leads to a method of summing a general geometric series.

The right way to do this change from a repeating decimal to a rational number is to call the repeating decimal x, multiply by the appropriate power of 10 to move the digits over one period, and then subtract to reduce to problem to a finite decimal expansion. Currently, there is another way which is being used in some textbooks. This is to use pattern matching. In the best version of this

$$\frac{1}{9} = .\bar{1}$$

and

$$\frac{1}{99} = .\overline{01}$$

are extended to

$$\frac{1}{99\cdots9} = .\overline{0\cdots01},$$

and so get

$$.\overline{45} = 45(.\overline{01}) = \frac{45}{99}.$$

This argument does not work directly when the repeating part does not start immediately, but there are ways around this problem. See [1], [13] for an illustration of what is currently being written. The real problem with this method is not that it is usually just given as a rule to follow without any reasons given, but that it is a dead-end method.

In middle school, after exponents have been introduced, students should be given the problem of how many grains of rice would be given if one is given for the first square of a chessboard, 2 for the second square, 4 for the third, and doubling each time until 2^{63} grains are given for the 64th square. The classical story is adapted nicely by David Barry [5]. On the last page, he illustrates approximately what the number of grains of rice would fill, or be equivalent to. He starts with 2 grains on the first square. 2^8 fills a teaspoon, 2^{16} - a bowl, 2^{24} - a wheelbarrow, 2^{32} - a festival hall full to the roof, 2^{40} - a palace with 256 rooms, 2^{48} - the World Trade Center, 2^{54} would cover Manhattan island 7 seven stories deep in rice, and 2^{64} would make a mound as large as Mt. Kilimanjaro. He concludes by saying that all 64 squares together would cover all of India knee deep in rice.

Students should be asked how much rice that is, i.e. sum the series $2 + 2^2 + \cdots + 2^{64}$. The use of different measurements for the amount of rice is nice, and needed in the United States. In the Third International Math. and Science Study, and in the Second International Math. Study, our eighth grade students did most poorly on measurement problems, as measured by how well our students did relative to the international average. We have to teach our prospective elementary and middle school teachers both the exact calculations which lead to $2^{65} - 2$, and the various approximate measures, if we expect them to teach this to our elementary and middle school children.

Another delightful children's book which can be used with late elementary and middle school children and in some of our college courses is Anno's "Socrates and the Three Little Pigs", [2]. I have used this successfully in a summer school for students between fourth and fifth grade, and in both our arithmetic course for prospective elementary school teachers and our undergraduate combinatorics course.

In high school, students should learn how to divide 1 by $1 - r$, getting

$$\frac{1}{1-r} = 1 + r + \cdots + r^n + \frac{r^{n+1}}{1-r}$$

and how to derive

$$1 + r + \cdots + r^n = \frac{1 - r^{n+1}}{1 - r}$$

by giving the left hand side a name, multiplying by r and subtracting. This should then be done formally for the infinite series, and then tied up with the repeating decimals from years earlier.

At present, since few calculus students know how to sum a geometric series, this needs to be done in calculus. I do it early, and start with repeating decimals. Unfortunately, even this is unknown by a large majority of our students. An increasing number of those who know that $.\overline{47}$ is 47/99 only know it by the rule mentioned earlier of taking 47 and dividing it by 99. A good way to check if this is all they know is to ask them what rational number corresponds to

$$.5\overline{47}.$$

To show that the geometric series is useful, I use it to differentiate x^n, $n = 2, 3, \cdots$, and later x^r where r is rational. I also use it to integrate x^n from 0 to a, first when n is a positive integer, and later when n is a positive rational number. See [3] for this and some references.

When students have seen special cases of the geometric series for abut six years before they see the general case, and have seen a number of different applications of it, they will have an easier time understanding the ratio test for convergence of series, and be able to use the geometric series to find error estimates for the remainder of some important power series.

A delightful problem on the geometric series was given by Gelfand and Shen [10, §45]. They introduce infinite geometric series by means of the paradox of Achilles and the turtle, with Achilles running 10 times as fast as the turtle. After giving four methods to sum this series, and getting the general formula

$$1 + q + q^2 + q^3 + \cdots = \frac{1}{1 - q},$$

they take the case when Achilles runs ten times slower than the turtle. This gives the absurd answer

$$1 + 10 + 100 + 1000 + \cdots = \frac{1}{1 - 10} = -\frac{1}{9}.$$

Then they give the following problem.

> Problem. Is it possible to give a reasonable interpretation
> of the (absurd) statement "Achilles will meet the turtle
> after running $-1/9$ meters"?
> Hint. Yes, it is.

4. Large lectures and teaching assistants. There are problems which arise when teaching a large lecture with a number of teaching assistants. Krantz has mentioned many of them. At the University of Wisconsin-Madison, first year calculus lectures meet three days a week, and there are discussion sections which meet twice a week. The size of the discussion sections is bounded, so the more students in the lecture, the more teaching assistants. This means that it is possible to give and grade exams where students have to write complete solutions. I always do this. I write the first exam. The remaining exams are written by the teaching assistants, usually in pairs. During our weekly meetings, we discuss the exams after a final draft has been written, and worked by the writer(s) of the exam. Working the exam in detail is important, both to catch little problems which show up when complete solutions are written, and so that the teaching assistants get some idea how long it takes them to work an exam in comparison to how long students will take. All of us have a chance to make comments and suggest changes. I try to say as little as possible, since this is very good experience for the teaching assistants. However, if the exam is too long, which is a common problem the first time an exam is written, I am able to get it modified appropriately. One question which regularly comes up is what is a specific question really asking, and is this the best way to ask it? I try to mention something about this when the first exam is being discussed, and that leads others to ask this question later.

A few of my colleagues have had one teaching assistant who essentially told the students what the questions are, and so do not let any of the assistants see the exam before it is given. I have not had this problem, probably because the teaching assistants know they will be writing one of the exams and do not want this to happen to an exam they help write.

I disagree strongly with one of Krantz's recommendations. He wrote that a teacher should *never* penalize a student for being honest, so that a student who comes and says the points were added incorrectly should just be sent home with a little praise for being so perceptive. Consider what happens when you are in a small grocery store without a fancy check-out machine, and the clerk gives you incorrect change. If you are not given enough, you tell the clerk. I hope you also tell the clerk when you are given too much. I would be insulted if the clerk said to forget it and praised me for being honest. In forty years of teaching, I can remember three students who came and said a problem was misgraded, and too many points were given, I regraded the problem, and gave the correct grade. I also remembered who had come, and made sure that the few points did not change the final grade. There have been a few other instances where scores were added incorrectly. Those were also changed.

When a student in a large lecture comes to ask about possible misgrading, my answer is that I will look at the grading of the full exam, not just that question.

My solution to the problem of a fixed formula for adding grades to get a course grade is to say that any one exam which is out of line with the others counts less, unless it is a higher grade on the final. Doing this both ways can be important in the first semester calculus with students who took a calculus course in high school, and think they know more than they do. They may do reasonably well on the first exam, even without studying, but after that do worse, and on the final show they really did not have a firm command of the early material in the sense of being able to use it in more complicated settings. They probably should fail the course even if the first exam pulls an otherwise failing grade up to D.

Both of the last two examples are illustrations of maintaining standards. Here is another instance.

A few years ago, after an eight hour committee meeting on Saturday, I went to an afternoon party at a local jewelry store. I was tired and a bit surprised when a man approached and asked if I were Professor Askey. After answering "yes", he said he had me for calculus. Bored, I said: "You and a few thousand others". He said I kicked him out of class. My reply was: "That was a while ago". He said I kicked him out for smoking. That was a long time ago. He continued, saying that he was just back from Vietnam and had a chip on his shoulder. I had started the first lecture by saying that smoking was not allowed in class. He decided to challenge me, and lit up a cigarette. I kicked him out. He went back to his room, and started to think what he was doing with his life. He decided he did not like what he was doing. He liked mathematics, so became a mathematics major. He changed to philosophy in his last year, now runs a jewelry store in another city. He said he had wanted to thank me for changing his life. We all hope to have an impact on some students, but I had not expected this to come directly from kicking a student out of class. Keeping standards is important.

References

[1] D. Anderson, Joey's shortcut, Math. Teacher, 89 (1996), 291.

[2] M. Anno and T. Mori, *Socrates and the Three Little Pigs*, Philomel Books, New York, 1986.

[3] R. Askey, What do we do about calculus? First, do not harm, *Amer. Math. Monthly*, 104 (1997), 738–743.

[4] R. Askey, On not identifying equations and identities, FFF#122, *College Math. J.* 28 (1997), 377–379.

[5] D. Barry, *The Rajah's Rice*, Scientific American Books for Young Readers, Freeman, New York, 1994.

[6] R. P. Boas, Inequalities for derivatives of polynomials, Math. Mag. 42 (1969), 165–174.

[7] R. Fischer and A. Ziebur, *Integrated Algebra and Trigonometry; with Algebraic Geometry*, Prentice-Hall, Englewood Cliffs, NJ 1967.

[8] T. Gardiner, *Recurring Themes in School Mathematics*, UK SMC, School of Mathematics, University of Birmingham, B15 2TT.

[9] I. M. Gelfand and M. Saul, *Trigonometry*, Birkhäuser, Boston, to appear.

[10] I. M. Gelfand and A. Shen, *Algebra*, Birkhäuser, Boston, second printing, 1995.

[11] J. Kilpatrick, Confronting reform, *Amer. Math. Monthly* 104 (1998), 955–962.

[12] J. Pelech and J. Parker, The graphing calculator and division of fractions, *Math. Teacher*, 89 (1996), 304–305.

[13] Reader's Reflections, Five letters on Joey's shortcut, Math. Teacher, 89 (1996), 544–545.

[14] J. Rotman, *Journal into Mathematics*, Prentice Hall, Upper Saddle River, NJ, 1998.

[15] J. H. Silverman and J. Tate, *Rational Points on Elliptic Curves*, Springer-Verlag, New York, 1992.

[16] H. Wu, The mathematics education reform: Why you should be concerned and what you can do, *Amer. Math. Monthly* 104 (1997), 946–954.

Department of Mathematics
480 Lincoln Drive
University of Wisconsin–Madison
Madison, WI 53706
`askeymath.wisc.edu`

Personal Thoughts on Mature Teaching

David M. Bressoud

Macalester College

This book is directed toward new teachers, but much of the advice can be summarized in two fundamental attitudes that are essential for all teachers: enthusiasm for the subject and concern for the students. These attitudes got me through all manner of pedagogical mistakes in my early years as a teacher. As I have matured, I have found that my early enthusiasm and concern are still important, but they are not enough. These are my thoughts about what comes next for those who have been teaching for several years, who know the basics of what to do and what not to do in the classroom, who already are good teachers.

I have always known that enthusiasm is critical. This intuition was confirmed in "The Quest for Excellence in University Teaching" [1] which lists enthusiasm as the most frequently cited attribute of excellent teachers, whether one asks students, colleagues, or administrators. This fact echoes what I have experienced in my own classes and those of others. Enthusiasm is infectious. Even when my students do not share it, they respect it.

My first responsibility in preparing for any class is to find something that I consider interesting and exciting, that I want to show and to share. The fact that I find the material extremely elementary is no excuse, nor that I have taught this course twenty times before, nor even that I was not involved in constructing the syllabus which to me seems disjointed and pointless. As a good teacher, I look for new and interesting examples and illustrations, even in the most elementary mathematics. I search for fresh insights and approaches to what I have taught before and never rely on last year's notes. I seek the threads that tie together the pieces of the syllabus.

One danger of enthusiasm is to become overly ambitious for your class. This is an error that I committed frequently in my early years of teaching; many new teachers do. Even after more than twenty years at the head of the classroom, I still make this mistake more than I would like to admit. It is a forgivable error that has the potential for good—I have seen it stretch students far beyond what they thought they were capable of—provided it is accompanied by the second critical attitude, concern for the student.

Steve Krantz has said a lot of good things about the basic components of concern for your students: Respect them and their questions; get to know at least some of them as people; honor the commitment to have times when you are available to them; learn to be sympathetic, receptive, and patient; be aware of your audience; care. It is common for new instructors to have a rough time the first semester they are in charge of their own class. The assignments and exams are often much too hard. They do not yet have an adequate repertoire of examples and alternative explanations to fall back on when difficulties are encountered. New instructors can be oblivious to some of the pitfalls that students will encounter. But concern for the students will overcome these. I have been privileged to see the student evaluations of several teachers who before that semester had never been responsible for their own class but who had both enthusiasm and concern. Many of the evaluations were critical, describing the things that had gone wrong during the semester. But all of the critical evaluations ended with words to the effect "but she is going to be a really good teacher." They were right.

There comes a time when the naive enthusiasm and concern of the new teacher is not enough. For me it arrived after I had developed confidence in my own ability to teach. I knew the difficulties that students would have and I carefully set out my warnings. I had laid up a stock of illuminating examples that I had seen work for students who were confused. I knew the important concepts that must be stressed and how to tie them together effectively. I was writing assignments that stretched my students, but not beyond their abilities. Exam results were coming in that were not quite what I wanted, but that were adequate. And so I dared to probe a little deeper, to give the students an opportunity to show off their understanding. What I discovered horrified me.

I believe that this moment comes to most of us. It might happen when talking with a student in your office. It might be the result of asking something on a test that goes beyond formulaic responses and requires students to draw on their knowledge to synthesize an answer to a question that is not quite like any they have seen before. It might arrive in a written report in which, for the first time, they are required not just to give the answer—which you know they can find—but to explain how they got that answer. Where you had thought there was understanding, you discover confusion. It seems that they have learned nothing and that your only recourse is to go back to the beginning and start all over again.

This is a moment of crisis—in its literal sense of decision—for the good teacher. I have known those who have resolved it by blaming the students, either their unwillingness to work or the poor preparation that they received from others. I have known those who have resolved it by resigning themselves to the belief that what happens in their classrooms for the majority of students will never be more than a superficial and temporary acquisition of what is needed to pass the course. They decide to focus on those few students who are most like themselves.

Either of these responses marks the beginning of a loss of concern. The concerned response is to seize this opportunity to recognize the depths of our own ignorance about what actually happens in the classroom and how students learn. If we are good teachers, we will force ourselves to begin exploring the murky waters of pedagogical theory and educational psychology.

This is our opportunity to honestly face what does and does not work in our own teaching and to discover what others have done, to experiment with their ideas and techniques. We have the opportunity to build up a body of insights and ways of capitalizing on those insights. It is my belief that this is a dynamic process that never ends. The truths that I now know about what happens in my classroom are subject to articulation, refinement, and elaboration. Priorities will shift. Techniques for achieving those priorities will always be under trial. Being a good teacher has become a highly personal quest.

Ironically, the great truths that I and others discover are never new. Reading through the literature of education, we find that each hard-won truth was known to the previous generation and to generations before that. But that does not eliminate the need to rediscover these truths for ourselves, individually. We are now seekers after wisdom, and there is no quick route to it. We can learn from the wisdom that others have accumulated, but it is not our own wisdom until we have sifted it, found our own ways of applying it to our teaching, watched it succeed and watched it fail, and made its lessons part of our own story.

I would like to offer some of my own truths. They are neither new nor will they solve your problems. These are not answers. These are places to begin asking your own questions.

Teach the Students You Have

The temptation to blame the attitudes of our students or their preparation from previous teachers is very real because it often seems justified. Working to improve teacher training or what happens in our K–12 schools is an important and worthy contribution, but such efforts do not absolve us from the responsibility to face the students before us with all of their imperfections. We must begin by engaging them where they are. This has nothing to do with abandoning standards or watering down the curriculum. On the contrary, nothing will be accomplished unless we are optimistically realistic about our students and their capabilities.

The situation is never as bad as the cynics and pessimists paint it. I start each semester by asking each student about expectations for the class and what they want to get from it. Almost universally, they want to "understand" the mathematics in this course, and they expect to have to work hard. But they are busy people with many demands on their time, and they are uncertain about what it means to understand mathematics or how to go about accomplishing this. I must provide the structure to help them achieve these goals. Also, no matter how well I may think I know the abilities of the students who will be in my class, there is no substitute for diagnostic tests or assignments.

Establish and Communicate Clear Goals

If you do not know where you want the students to be by the end of the semester, there is little chance that they will get there. As you set your goals, you need to be optimistic. I have found that students respond well to the challenge to work hard provided that they believe that your goals are attainable and that the necessary support mechanisms are in place. Unrealistic goals can be dangerous. Once students begin to feel overwhelmed, they adopt a strategy that David Tall [5] has called *disjunctive generalization.* They jettison the search for understanding and switch to memorization, regardless of contradictions or inconsistencies, as the safest route to a passing grade.

Writing down your goals for the course is a good exercise for you. Think about the kinds of problems you want students to be able to solve by the mid-point, by the end of the course, a year from now. How important are the ability to apply this mathematics to novel situations? to analyze the components of the theory they have learned? to creatively synthesize these components into creative problem solving? What are the big ideas that they need to be conversant with? What are the connections that they have to be able to make?

If you want to share this list of goals with your students, that is fine. It will help relieve some of their start-of-semester anxiety. But you communicate these goals in how you choose to assess their performance. If you do not measure and reward their progress, then few students will make those goals their own.

Use Assessment Effectively

It is common to base most of the course grade on student performance in two midterm exams and a final. Students are accustomed to exam questions that are a mixture of routine straight-out-of-the-book exercises and a few more challenging problems. They know that they can get a good grade by practicing homework problems until they have the basic patterns memorized. They do not have to be able to answer the most difficult problems. We may say that our goal is for our students to understand the mathematics, but if we use this standard format to assess their knowledge, then we are telling them that what we really care about is whether they can mimic the template solutions quickly and accurately. We are allowing them to bypass understanding.

We want our students to be able to do basic calculations, but we also want them to learn more than this. We would like to be able to replace the standard exam with one that consists of probing, challenging questions that give students a chance to show off what they really know. I know from my own experience that unless I have specifically prepared my students for such a test, the result is abysmal scores and vociferous complaints that the test was unfair; it was not the kind of test my students were expecting. If we have given a challenging test on which students have scored poorly, we are tempted to use a curve to determine grades. This undermines what we were trying to accomplish because it confirms the message that partial credit is good enough and real understanding is not important.

The standard assessment based on two midterms and a final is a trap that leads to one of three unsatisfactory outcomes: Either students see facility with certain well-defined procedures as the goal of the course, or they learn to rely on a curve and so know that they will not be held accountable for any real understanding, or most of the students fail, which not only makes life difficult for the math department but shows that you are not teaching *your* students.

There are other ways of approaching assessment.

Assessment is the carrot and stick that you can use to shape student attitudes and study habits and to communicate what you want students to learn from your course. If you want students to read the text, then give unannounced quizzes on the readings that they were to have done. If you want students to reflect on what they have learned and think about what is happening in the processes that they are mastering, then have them keep a journal that is periodically graded or have them write reports that require this reflective approach. If you want your students to be able to apply their knowledge to unfamiliar situations, then give them problems and projects that require this level of understanding. If you want your students to listen critically to your lectures and be able to distill the main points, then ask them to write these down at the end of the class and hand them in. If you want students to be able to use definitions and theorems correctly and unambiguously, then have them write assignments where this is required. If there are basic skills that you want students to master, then test these skills and set the bar for a passing grade as high as you feel is needed, whether that be 80%, 90%, or 100%.

Assess early; assess often. In my experience, students react positively to this. They appreciate the feedback and direction that it gives them. It reduces the stress of a major examination that counts for a third or more of their grade and for which they are not quite sure what will be expected. Students should never be surprised by what you expect of them. You should be shaping their approach to the class from day one. There is nothing wrong with putting probing, challenging questions on your examinations provided this is nothing new and that students have been given the means to tackle such questions. There lies the nub of the difficulty. Once you know where your students are as they begin the class and you have made clear to them what will be expected, how do you make it possible for them to achieve those goals?

Put Supports in Place

Take your big goals and look at the pieces that have to be in place in order to achieve them. You want your students to be able to tackle and solve an unfamiliar problem. It requires going back through all that they have seen this semester, picking out appropriate ideas, and putting them together in what is for them an original configuration. Do they need to be conversant with the big ideas of the class? Do they need to be able to go back and reread sections of the text? Do they need to be certain about the meaning and significance of certain theorems? Do they need to be able to express their own understandings clearly? If so, then these are things that you should have been emphasizing and assessing all along.

You may need to devote time to how to read a math book. Most students believe that this is not possible, so just testing them on how well they read will not be enough. You may need to spend time talking about what you expect in technical reports. You may need to talk about techniques of problem solving. If students know that these are skills for which they will be evaluated, they will pay attention. If your syllabus is so crammed that there is no space to work on the skills you consider to be important, then there is something wrong with your syllabus.

The most important support that can be put in place is the opportunity to practice the high level skills that you will demand. This should include critical feedback and provision for students to redo the assignment. One of the great drawbacks of the traditional "two midterms and a final" model of assessment is that it re-enforces the perception that the way to get through classes is to take the week before an exam, focus all energy on that course, clear that hurdle, and then forget all that has been learned to clear memory space for the next hurdle. Traditional assessment re-enforces the attitude that success in mathematics is the result of natural aptitude and not something that can be cultivated and developed. We know that mastery comes through the process of attempting and failing and then going back to find the roots of that failure and correcting them. This is an attitude toward learning that we would hope that our students know before they arrive in our classes, but they usually do not. It is our responsibility to lead them through this process until it becomes their own.

I do this through projects and the reports that students write. I have a clear set of criteria about the level of precision and clarity that I expect. Incorrect mathematics, inadequate explanations, and poor writing are all critiqued in the first submitted version which is returned to the students for reworking. On midterm examinations, after I grade and return the test, my students have several days in which to correct the answers that they missed. They can regain some of the points that they lost if they can show that they can now solve that problem. The only complaint I ever received about this policy was from a student who did not want to be forced to think about the test questions he had missed after the exam was over. That complaint has become one of my primary justifications for this policy.

Make Students Active Participants

At one of my first national meetings, I picked up a button that proclaimed, "Math is not a spectator sport." We all learn by doing. I do not know any responsible teachers who do not want students to wrestle with the mathematics that we have explained to them and to practice the higher level thinking skills in which we want them to excel. The problem is that most of our students have no idea how to begin to interact with mathematics in this way. For them, practicing mathematics outside of class is doing the even exercises numbered 2 through 20 at the end of the section, and nothing more. Most of our students come into our classes with no conception of how to begin tackling an unfamiliar problem, or what to do if the first line of attack fails. The answer is not to despair of who they are, but to help them become the students that we want them to be. Our opportunity for shaping the behaviors that we want them to adopt is in the classroom.

Some of the most effective learning I have witnessed has been in group situations in the classroom where a small group of two or three or four students is tackling a challenging and unfamiliar problem, a problem that they will then carry out of the classroom to continue working on. The group dynamics are important. Students working on their own are more likely to freeze and try nothing if the correct approach is not immediately apparent. They are more likely to stick to an unproductive strategy despite its futility. They are less likely to see alternate procedures that might simplify or simply clarify what has worked. As I have seen repeatedly, a small group of students collectively can solve a problem that none of them individually could have worked out. The result is increased confidence and experience in lateral thinking.

Lecture is still one of my tools for teaching, but I have learned that it is most effective when broken up by opportunities for students to actively engage the topic that I am explaining. This includes asking probing questions and giving students time to think about or work on the answers. When there is doubt or hesitancy about the answer, I ask several students to put the answer into their own words. Where there is divided opinion about the correct answer, it is helpful to stop the class and have students discuss it with those around them. A tremendous amount of learning transpires when a student has to explain his or her own understanding to someone else.

Encourage Group Work

Few of the catchwords associated with the reform movement in mathematics instruction have been as controversial as collaborative or cooperative learning. I am a believer for the reasons given above, because I have had students tell me that this is where they learned how to analyze unfamiliar problems, and because of the role that it plays in developing study groups and support networks.

I have held exit interviews with graduating seniors. One of the consistent factors cited as important for their success in college was learning to form and make use of study groups. This was not something that they knew would be helpful when they came to college. Their first study groups grew out of the interactions with classmates that were created by group projects.

There are many approaches to collaborative or cooperative learning. My advice is to talk with others who have experience and then experiment with what feels comfortable to you. The fact that students will brainstorm a problem collectively does not mean that they have to receive a common grade. I have often had each member of the group write up his or her own report. But group grades are useful early in the semester to force students to learn to work together.

Use Technology as Appropriate

Technology is the other controversial catchword of the reform movement. This is the one place where there really is something new under the sun, but reformers are far from unanimous in their understanding of what it means for teaching or how it should be used. My own advice is to stay informed and be willing to experiment with the ideas and approaches that make sense to you. There have been many failures. There also have been many successes. We are still in the early stages of learning how to use this tool.

Advocates of technology say that it enables students to focus on the ideas rather than rote manipulations. Critics assert that it becomes a crutch and that students who are not fluent with fundamental processes are handicapped when approaching higher level problems. My own experience is that both sides are correct, and the important and difficult question is where to draw the line. At what point do we introduce, permit, or encourage the use of technological tools?

The answer is highly dependent on our immediate objective. When I want my calculus students to discover the orthogonality of the sine and cosine functions of various frequencies, it is essential to use a computer algebra system to ensure speed and accuracy in performing integration by parts. But before we begin such an exercise, my students will have done a lot of integration by parts problems by hand because I believe that it is a fundamental skill. Its mastery is essential to the appreciation of its consequences.

There are many different ways in which computers or graphing calculators can be incorporated into classes. They can be used to prepare demonstrations, but be wary of presentations that are too slick. Remember that your job is not to entertain but to get students to think. Many math classes now incorporate labs where computers are used for numerical calculations, or to aid in visualization, or for work at the symbolic level. What I have found through painful personal experience is that any laboratory must be tightly integrated into what is happening in the classroom, reinforcing what happened in the previous class and preparing for the next. When the laboratory experience is well thought through, it can provide powerful reinforcement of the lessons that you want to communicate. When it is poorly conceived, it is nothing more than a frustrating waste of time for you and your students.

Be Open to Curricular Reform

When I have thought long and hard about a course—what works and what does not work for my students—I find myself dissatisfied with the traditional syllabus and the available textbooks. The fact that I am not alone is reflected in the myriad reform curricula and textbooks. This is not a monolithic movement. It is characterized by wide diversity. In sorting through what is available, you need to be aware of the goals and priorities of those who developed the materials.

My own primary criterion in choosing or developing curricular materials is to have a driving theme that generates questions that will puzzle, discomfort, and challenge my students. I have come to appreciate that the pure Euclidean ideal of finished mathematics is not appropriate for most teaching. It is a polished surface too slippery for most students to grasp. Imre Lakatos [**3**, p. 140] has gone so far as to accuse Euclid of being "the evil genius particularly for the history of mathematics and for the teaching of mathematics." A kinder assessment is given by David Tall in his "Reflections" on *Advanced Mathematical Thinking* [**6**],

> This does not remove the need to pass on information in the theorem-proof-application mode, for this is the crowning glory of advanced mathematics. But students need to be assisted through a transition to a stage where they see the necessity and economy of such an approach.

Some of my own thoughts on how to assist students through this transition are expressed in my review of Serge Lang's *Undergraduate Analysis* [**1**] and in "True Grit in Real Analysis" [**2**].

If we want to change what students take from our courses, then we must change what we do. If nothing changes, then nothing changes. We must be realistic about where we and our students start, clear about what we want to accomplish, knowledgeable about how our students learn, and willing to experiment with our teaching to make it as effective as possible.

References

[1] David M. Bressoud, Review of *Undergraduate Analysis: Second Edition*, *The Mathematical Intelligencer*, **20** (1998), 76–77.

[2] ———, True Grit in Real Analysis, submitted to *The American Mathematical Monthly*.

[3] Imre Lakatos, *Proofs and Refutations: The Logic of Mathematical Discovery*, Cambridge University Press, 1976.

[4] Thomas M. Sherman et al, The Quest for Excellence in University Teaching, *J. of Higher Education*, **48** (1987), 66–84.

[5] David Tall, The Psychology of Advanced Mathematical Thinking, pages 3–21 in *Advanced Mathematical Thinking*, David Tall ed., Mathematics Education Library, vol. 11, Kluwer Academic Publishers, Dordrecht, 1991.

[6] ———, Reflections, pages 251–259 in *Advanced Mathematical Thinking*, David Tall ed., Mathematics Education Library, vol. 11, Kluwer Academic Publishers, Dordrecht, 1991.

Thanks to Danny Kaplan, Karen Saxe, Stan Wagon, and my wife, Jan, for thoughtful comments and suggestions.

Remember the Students

William J. Davis

The Ohio State University

I didn't like Steve Krantz's first edition of *How to Teach Mathematics* because of its narrow focus on traditional methods.[1] As strange as it may seem, even though he has spent a lot of time talking and learning about different modes of teaching, and even though he has tried to incorporate much of that new knowledge into this book, I think I like this version even less. It seems to me that now he's patronizing people who want to make basic changes in how math is taught.

I think a good place to start to see what I dislike in the book is in Steve Krantz's section 1.6 on lectures. Here are some snippets from that section.

> Those who say that "the use of the lecture as an educational device is outmoded" rationalize their stance, at least in part, by noting that we are dealing with a generation raised on television and computers.
> *SK, p.14*

I don't think the lecture is outmoded. I don't really think it was ever "moded" in the first place. The people who succeed in learning math in a typical lecture-based course are probably capable of learning math without the lecture. After all, most of the learning happens when the student grapples with the ideas for himself after class. Professor Krantz seems to agree with this point.

> To be specific, the old-fashioned paradigm for student learning was that the student would sit in class for an hour and take note; then he would go home and spend three to five hours deciphering the notes, filling in the gaps, and doing the homework.
> *SK, p.16*

Later in the same section, he proceeds.

> The jury is still out on the question of whether students taught with reform methods or students taught with traditional (lecture) methods derive the most from their education. Which students learn more? Which retain more? Which have greater self-esteem? Which have greater interest in the learning process? Which teaching method encourages more students to become math majors? Frankly, we don't know.*SK, p.17*

I leave it to you. Should that jury stay out? Can we do anything to answer those questions? What's that *self-esteem* part about, anyway? Who's talking about self-esteem? I think most of us agree that the only true self-esteem comes from real accomplishment.

---

[1] I thank Professor Krantz for inviting some of us who have very different views of how to teach math to write short pieces for his book. It's a generous act of a person sincerely interested in the subject of teaching and learning.

I have heard the charges over and over that people trying to change the way we teach math don't give evidence that what is being done is an improvement over the traditional teaching methods. There *are* studies that indicate improvements in the items listed above. Why are such studies neglected, in particular, here? Steve Krantz should know about at least some of the studies.

Steve Krantz has inadvertently given us some guidance in this direction.

> We make no statement unless we can prove it.
> *SK, p.27*

Of course Professor Krantz was talking about our mathematical craft when he wrote that, and not about the topic of this book. It gives me an opening to complain, though, so I'll take it. We mathematicians seem to leave the tools of our trade behind us when we venture away from mathematics. We tend to abandon clear, supported argument. We all serve on committees, so you know what I'm talking about. I think what Steve Krantz said about the *new methodology* is one such example. There *is* evidence. There's not enough. There's not, as far as I know, a standard set by our learned societies against which we can measure success and failure. Why not? Other groups, including physicists, *have* set such standards against which they measure the effectiveness of different approaches to teaching. We could certainly decide to stop bickering and do the same.

I believe students can learn, but that we can't teach in the way that Steve Krantz believes we can. Lecturing is telling, and telling just doesn't do the job.

> My teacher knows a lot about the subject. He just doesn't know how to tell it to me.
> *A freshman honors student at OSU*

Steve Krantz's *How to Teach Mathematics* is about teachers and lecturing, but not much about teaching or about learning. The student said it all above. The teacher is telling, and the student is assumed to be absorbing, interpreting and internalizing. That may be what a very small percentage of our audience does, and that's probably as it always has been. We mathematicians come from that group. We tend to believe that you either come from that group, or you can't learn and do math. There's a natural tendency to call it the top 10 percent, or something like that. The group I think is really different from the way they were years ago is the other 90 percent. I believe, but can't prove, that the skills people in that majority bring to the task of learning mathematics have dropped significantly since I began teaching in the mid '60s. That's a topic for a different note.

Can we take the other 90 percent and teach them mathematics? I believe we can for a large fraction of that group. If we are to accomplish that, we must become more effective teachers. Effective teaching must include getting students involved and learning. This book essentially ignores students' learning processes. Steve Krantz seems to assume that his audience is full of copies of people who learned the way we did. We were capable of going to class (or not), and then doing the work we had to do to master the material in the text and the assignments just as it was presented. I believe that can happen only when the objects being studied have some meaning for the student, and I believe that most of the students we see in our classes have attached meaning to very few of our objects.

When I started to write this appendix, I went through the Krantz manuscript looking for a statement saying something like, "Lecturing is a good and effective way to teach mathematics." It must be there, but I couldn't find it. Here's one, and it bothers me.

> Turn on your television and watch a self-help program, or a television evangelist, or a get-rich-quick real estate huckster. These people are not using overhead projectors, or computer simulations, or Mathematica. In their own way they are lecturing, and em very effectively. They can convince people to donate money, to change religions, or to join their cause. Of course you calculus lecture should not literally emulate the methods of any of these television personalities. But these people and their methods are living proof that the lecture is not dead, and that the traditional techniques of Aristotelian rhetoric are as effective as ever.
> *SK, p.15*

Steve, whatever is going on in these TV shows isn't teaching. It isn't learning. It's largely emotional response to *convincing* stimuli. These purveyors have mastered the art of appeal to the reptilian brain (see below). It is clearly in the best interest of TV hucksters to keep their audiences from thinking.

The problem I have with the book is that it's about the teacher's teaching, not the students' learning. We know a lot about learning. We should use what we know.

Teaching and Learning

We live in a wonderful time. As we near the end of the 20th century, we have access to whole new fields of evidence about how people learn. We know a lot about how the brain functions, about the structure of the brain, about organization of information, about the differences between information and knowledge, and about the processing of all of that. How can someone write a book about teaching and ignore what we know?

I'll take one simplified model of the brain as a learning machine, explain it briefly, and try to indicate how I believe that model should influence what we do for and with our students. A friendly place to read about this model and its consequences is in the book by Caine and Caine [3]. A more detailed description of the functions described here appears in the recent dissertation of Lee Wayand [13].

The brain's job is to organize itself and the rest of the system it regulates. The human brain consists of three parts; the R-complex (or reptilian brain), the limbic system which contains the hippocampus and amygdala, and the neocortex. That reptilian brain, or lizard brain as Steve Krantz now calls it in private conversation, is basically the brain stem. The next layer is the limbic system and that large surrounding pudding is the neocortex. For purposes of our discussion, the R-complex performs the base level survival functions of the body, from breathing and blood flow, to basic survival skills like fight or flight responses to threat. The hippocampus mediates all sorts of activity, including some information processing and learning. The amygdala is likely a center for mediation of emotion.[2] High level thinking and learning goes on in the neocortex. If you want people to understand and learn, you want them functioning in the neocortex as much as possible. For the neocortex to be in charge, the R-complex and limbic system must cooperate.

There are two basic kinds of learning [1]; facts and understanding are good enough names for my purposes. Facts are what you think they are. They consist of specific instances of features of our surroundings, like specific faces of friends and acquaintances, characteristics of objects we encounter outside ourselves, and, for mathematicians, numbers, formulas and processes. Understanding is the collection of maps and templates we have at our disposal to fit facts into. These templates and maps are the places that facts are connected to gain understanding.

On the next page is a schematic representation of the system needed to get facts into a person's memory.

On the very left we find our senses, and layers of filters one's brain uses to avoid information overload. Most input is ignored.[3] The chamber immediately after the input and filters is short-term memory. It seems to be able to store small numbers (say 7) of pieces of information for only short time. Some of those pieces can be kept alive in short-term memory by forced repetition. Some can be passed on into long term memory by repetition and memorization. Information gets into short term memory and hangs around for a very short time before it either passes through that upper channel to long term memory, or until it is flushed into oblivion through that drain because the brain didn't have any use for it. That flushing is a physical process. It appears that much of the mediation of short-term retention, passage to long term memory, and that flushing is done in the hippocampus.

Take special note of the circles. Those circles around the channel between short and long term memory are sphincters, which can be shut down either by the reptilian brain or the limbic system, and which can prevent the transfer of information either into or out of long term memory. Understanding what activates and what relaxes those sphincters is critical to good teaching.

Information doesn't stay in short-term memory for very long. It seems that it can be kept there longer by repetition, but it won't stay forever in any event. There seem to be two ways things pass from short-term memory into long-term memory. The first and commonest way we build our "facts" database is through repetition or practice. Building this database can be linked to external motivation. Storing facts without understanding almost certainly explains how

---

[2]Separating emotion from logical thought and learning is impossible [4].

[3]Think about it. Most of the things you hear and see are not important enough to be remembered. They may be important to your understanding your current environment, but they aren't worthy of their own permanent place in your brain.

students can successfully cram for tests and then immediately lose the ability to use those facts again after the test is over. Recall is more difficult. Even if facts reside in the database that is your brain, your internal search mechanisms must have ways of finding them when they are needed. If they aren't attached to objects that carry some meaning (understanding), they may be lost forever. Another way information can pass from short-term memory to long term memory is that it finds an existing hook to attach to. That can be a place that says, "This is like something else I know," for example. Frequently it means that there are slots in various templates for the new information. Of course that means that it is like something else that's in there, because otherwise the templates wouldn't have such slots.

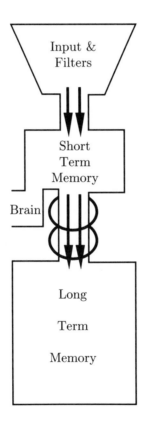

How do those maps and templates work? Here's a simple example.

Imagine walking into a strange airport here in the United States. You see counters at which agents are selling tickets and checking people in, and you see signs pointing you to various gates. You don't have to be told which is which, and you instantly are looking only to see the specific differences that might be important to you. You look for the names of airlines, monitors with flight information, and the like. You don't have to be told what each part of the terminal is about. You know. You can also fill in specific information quickly and accurately, and keep it working for you as long as you need it. No one had to tell you where things were, their function, or the reason for their existence.

This is an example of how templates are frequently spatially organized. Spatial organization can work against our data search as well. To see that, think of running into an acquaintance from work at a clearly independent place, like a beach or mall. You will frequently know you know the person, but won't be able to attach simple information such as a name, because that person is simply out of place.

Our hooks and templates aren't all spatially grounded, of course. Early in the development of the species, they probably were because of our more intimate relationship with our surroundings. By now we have learned to build templates for conceptual structures. We mathematicians have developed elaborate systems of such templates. (In fact, we frequently give them a pseudo-spatial meaning, don't we?) These templates are at the center of concept understanding. Good templates make addition of new facts easier, and make retrieval of relevant knowledge possible. Imagine your own 'function' templates. When, for example, you encounter an expression like $au^2 + bu + c$, and when your hippocampus asks for templates that might have facts about the expression, several things might happen. You would very likely recognize the pattern, and not the actual names of the constants and variables. You would likely think of it as a generic function whose variables might be real or complex numbers or quaternions or matrices or even other functions. You might associate a graph of some sort with it. You would probably not think of this expression as just something to stick into the quadratic formula. (That is what the majority of undergrads do with such an expression once they manage to translate that u to x.)

Let's get on to the process of teaching and learning. Teaching is simply helping students learn. The first step that must be taken is to get the students involved in the process. They are ready to learn, and their minds have learning as their primary occupation. It seems that this makes several jobs for us teachers. One is to keep those channels open between short and long term memory so facts can flow in and knowledge can flow out. A second is to help organize that database of facts into meaningful pieces of knowledge. A third is to help students find the relevant maps and templates that might help them learn and solve their problems. When the templates and maps don't exist, we must help the students build the new ones they need. That comes most frequently from modifying and expanding existing ones.

Those of us who are interested in reform are frequently charged with trying to make learning math simple, or pleasant, or more in tune with students' life styles and experiences. I don't think that's true. The things students do that annoy us as we try to help them learn are probably emanating from inside their reptilian brains. That's certainly true about their hormonal activities, fear and apprehension of math classes, love of MTV, and quest for beer. Those activities may drive a large part of their lives, but I certainly can't get into their learning systems in the neocortex by pandering to the primitive being. If I try to take advantage of, say, students' intimate involvement with MTV or computer gaming, I'm going to be conversing with the wrong part of the brain and real learning won't take place.

Finding the places to connect ideas and facts in students' minds has been recognized as a primary role of the teacher for at least the 2500 years since Socrates. In our lifetime, and our discipline, we have the legacy of George Polya [11] to remind us that our job in teaching as well as in research is precisely that of finding, building and using the appropriate templates. In slightly more recent time, we have the constructivist approach to learning with its primary proponent, Piaget. These days we have the ongoing work of E. Dubinsky and others following the serious ideas of Piaget into real experimentation and research in teaching mathematics.

Many of us believe that the primary tool for the teacher for keeping the channels open, and for helping students find connections in their current knowledge is Socratic dialog. What do we mean by a Socratic dialog? Simply put, the teacher doesn't tell, the teacher guides. The student usually initiates the dialog by asking for help on a problem. The teacher then asks questions about the problem at hand. "What was the original problem? Is this like something you've seen before? What have you done so far, and how did you decide to do that?" The questioning needs to be non-threatening so those channels stay open. The teacher really needs to listen carefully, because in the majority of cases, the difficulty the student has is not expressed in the first few questions and responses. For example, in setting up equations for a simple polynomial interpolation, the difficult part may go back to the student's not understanding that x^2 is not the most general quadratic expression. Once the student decides that the generic quadratic is $ax^2 + bx + c$, a major part of the battle is won. The job of the teacher is to get the student to that point.

Many communication skills are used in Socratic encounters. They are relatively simple to understand, but difficult to use because we aren't used to using them. One example of such a skill is timing. Just as in good jazz music, the spaces and quiet times are frequently more important than the notes. There is a whole craft built up around effective communication of this kind. My familiarity was derived from the community of people involved in what we now call conflict management. The skills for teachers are very similar to those required of a dispute resolution mediator.

Students react in interesting ways to attempts at Socratic dialog. The following is typical.

> He doesn't just answer our questions. He asks more questions until
> I begin to see where my real questions are.
> *Anonymous OSU Student*

Here's my recent favorite quotation from a student who was a participant in this sort of discussion for several months. It represents an extreme view from a student's perspective of what's happening during such encounters.

> They have this thing. You ask them a question and they just stare
> at you, and suddenly you know the answer. It's really annoying.
> (Now they are trying to teach us how to stare at students over the
> Internet.)
> *An OSU Student and Distance Mentor*

I believe that Socratic dialog between students or groups of students and instructors is where learning begins. I believe that very few students can sit and assimilate facts and turn descriptions into knowledge and processes if I mostly just stand in front of them announcing new ideas. I'm sure they can't do it *during* a lecture, so in Calculus&*Mathematica* courses[4] [5], we present students with new ideas in Mathematica notebooks. They are encouraged to explore the ideas by experimenting, discuss their ideas about what is happening, and build their own templates for understanding. Obviously students can't be expected to invent all of the ideas.[5] What they can do is become familiar with processes and ideas before they are formalized by the teacher. Only then will the ideas have some meaning attached (templates built).

Obviously class time is not where the bulk of learning occurs. Our C&*M* students spend long hours working at home and in the lab. In fact, if one is to believe their estimates of time spent on learning math, our students are involved for many more hours than the typical student is. Of course, they aren't usually spending Steve Krantz's recommended 3 to 5 hours outside class for each hour in class. The last time we tried to estimate the time, the average was somewhere between 15 and 17 hours per week per student in the lab for our 5-hour courses. Class time is spent helping students strategize for learning.

To reiterate, I see my role as that of questioner. My job is to help students find the hooks that might accommodate new facts. My job might be as simple as getting the student to admit that $au^2 + bu + c$ is simply an alternate spelling of $ax^2 + bx + c$. I keep asking questions as they search their memories. All the time these discussions are going on, it is my job to keep those learning channels open. I can't tell students how far back into their memories they may need to go to find a hook they need for the task at hand. They must find that for themselves. I can only guide.

At this point, all of this probably sounds like an impossible task. I am asking you to sit with students patiently guiding their thinking back to places in their knowledge that impinge on the problem they are working on. Well, it is difficult. On the other hand, it is probably not as daunting a task as it appears at first. For one thing, you should be talking to small groups of students most of the time rather than individual students. For another, you'll soon find that the paths to the correct hooks and templates are very similar for the majority of students. You'll also find that giving students related tasks to perform can shorten the process. Finally, the real experts at helping students find the way are other students. Once a path has been found, students are happy to share what they have found. Surprisingly enough, once they are used to the Socratic business, they also don't simply tell other students what they have discovered. They try to help their peers find their own way.

So what about lectures in my courses? At the end of each Calculus&-

---

[4]Most of my experience with using the ideas in this note come from my involvement with Calculus&*Mathematica* classes. I'll refer to them throughout the rest of this note.

[5]S.K. (p.104) includes as one of the hallmarks of the reform movement, in a bulleted list, "Students should discover mathematical facts for themselves." I'll wager that Steve Krantz and I read completely different meaning into this sentence. I'll bet Thomas Kuhn [9] would say that Professor Krantz and I are encountering a paradigm shift.

Mathematica lesson, as the students try to summarize and make sense of what they encountered, we have a review session over the ideas the students wrestled with as they worked. This is the only time we do anything like a lecture in the class.[6] I don't start new lessons by telling the students what I want them to do.

This is difficult stuff for mathematicians to cope with. Occasionally, you must suspend disbelief as you listen to the places students can take you in their quest for hooks. Here's an example. Early in my differential equations course, I ask students to check on whether or not various functions are solutions to differential equations and some initial value problems. They have trouble, probably because checking that one function equals another function isn't something they've done much of. In order to clarify what I'm asking, I might start a discussion with, "Check and tell me if $x = 7$ is a solution to $x^2 + 6x - 7 = 0$." Most of the time, a sizable number of students in the classroom will say, "No. The solutions to that are $x = -7$ and $x = 1$." When I ask what they did, they tell me that they dug out the quadratic formula, plugged in the numbers, and got those answers. When I ask if anyone did it differently, I rarely get anyone saying they just plugged 7 in and didn't get 0. Why is it worrisome? It indicates to me that they have only programmed responses to much of the mathematics they encounter. The processes they bring usually don't come with any understanding; they are rote procedures. We have trained them to react rather than analyze. The programmed response to a quadratic expression is plugging something into the quadratic formula. I change the problem to something they don't know how to solve, like $x^4 - 28x - 701 = 0$. It takes rather a long time to get them to plug the 7 in for x and see what they get. Then we get back to the problem of whether or not $y(x) = sin(2x)$ is a solution of $y'(x) = 2y(x)$. Following a great deal more discussion, they frequently say that it happens at lots of points, and give me as evidence a graph of $2cos(2x)$ plotted together with $2sin(2x)$. They point to the crossings, and say, "Here." It turns out that we are still a long way from having them internalize the notion of functions being equal, and checking that some expression is a solution to a given differential equation.

One more time, here's the challenge. If it's true that students can only build new knowledge upon their existing knowledge, we should know what that knowledge base is. George Polya [11] opened that can of worms for us mathematicians a long time ago. What do our students know? What does know mean? What is knowledge? A group of my students announced in a discussion we had this spring that knowledge is information that's usable. Postmodern philosophers have made knowledge into a deep and complicated activity. At virtually any level, though, we need to know what usable knowledge our students have at their disposal. That, to me, doesn't mean, "What formulas do they remember?" or, "What formulas don't they remember?" I don't know how to answer the question about what students know and don't know. I know even less about how I might assess what they are capable of doing with what they remember. Look at the previous paragraph. What are the students struggling with? Is something missing in their 'facts' database, or so they have rusty templates for

---

[6]There is a frequently encountered exception to that. If I spot something that needs a quick redirection for several groups of students, we'll take a short trip to the board for a discussion of the particular sticking point.

interpreting 'equals', or what? The best I can do for now is to question students as the need for help arises, and hope that together we can find a way to help them understand the question.

We mathematicians tend to talk to the top 10 percent of the student body: that part capable of learning mathematics well, and as we learned it. Perhaps we are simply so proud of ourselves that we have failed to notice that other people have skills and abilities that can make their understanding of mathematics as real as our own even though it may look different to us. We have always known that some mathematicians deal best with geometric and spatial objects while others deal most fluently with symbolic representations. Our teaching, though, seems to concentrate most strongly on the latter, and reward for hard work comes to those who do well with the symbols and the formal argument. In other words, success in mathematics continues to be, for the most part, limited to people who think and learn the way we did. Are there other ways? I think so.

If our brains develop according to our life experiences, our environment, and our inherited family traits, it makes sense that each of us has a different cognitive structure to deal with observation and learning. When I was growing up, the preferred measure of intelligence was IQ. Tests were given which measured linguistic and mathematical propensities, and declared intelligence or lack of it on the basis of the results. That seems to be a rather narrow view of intelligence. Don't you agree? What about other 'intelligences'? Going back to Dewey [6] and Jung [8], and moving forward to Gardner [7] and many others[7], we are forced by reason and observation to consider other capacities. Among these are the ones Howard Gardner added to the Verbal, and Logical/Mathematical[8] upon which IQ testing rested: Kinesthetic, Spatial, Interpersonal, Intrapersonal, Musical, and a newer addition, Natural. We should be prepared to take advantage of the different learning styles our students bring to us.

Most of the reform teaching projects expand the approach to teaching and learning to accommodate a broader learning style base. The first clear statement of that came from the Harvard consortium's rule of three (now four): Symbolic, Numeric, Graphical and Verbal. Quite simply put, a student should first be able to approach a new mathematical idea through that student's own strengths. After that, the concepts should be viewed through each of the other lenses, and the student should be able to put real flesh on the bones of the concepts by making the connections between the four perspectives. Is anything different from what we have always done in our classes? Probably. For one thing, a student should be able to approach a concept from each of the different points of view. For another, we all expect students to be able to give clear and concise explanations of what they do, observe or calculate orally or in written form. In Calculus&*Mathematica*, we expect such descriptive writing in each problem solution.

Go back to the illustration of the learning system above for a minute. That

---

[7]One would do well to look at the ideas of Myers and McCaulley [10], and Gregorc [12], [2]. A synthesis of many of these ideas is emerging from Bernice McCarthy's 4MAT model (http://www.excel.com).

[8]I think this is a terrible name. It certainly doesn't encompass the wide variety of skills that might describe success in mathematical adventures.

simplistic cognitive model can also let you see that there are different ways of interpreting and processing information and experience. One interpretation, from the work of Gregorc, suggests that one perceives new information best from either a concrete or abstract presentation, and then processes that information internally either sequentially or randomly. One can simply read the descriptions of what these terms mean and make good guesses about who will succeed and who will fail in our traditional classrooms. We can accommodate more people if we simply make our teaching more flexible than it currently is.

There's one more book that's a bit off the beaten path that I recommend to you now. It's *In The Mind's Eye* by Thomas West [14]. The premise is straightforward. Too little attention has been paid in the past to the visual learner. The contention is that we live in a visual age, and that there are many people who are labeled as being dyslexic, or learning disabled, whose ability to learn in standard modes is less than what our educational system expects.[9] These people generally have very strong spatial abilities. According to Tom West, we are also entering a far more visual age. It may be that the pendulum will swing, and that the greater value may soon be placed on people who excel in this mode of information processing.

There's not room to do any of the ideas introduced above a fair hearing here. Why don't you talk to some people, visit a bookstore and read some more about all of this?

Epilog

Those of us working toward change are frequently accused of trying to simply make math fun, or make math easy. I don't agree, of course, with that assessment. I expect my students to work very hard, and I expect a lot from them. One C&M instructor at Ohio State this past year decided that the text and electronic lessons were deficient, that they didn't make the students master enough facts. He proceeded to fix all of that by taking the students from the lab for regular sessions on the topics left out of the course. He made his expectations very clear, just as we all usually do in a traditional classroom. He expected the students to be able to perform a clearly defined list of manipulative exercises, the kind one finds on most calculus exams at our colleges and universities. We had a protracted conversation in e-mail. Here's one thing he said.

> I can show you my student teacher evaluation forms if you'd like which are in general very positive and as a matter of fact there's one which I'm looking at right now which explicitly thanks me for taking the students out of the classroom once a week to explain/review the material.
> *Calculus Instructor*

I'm not surprised, are you? This instructor gave the students clearly defined tasks of the sort they had experienced for their previous twelve or thirteen years

---

[9]Here I go again. This is yet another topic for discussion.

of math. He took them to a place where he could explain clearly how they should answer each of those questions, and then tested them on what he showed them. What could be easier for a student? A student has a success template laid out in front of her. She works hard at practicing the skills the instructor recommends, and if she does that well, she'll succeed. There's no need to impose any internal meaning on the processes, and there's no reason to think that much, if any, of the material will ever come back to haunt her. This student can get a good grade in this course without having anything she learned be transferable even to the next course.

Ask yourselves again, "Who is trying to make it easy for the student to succeed?"

References

[1] Anderson, J., *Cognitive Psychology And Its Implications*, W.H. Freeman, New York, 1990

[2] Butler, K., *It's All In Your Mind: A Student's Guide to Learning Styles. CT: The Learner's Dimension. Cognitive psychology and its implications*, W. H. Freeman, New York, 1988

[3] Caine, R. and Caine, G., *Making Connections*, Addison-Wesley Publishing Company, Reading MA, 1994

[4] Damasio, A. R., *Descartes' Error: Emotion Reason, and the Human Brain*, Avon Books, New York, 1994

[5] Davis, W.,Porta, H., and Uhl, J. J., *Calculus&Mathematica*, Addison Wesley Publishing, Reading, 1994

[6] Dewey, J., *How We Think: A restatement of the relation of reflective thinking to the educative process*, D.C. Heath, New York, 1933

[7] Gardner, H., *Multiple Intelligences : The Theory In Practice*, Basic Books, New York, 1993

[8] Jung, C. G., *Psychological Types*, Princeton University Press, Princeton, 1971

[9] Kuhn, Thomas, *The Structure Of Scientific Revolutions*, University of Chicago Press, Chicago, 1970

[10] Myers, I. B. and McCaulley, M. H., *A Guide to the Development and Use of the Myers-Briggs Type Indicator*, California: Consulting Psychologists Press, California, 1985

[11] Polya, George, *How To Solve It*, Princeton University Press, Princeton, 1988

[12] Tobias, C. U., *The Way They Learn*, Focus on the Family Press, Colorado Springs, 1994

[13] Wayand, Lee, *Identifying Communication Obstacles That Arise When Translating the Modern Mathematics Classroom to Distance*, Dissertation, The Ohio State University, 1998

[14] West, T., *In The Mind's Eye*, Prometheus Press, New York, 1991

Reflections on Krantz's *How to Teach Mathematics*
A Different View

Ed Dubinsky

Georgia State University

1. Introduction

When Steve Krantz asked me to write, for the appendix of this book, an essay expressing my views of teaching, I thought it would be useful to do this in relation to the book itself, rather than just give my views in isolation. So my first step was to figure out what the book was all about. I began, naturally enough, by reading it. And I discussed a lot of things with Steve. And I thought about it. Here is what I think the book is.

Krantz has produced a serious reflection on a traditional approach to teaching and learning. He acknowledges the existence of views other than his, but although the book goes far beyond a mere reference to these alternatives, it is considerably less than a balanced presentation of a variety of different approaches to collegiate mathematics education. Even with these appendix essays, the book retains its strong reflection of the very personal views of the author, based mainly on his many years of experience in the classroom together with some, not overly extensive, awareness of past, present, and future investigations into the teaching and learning enterprise.

We get two things from what Krantz has written and both are welcome. First, the text, together with the appendices (some of which I have seen as of this writing) contribute to moving our considerations of teaching and learning collegiate mathematics from an argument in which we are all fighting to defend our positions and attack those with whom we disagree, to a set of reasoned discussions in which we clarify points of agreement and disagreement, provide justifications for both, and ultimately search together for the syntheses that alone can make significant improvement in the breadth and depth of what mathematics our students learn.

Second, we learn from this book that although there really are identifiable categories of viewpoints such as traditionalists and reformers, it is not the case that a member of a category differs in all of her or his ideas from any member of a different category, nor do all members of a category agree with each other on every issue. What this realization helps us with is the need to focus on ideas and not on those who hold them. When we do this, we find many points of unexpected agreements and disagreements varying with the issue and the individual. I find that the most important precondition for me, in the words of Secretary Riley, to stop fighting and get on with the job of improving education, is to understand that Andrews, Askey, Hughes Hallett, Krantz, Uhl and everyone else might, on any given issue, agree or disagree with me.

So, in the spirit of this latter point and in hopes of contributing to the former, I will try in this essay to reflect on a number of issues that arose for me in reading Krantz's book. My goal is, in each case, to explain and perhaps justify my position, to express my agreement or clarify and possibly justify my disagreement. The topics I will consider are: beliefs and theories about the nature of learning; teaching methodologies; the use of technology in mathematics education; applications and pedagogy; and some specific pedagogical issues.

2. Beliefs and Theories about the Nature of Learning

In this section I would like to talk about traditional vs. reform approaches to mathematics education, the particular approach called constructivism and its relation to discovery learning, the relation between knowing how to calculate and understanding concepts, the intrinsic difficulty of undergraduate mathematics topics, and how we can try to decide what are the effects of a particular pedagogical method.

2.1. You @!*#?!** Traditionalist!

Krantz believes that mathematical knowledge is something, in some sense or other, very definite and concrete, that exists in the minds of some people (usually called mathematicians) and that it is possible to take this entity and transfer it to the minds of other people (usually called students) using media like speaking, or writing, or showing pictures. I believe, on the other hand, that mathematical knowledge is much more elusive, being a characteristic of the behavior and thought, external and internal, active and reflective, of *all* people, in varying degrees of quality and sophistication, and that it is built by individuals, acting in social contexts as they try to make sense of certain situations in which they find themselves.

Something like Krantz's beliefs have been held by a large number of people for a very long time. My position is relatively new and although it has a name that is becoming, unfortunately, a buzzword (constructivism — there I said it, I even carry a card!), it is not a position really held by very many people. It is in this sense that I say Krantz is a traditionalist and I am a reformist. It only means that he is trying to hold on to what is best about a going concern, and I am trying to break some shells and make a new omelet.

Now you might think it the epitome of arrogance for *me* to say what *Krantz* thinks. Indeed, doing this contradicts one of my most deeply held beliefs (e.g., that one person can never really know what is in the mind of another person). But I only put it that way to catch your attention. What I really mean is that what I just wrote is one way that one might interpret many statements, such as the following, by Krantz in this book (the comments in square brackets are mine.)

> "...you [the teacher] must be conscious...of what are the key ideas you are trying to communicate to the students..."

> "If you [several bad things some teachers do sometimes] then you will not successfully convey the information."

> "...the teacher who [several good things some teachers do sometimes] conducts a good class."

> "Remember that you are delivering a product."

If we were to schematize the situation suggested by such comments, it is that of an active teacher trying to do something to passive students.

2.2. I Admit It, I'm a Constructivist!

I am convinced that learning does not happen in the way suggested by Krantz's comments. I believe that a person learns by actively trying to do something, or make sense of something and must, almost consciously, make fundamental changes in the make-up of that vague entity called her or his mind. It is making mental changes that I call constructing and it is my belief that the teacher can only act indirectly to get the students to construct these changes. That makes me a constructivist in my approach to mathematics education.

So what does a constructivist do by way of research in teaching and learning together with curriculum development? There are many people engaged in this enterprise and they have many different answers to these two questions. One pair of answers that a number of people have adopted in their work is this. Research means theoretical and empirical studies to understand the nature of the specific mental constructions that a person might make in order to understand a particular mathematical concept. These studies should point to pedagogical strategies that might help students make the mental constructions that are proposed as a result of this research. Then, of course, curriculum development involves the design and implementation of courses in which this pedagogy takes place. The instructional treatments should focus on getting students to make the specific constructions and it should also, guided by what we can learn from research, make use of innovative pedagogical approaches such as cooperative learning and writing computer programs as a way of making mental constructions.

Consider, for example, mathematical induction. In some research that was done over a decade ago, an important difference between students who were successful and those who were not with making proofs by mathematical induction appeared to emerge. It seemed that the successful students were constructing a mental process — or function — that took a positive integer and returned a value of true or false depending on some operation. For instance, if the problem was to show that for n sufficiently large, a casino with only \$3 and \$5 chips could accommodate any number n of dollars, then successful students seemed to be thinking of a function that acted on a positive integer n and returned the truth value of the proposition: There exist positive integers k, j such that $n = 3k + 5j$.

For many students this process conception represented serious progress from being restricted to thinking about a function in terms of plugging numbers into algebraic expression, which we would call an action. It was even harder for students to be conscious of an operation *on a function* which converted such a boolean valued function P of the positive integers into another such, say Q, in which $Q(n)$ is the truth value of the implication $P(n) \implies P(n+1)$. This requires thinking of a function as an object to which higher level actions can be performed.

A somewhat more detailed and extensive version of these comments came out of a research study that combined theoretical analysis with empirical results to obtain what was called a "genetic decomposition" or "cognitive model" of the concept of mathematical induction. The next stage of the study was to develop and implement an instructional treatment that helps students make such constructions. For example, to construct the above function P, we had them write and use the following computer program.

```
P := func(n);
        return exists k,j in [1..n] | n = 3*k + 5*j;
     end;
```

For converting the implication we first had them build a "machine" for doing this in general:

```
convert := func(F);
               return func(n);
                          return F(n) impl F(n+1);
                      end;
           end;
```

And then they could just do

```
Q := convert(P);
```

This code is written in the computer language **ISETL**, which is what we call a *mathematical programming language*. Its syntax is very similar to standard mathematical notation and it can express almost every finite mathematical construct (such as quantifications, functions as sets of ordered pairs, sets and subsets defined by conditions, finite sequences, etc.) in the language of mathematics. The idea is that by making mathematical constructions on the computer the student tends to make corresponding constructions in her or his mind. See [5] for a discussion of **ISETL** and its use in mathematics education.

The study found that using such an idea — together with some other pedagogical strategies, students developed what appeared to a surprisingly deep understanding of making proofs by mathematical induction — and they could make such proofs, even when the problems were somewhat different from what they had been practicing on!

You can see the details of this study in [3], [4].

This kind of approach has been applied to a fairly large number of concepts in undergraduate mathematics and courses (including textbooks, lesson plans, sample exams, etc.) have been developed in discrete mathematics, precalculus, calculus and abstract algebra. It is a general research program that I, together with several others, are trying to implement. An overall description of the research we are doing can be found in [2] and you can find links to specific published research reports and textbooks at the following WEB addresses:

http://rumec.cs.gsu.edu/

http://www.cs.gsu.edu/ matjbkx/edd/mypapers.html

Finally, since I am talking about constructivism, I think I need to say something about discovery learning. There seems to be some confusion here about what is being advocated. Krantz says that the "reform school of thought" favors discovery and that one of the "hallmarks of the new methodology" is that students should discover mathematical facts for themselves."

Not exactly. One does not have to be an educational traditionalist to realize that what took hundreds, if not thousands of years for mathematicians to discover is not likely to be figured out by very many students — even if they do have lots of giant shoulders to stand on, powerful computers to work for them, and intelligent, sensitive teachers (all up to speed on existing research) to guide them. It is almost insulting to hear so many people suggest that reformers are so misguided as to not realize this. If we set before our students mathematical task to perform *before* they have learned the mathematics then they are not likely to invent a proof of the fundamental theorem of calculus.

So why do we do it? Why do we try to get students to work on mathematical problem situations that use mathematical concepts and methods they have not yet learned. It is not that, given the appropriate situations, they *are likely* to realize on their own that there is some kind of inverse relationship between velocity and area, between derivatives and limits of Riemann sums; or that they *can* (and I can tell you that many do) invent the chain rule after reflecting on some at first mysterious examples. All of these things happen in many reform courses and they happen with relatively small loss of "coverage". But they are not the most important reason for giving students an opportunity to discover various things in mathematics.

The real reason is the word I just used—opportunity—and it is an idea that goes back to Piaget. If you give your students the *opportunity* to discover some bit of mathematics, then whether they succeed or not, it seems to be a psychological fact that much of the attention and mental energy they put into figuring something out is directly transferred to understanding the explanation when it is given by a member of the student's group, or another group, or the teacher. And since you are not planning on waiting until everyone (anyone?) has made the discovery, you can cut it off and move on whenever you decide, thereby controlling the amount of time spent on *trying* to discover something.

2.3. A Mathematician? You Must be Able to Add up Really Long Lists of Numbers.

We all know what happens at cocktail parties when we own up to being mathematicians. People think that all we do is arithmetic operations with numbers— lots of numbers and big ones, too. Traditionalists accuse reformers of ignoring calculations and reformers confuse traditionalists with mindless guests at a party. Both are wrong.

On one level, I am somewhat bemused that there is an argument here. How can anyone in the mathematics enterprise, from high school teacher, to industrial user, to researcher, even conceive of mathematics without lots of heavy computations? On the other hand, how can anyone who actually understands mathematics think that calculation is the whole story. To be honest, I don't think I know any traditionalists or reformers who hold to either of these views. I think we all believe that understanding mathematics means that you are able to perform calculations and that you have some understanding of what those calculations are about.

You don't have to be a whiz at calculations, but you do have to be able to do them, however slowly and however many times you have to repeat them to avoid errors. Technological tools can and should be used—but as aids, not to replace the ability of the person who uses them. To be specific, you have to *be able* to take the derivative of any composition of elementary functions, even though you may often choose to use a computer algebra system to do the job, or to check your work.

As to understanding, you have to have the proverbial "feel" for calculations so as to have a rough idea of how they are going to come out without necessarily doing them, you have to be conscious of patterns, you have to know what the calculations are useful for and how they relate to other situations, mathematical and otherwise, and you have to know *why* particular calculations are made in particular ways.

The fact is that both calculations and conceptual understanding are essential parts of mathematics and they are completely dependent on each other. Differentiation is an incredibly useful tool for solving a multitude of problems from the everyday world. It is also an operator on certain function algebras that is linear but not exactly multiplicative—and it transforms composition to multiplication. I think it is important to understand these abstract properties of differentiation, but I can't imagine a very deep version of such understanding that is not based on a thorough knowledge of the rules for calculating derivatives.

Surely everyone agrees with what I have just written. So why is there a controversy? The controversy comes because there is a very large number of students who have difficulty with either calculations or conceptual understanding or both. Our differences here are about how to deal with this. I think it was a wonderful idea that maybe, if you could use a computer to do the calculations, the student would be freed to spend more time and energy understanding the concepts. Either that would be enough or it might turn out that going back later, the student could better learn how to calculate. So maybe we should stop insisting the students learn the calculations, let them use the technology and we can get on with the concepts

Unfortunately, it seems fairly clear to me that this does not work. Krantz is right when he points out that "...there is no evidence ...that a person who is unable to use the quadratic formula" by hand but can do so with a machine will "...be able instead to analyze conceptual problems". My experience suggests that calculating derivatives is hard for many students but understanding differentiation as an operator on functions is much harder. There is simply no way around the pedagogical problem that we must find way to get our students to be better at both calculations and conceptual understanding.

I think, incidentally, there are ways to use technology to help do this and there are several references given by Krantz (for example, the paper by Heid and my own, joint, work in calculus with several others.)

An important point for teachers who plan to think about their students' work with calculations and conceptual understanding is my observation that the strongest resistance to de-emphasizing calculations can come from the students who (with considerable justification) feel extremely insecure when asked to reflect on calculations they can do by hand only haltingly. In the context of a theory I work with, there has to be a strong base ability to perform actions (calculations) before an individual is ready to move on to thinking about processes (reflection on the actions).

2.4. Yes, Barbie, Math is Tough!

I think Krantz is wrong when he says: "There is no topic in the [calculus] course that is intrinsically difficult. We merely need to train our students to do it." Aside from the fact that his second sentence here is more of a focus on actions as opposed to reflection than I think wise, I am not sure what could possibly be meant by the term "intrinsically difficult". Given any topic (in calculus or any mathematics subject), any students, and any point in their development, there will be some for whom the topic is hard and some for whom it is easy. Both my experience and my research suggests very strongly that the idea of a function as an object, so that the derivative of a function is a function, or the solution to a problem can be a function, is hard. It was hard for entire societies and required a long historical development. I think that almost everybody has to make some serious mental changes to move from thinking of a function as something that does something (process) to thinking of it as something to which something is done (action on an object). Whatever the phrase means, I think that developing the ability to understand a function as an object is "intrinsically difficult".

The point is important because we really have to decide, regarding such examples, whether we can get by with pedagogy focused on just training our students to do it, or if we have to develop substantial pedagogical strategies to help our students overcome a major obstacle.

2.5. So Who's Right?

I mean who is right in all controversies about how people learn and what teachers can do to help? I think Krantz's answer to this question expresses the situation exquisitely in four words: "Frankly, we don't know".

But we had better find out! I think that research is one way to do so, but we cannot expect the kind of research that makes a small number of studies (or even a single one) and allows us to say that this or that pedagogical approach is best—or even better. Our domain is much too difficult, compared for example, with medical research and even there we have extensive studies over long periods of time that tell us things like thalidomide is relatively harmless and watching atomic explosions can be done with impunity.

The kind of research we need is basic, long term and must relate to theoretical analyses in addition to the gathering of data. Research can begin to tell us a few things, such as that it helps students understand and succeed with proofs by induction if they are able to think in terms of functions whose domains are the set of positive integers and whose ranges are the set of two boolean values. But we need to integrate this research gradually and intelligently with our teaching and this is one reason I hope that the fledgling field of research in undergraduate mathematics education becomes accepted as one of the mathematical sciences.

But research will not be enough. We need discussions in which people, no matter how much they disagree, are working together to find solutions to our educational problems and not just score points off each other. We need working together in modes such as this book and its appendices. Indeed, one of the most important contributions of Krantz's book is those four words above, together with the clear thrust of doing something to find out. Perhaps here is a place to point out, as I have done elsewhere, that Piaget (who is a great role model for me) habitually dealt with major figures who disagreed with him by inviting them for a year or so to Geneva to work together with him on a project—not to decide who is right, but to synthesize their opposing views (the book by Beth and Piaget listed in Krantz's bibliography is one product of this custom). As far as I can tell, in the field of collegiate mathematics education, the only person who has really done anything that even moves in that direction is Krantz in writing this book, in inviting these appendices, and discussing with many people what are their objections to what he is writing.

3. Teaching Methodologies

Throw out the old and bring in the new! Well, perhaps, but let's be careful. I would like to talk about why I don't think the lecture method is a very good idea. I would also like to describe some possible alternatives. This brings up the question of how to decide what kind if teaching one should do? Having made that decision, it turns out that the work is not over, it has just begun. Learning how to use a particular teaching methodology is not automatic. We all went through many years (decades for some of us!) being subjected to and using what very roughly might be called a traditional pedagogy. So we will be prepared to use that if it is our choice. I hope it won't be. But that means that there is a lot of work to be done in developing the ability to use a pedagogical strategy on your students that is different from what you experienced.

3.1. Do Lectures Work?

Krantz says that he is not ready to give up on lectures because:

> "They have worked for thousands of years, in many different societies,
> and in many different contexts. And they have worked for me."

Is this really true? Forget about "working" for a moment, and let me ask if lecturing as we know it has been happening for "thousands of years". Did Socrates or Plato lecture? Is this what the mathematical monks of the Middle Ages were doing? What was Fermat's classroom presence like? Was Newton effective in his presentations to students? How large a class size did Galois have to deal with? Was lecturing the way European students of the 18th and 19th learned mathematics? Or maybe Krantz is referring to thousands of years of lectures on mathematics in China?

I am not an historian and I don't know. From what little I have heard, however, I think that the practice of lecturing as the main teaching methodology for university mathematics is much more recent. I suspect that instead of "thousands of years", we may be talking about hundreds of years, or perhaps even decades. Whatever is the extent to which lectures have worked, and whatever is meant here by "work", I am not so sure there is an extensive history supporting this particular strategy.

What cannot be doubted, of course, is Krantz's claim that they have worked for him. Well, something worked for Krantz and the multitude of other successful research mathematicians we have produced, let us say in this century. It is possible that for these particular individuals, just about anything would have worked (even cooperative learning!) Even they did learn mathematics by attending lectures, this collection of individuals is hardly typical and certainly not very representative of the student body we are working with in our society's great experiment with mass education at the post-secondary level.

What does seem clear now and for this student body is that lecturing is not working. This is attested to by reports of minuscule attendance at the lectures, poor performance in tests based on those lectures, dissatisfaction on the part of teachers of these students in subsequent courses (in mathematics and courses for which mathematics is a prerequisite) and what seems to be a decline in the rate of successful completion of mathematics courses.

If we can all agree that this problem exists, there are certainly very different views about its causes, and what to do about it. Krantz takes the position that lecturing would be sufficiently effective if we were better at it. I differ with that on purely personal grounds. I think I am an excellent lecturer—but I don't think my students got nearly as much out of my lecture courses as they do today in my courses which use other pedagogies.

Others argue that it is the students' fault. The background they obtained in high school (where they were taught by teachers that are largely products of lecture-pedagogy) is too weak, or their attitudes are all wrong, or the conditions under which we teach are inadequate. Be all that as it may, if we have a huge enterprise (collegiate mathematics education) which is not working, then it is pretty unlikely that all of the fault lies in places other than the methods teachers use in that enterprise. I think that as mathematicians we are responsible for working to make changes in the overall system to help deal with our problems, but that is a long-term operation. In the meantime, we must figure out how to do the best we can with the students we have and the conditions under which we work. In my view, that means we must look for alternatives to the lecture method.

3.2. What Are the Alternatives to Traditional Pedagogical Methods?

There is a multitude of new pedagogical strategies in collegiate mathematics education that many people have been developing and implementing over the last decade. They include various ways in which technology can be used, cooperative learning, replacing some lecturing with methods in which students are more active, writing, and the use of history.

It would be very nice if I could point to a place where all of these methods are described. Now is the wrong time for this. We are too much in a period of new ideas, revising first attempts, discarding some approaches and pursuing others that seem more promising. Krantz gives some information on what is available and MAA OnLine is another source. But I am afraid that the interested faculty member must do a lot of digging in the library, read publications of the MAA and AMS, attend sectional and national meetings where some of these approaches are discussed, and generally be on the lookout for new ideas as they emerge and are reported.

There is some discussion above in my section on constructivism where I talk about some of the pedagogical approaches I have been working with. This is in the context of pedagogy that supports a constructivist view of how learning takes place.

At some point in the future, we will have to think about producing compendia of new pedagogies. Amongst other issues this will raise is the question of effectiveness. What do we have to say to the working teacher about the relative effectiveness of these new methods?

3.3. How Does a Teacher Decide on What Pedagogical Approach to Use?

There is not a lot that can be said about this today. As I commented earlier in agreement with Krantz, we simply do not know how to decide with any great degree of certainty on the effectiveness of a particular teaching method. It is very much like parenting. There are lots of views and many things to read, but in the end, each teacher, like each parent, must decide as best he or she can what pedagogical approaches to adopt.

I must, however, insert a word of caution here. Some people might interpret the previous paragraph as stating that, for a given teacher, whatever "works" for her or him is the method that should be adopted. The danger here is in restricting the concern to the teacher. As Krantz puts it, "...we should each choose those methods that work for us and for our students."

This is not so easy as it sounds. Aside from the difficulty in deciding what works for our students, we must acknowledge that sometimes, what works best for the teacher, may not work very well for the students. To put it extremely, the teacher who faces the blackboard speaks in a monotone, writes out the text on the board, works a few illustrative exercises, and assigns homework, may be using a methodology that "works best" for her or him (in the sense, for example, of minimizing the distractions from other issues in the teacher's life). But I hope we can all agree that this is not likely to be an approach that works very well, much less best for the students involved. At the very least, it is not at all clear that a teacher alone can always determine how effective is her or his approach to teaching.

3.4. What about Implementing a New Pedagogy?

Here, I agree with Krantz that, as hard as it is to decide to implement a new pedagogical approach, this is only the beginning. As Finkel and Monk [7] point out, it is very difficult for a mathematician, with no background in educational methodology, only the experiences he or she had as a student, and in some cases, long years of practice with a method he or she has decided to replace, to actually change the way he or she teaches.

First, one has to make the decision. Then you need to learn about the method you have chosen to implement—how it works, what you can use from previous methods, what changes you need to make. For most people, this is still not enough. You need some kind of mentored experience in actually using the new methodology.

There are some indications that mathematics departments are beginning to introduce courses in pedagogical methods for graduate students who will have a career in college teaching. There are also a number of mini-courses and work-shops organized by the professional societies. Krantz has mentioned MAA's Project CLUME and, in fact, MAA has a program of Professional Development. Finally, those of us working in educational reform produce a great deal of written material that can be helpful and Krantz has referred to these.

All of this is a good start, but I am afraid it is still not enough to make systemic change in teaching and learning collegiate mathematics. I hope that readers of this book will not only pursue the opportunities for professional development that do exist, but also will push for an extension of these activities to a sufficient level.

4. The Use of Technology in Mathematics Education

The use (or not) of technology is one of the most controversial issues in mathematics education today. We appear to be totally polarized—there are those for and those against. To me, this is completely ridiculous because there are many ways in which technology can be used in mathematics education and I differ with some people who advocate some of those ways at least as much as I disagree with those who are more or less against any significant use of computers.

For example, both Krantz and I deprecate what is called programmed learning, or Interactive Tutoring Systems. Krantz worries, correctly in my view, that some of these systems may not provide enough opportunity for the student to ask questions and the teacher to respond. I am also concerned about the ways in which these systems try to get students to think about concepts. I do not believe that mathematics can be reduced to a collection of goals and subgoals together with not very rich ways of connecting them.

I also have a lot of concerns about the use of today's sophisticated calculators. In earlier times, I did not hear students say that the limit of a function at a point is the value of that function at a nearby point. Is this new misconception due to certain ways in which calculators tend to be used? I have similar troubles with the graphing capabilities of hand-held calculators because I think they can focus the students' attention on the least interesting examples. But this particular issue will soon go away. Hand-held computers are rapidly approaching the functionality of desk-top computers so that soon we will have to look for something else to fight about.

So we have to talk about the ways in which technology can be used and I also want to explain the way I think is the most effective and why.

4.1. What are the Different Ways in which Technology Can be Used to Help Students Learn Mathematics?

I have written elsewhere [6] about the different ways that I think technology can be effective. They are, roughly: using graphics capabilities to *show* mathematical phenomena; using the computational abilities of a computer algebra system to *do* mathematics; and using the expressibility of a programming language to *construct* mathematical entities on the computer.

4.1.1. Using Technology to Show Mathematics

There is no question that we can produce incredibly wonderful pictures on a screen using today's technology. On the one hand, I think that makes many mathematical situations more real and accessible to students—at least as phenomena to be explained, manipulated, and perhaps understood. This can be very helpful but, in my view, entering an expression, pressing buttons and looking at a screen, or even manipulating the screen with a mouse is a little too passive in terms of the mathematics involved in producing the picture.

To take a simple example, consider entering an expression that defines a function and having the computer produce a graph. Suppose even that a table of values is produced, and that it is possible to manipulate one or more of these "representations" (expression, picture, table) and have the others change correspondingly. Even in such a sophisticated system, I do not see that the student is helped in any way to understand the connection between the various processes of plugging a number into an expression to get a result, looking down one column of a table for a number to see what is in the corresponding place on the other column, or locating a point on the horizontal axis and seeing how far up you have to move until you hit the curve. Conceptually these are all the same process and it is very important for students to understand that. I am not sure they do. Yes, students can learn from such technological systems that if you add a constant to the expression for the function the graph goes up. But what is it that helps them realize that the reason is that you are still computing the same values to place on the graph, but now every answer you get is increased by the constant, and that means higher up on the vertical axis?

My conclusion from all this is that using technology to show mathematics to students can be helpful, but it will be much more effective if used in conjunction with other activities.

4.1.2. Using a Computer Algebra System to do Mathematics

Using the power of **Maple** or **Mathematica**, students can learn to perform highly sophisticated and powerful algorithms. The idea is that making use of mathematics in this way is going to help the students learn elementary versions of the mathematics involved. Perhaps, but I am not yet convinced that using a computer algebra system to do applications that involve Padé approximations or Tchebycheff polynomials is going to help the first year calculus student understand how to compute the McLaurin series expansion of $\sin x$ and the issues that arise when you reflect on what relationship that series has to the sin function. It is possible, but there is absolutely no evidence in favor of this approach—even less so than evidence for other claims, both traditional and reform.

Some people take the very opposite view of using a computer algebra system to reduce the time and effort spent on learning standard manipulations. Please pay very careful attention to the fact that I said "reduce" and not eliminate. My remarks earlier about the importance of calculations still stand. What I am saying is that we can use a computer algebra system to do as well or better than in traditional classes—in less time. Here there is some research here for example, the study by Kathy Heid [8].

Yet another approach is to use a computer algebra system to provide data on the basis of which students can discover mathematical relationships. For example, before talking at all about the rules for differentiating combinations of functions, I ask students to use **Maple** to calculate about a page (closely spaced, I admit) of derivatives of elementary functions using **Maple**. Then, in class discussions, I can get most students to invent the rules for derivatives of sum, difference, scalar multiple; many will invent the product rule; and a few will come up with the chain rule.

Of course all of that is preliminary to a class discussion of why these rules hold and some elementary proofs. But the computer work helps them understand these rules and also seems to get them reasonably good at doing the manipulations by hand.

So I think that, like visual effects, using computers to do mathematics is a good educational strategy. But it is not the best way to use technology.

4.1.3. Using a Programming Language to Construct Mathematical Concepts

Let me begin with the biggest objection to having students write programs in order to learn mathematics. A couple of decades ago, this was a very popular idea and there was even a project (I believe its acronym was CRICISAM) to foster it. The trouble is that, in those days, programming was done in FORTRAN and there was so much extra effort in learning the programming language, dealing with bugs and other *non-mathematical* issues that any benefit that might accrue was canceled out.

I think this was a very accurate assessment. But that was then and this is now. The syntax of programming languages today can be quite simple and I hope that my illustrations in the section above on constructivism will suggest to the reader that rather sophisticated programs can be written with little syntactical difficulty. In fact, my experience with students is that although they continue to complain that the syntax gives them difficulty, I find that in just about every case the problem is either a mathematical concept that is not understood (like a function returning a function) or has to do with very inefficient work habits and methods of organizing material. My experience over the last decade or so has been that there are reasonable ways of using an appropriate computer language so that, in writing programs, syntax and system issues are minimized and the focus of the students can be almost entirely on the mathematics.

There are several reports of how we go about this and the reader can consult my WEB page at http://www.cs.gsu.edu/ẽdd/ for details. One thing I can say here is that our approach very much makes mathematics a laboratory course and, as Krantz suggests, we have the students meet, certain days of the week, in a computer lab where they work in cooperative groups on computer activities designed to foster the mental constructions we think they need to make in order to develop an understanding of the mathematics being studied. On the other days of the week, they meet in a classroom where the computer activities are discussed and work is focused on using the mental constructions they made in the computer lab to develop mathematical understandings.

4.2. Which is Best?

I think that all three of the above ways of using computers are effective in various ways, but I think that overall, the third is more effective than the other two. The most effective, however, is when all three approaches are used so that the student constructs mathematical concepts on the computer and then uses these constructs, or fancy versions of them found in a computer algebra system to do mathematics and to produce visualizations on a computer screen.

My argument for the effectiveness of writing programs in learning mathematics is two-fold. First of all, it relates to the theory in that specific mental constructs seem to arise out figuring out how to perform certain computer tasks. For example, if the student understands a certain mathematical procedure as an action or externally driven activity, then asking her or him to implement the procedure as a computer program tends to lead her or him to interiorize this action to a process. Moreover, I know of no more effective way of learning to encapsulate a process to an object than implementing the process as a computer program and then writing a program that uses that process as input and/or output. Thus writing a program that accepts two functions, constructs a program implementing the composition of those two functions and returns this program and then applying this tool in various situations helps a great deal in developing an understanding of functions as objects.

Again, there are reports of our research in which these effects are described and they can be found by consulting the above WEB page.

5. Applications and Pedagogy

Here we come to an issue on which I differ with my fellow reformers—and many traditionalists as well—perhaps more than on any other issue. Like Krantz, I feel that the wrong kinds of applications used in the wrong way can provide distractions to the mathematical issues on which we want the students to focus. Krantz refers to a problem about the destruction of trees in a tropical rain forest and suggests that a multitude of details about the situation tends to obscure the fact that what is needed here is to construct a function, take its derivative, set it equal to zero and solve. I think he is right and details that distract should be avoided, but I would go even further.

One argument is that is that a "real-world" context will make the problem more interesting for students and motivate them to use more energy in dealing with the situation. Unfortunately, one-person's real world is another's vague abstraction. This was brought home to me very sharply about 25 years ago when I was teaching (more or less traditionally) a unit on permutations and combinations. I asked the students how many starting line-ups could be made from a basketball squad with 12 members. I thought I was using a "real-world" example, but after class a student approached me and asked how many players there were on a starting line-up in basketball! It is true that this was at a "hockey school" and the student was a woman at a time when women were more or less excluded from basketball. Nevertheless, I realized that at least for this student, the application was not very helpful.

But even for students who do find a particular application context interesting, I question whether the resulting motivation really relates to the mathematics as opposed to the context—which can take them away from the mathematics! At the very least, I do not find any examples at all in the literature of research providing even a suggestion that using contexts that are interesting for students helps them understand the mathematics that *we* see in the context.

Given the lack of information that applications are helpful and the concern that the student will miss the mathematics in a context, or even be turned away from it, I think we should seriously reconsider our enthusiasm for the use of applications in helping students understand mathematics.

6. Some Specific Pedagogical Issues

Let me close with some brief comments about several issues that are, in fact, more at the heart of Krantz's book than perhaps are the matters I have been discussing at (probably too much) length. In spite of his very laudatory decision to relate his book to current issues of teaching methodology, research in learning and curriculum reform, Krantz has written a book on how to teach (traditionally). As such, he provides a lot of useful suggestions for beginners, and he presents his views on just about every educational issue one can imagine. In addition, he has provided a few of us with a platform on which to state our own views on some of these topics. Who can resist such an invitation? I will try to be brief.

Beginning with the most important, let me say that I agree with Krantz completely when he says that there is nothing essentially wrong with the content of any standard lower-division math course. I have a few quibbles such as too much emphasis on the analytic as opposed to the algebraic and geometric or the importance of adding courses that make mathematics a service course not only for the physical sciences, but the social and computer sciences. But for the most part, I think we have the content about right. This is my belief as a mathematician. I have also surveyed faculty who teach courses for which mathematics is a prerequisite and they tend to confirm this view. Perhaps the most telling argument is to take a look at the reform textbooks that are emerging and notice that the content is not really very different. After all, I am told that even the Babylonians asked their students about the rate of descent of the top of a ladder leaning against a building!

I think Krantz is right that students are not really ready for formal proofs until they have completed calculus and are taking one of the transition courses that have emerged in recent years. I think that forcing formalisms on students too early can contribute to our society's turning away from studying mathematics.

In my opinion, Krantz is too evenhanded about so-called objective examinations such as multiple-choice. No exam is really objective. Even a multiple choice examination makes a selection of material and what could be more subjective than making up the incorrect choices on a multiple choice exam? I think the overwhelming weight of arguments for such exams is the convenience of the person who grades the exam. A similar point can be made for timed exams.

Would we really use them if it were not inconvenient to give students as much time (within reason) as they need? Do we really want to know only what they know so well that they have it on their finger tips and can produce it in high-pressure, emotionally tense situations? Don't we also want to know what they say when they have a chance to relax and reflect on a mathematical issue?

I think Krantz is very wrong when he says that "hard copy textbooks, more or less of traditional form, work. My 42 years of teaching experience tells me that traditional textbooks are essentially unread by students who use them mainly to find template solutions for problems that will be assigned for homework and given on the test. Moreover, I find that the greatest unanimity in all of education is found in the community's reaction to any attempt to vary from this norm. Such attempts are resisted, if not rejected, by students, publishers, and the overwhelming majority of faculty. This has been my experience with the textbooks I have written that are designed to support the ideas I have expressed here and I hear the same story from other writers of "different" textbooks. It seems to me that if we are to have real improvement in teaching and learning mathematics, either this situation has to change, or alternatives to textbooks must be found.

There is an orthodoxy about class size neatly expressed by Krantz who says that "We all know, deep in our guts, that small classes are a much more effective venue for learning than large classes." I sometimes wonder if what we really know deep in our guts is how to count and the salutary effect small classes will have on the job market. I am not sure we can't do really wonderful things with large classes. Personally, I feel that the most effective teaching I have ever done was with a calculus class in which there were 74 students. Not very large, but not exactly small either. My personal opinion is that using the methods referred to in this essay, I think we could learn to teach mathematics as well to classes of 200 as 20. Indeed, certain groups in France report success with new pedagogical strategies they are using in classes this large [1]. This is a largely unexamined issue. The research is, at best, mixed on the effects of class size on learning. In spite of the economics of the situation, I think we should take a good look at it and see if we can find pedagogical strategies that are both learning-effective and cost-effective.

Well, what better topic to end these ramblings with than the authority of the teacher and what is its source. Krantz suggests it comes from how you dress, from maintaining a certain distance, and from not being too chummy with the students. I think he has this completely wrong. Authority in teaching as in anything else comes from a strong, secure knowledge of and satisfaction with who you are and how you want to be, to dress, to talk and to move. This varies with people and the ones who get the most respect from their students are the ones who remain completely true to themselves, their nature and their personality. I have run many academic programs and often I am asked about a dress code. My answer always is: You dress to please yourself and anyone else you feel like pleasing.

Thank you, Steve, for this opportunity to spout off. Let's get a cup of coffee.

References

[1] Alibert, D. & M. Thomas (1991), Research on Mathematical Proof, in D. Tall (Ed.), *Advanced Mathematical Thinking*, Dordrecht: Kluwer, 215-230.

[2] Asiala, M., A. Brown, A., D. DeVries, E. Dubinsky, D. Mathews, & K, Thomas, (1996), A framework for research and curriculum development in undergraduate mathematics education, Research in Collegiate Mathematics Education II, 1-32

[3] Dubinsky, E. (1987), On Teaching Mathematical Induction, I., Journal of Mathematical Behavior, 6(1), 305-317.

[4] Dubinsky, E. (1989), On Teaching Mathematical Induction, II, Journal of Mathematical Behavior, 8, 285-304.

[5] Dubinsky, E., (1995), *ISETL: A programming language for learning mathematics*. Communications in Pure and Applied Mathematics, 48, 1-25.

[6] Dubinsky, E. & R. Noss, (1996), Some Kinds of Computers for Some Kinds of Math Learning, Mathematical Intelligencer, 18, 1, pp. 17-20.

[7] Finkel, D. L., and G. S. Monk (1983), Teachers and Learning Groups: Dissolution of the Atlas Complex, In Learning in Groups, Jossey-Bass, 83-97.

[8] Heid, K. (1988), Resequencing Skills and Concepts in Applied Calculus Using the Computer as a Tool, Journal for Research in Mathematics Education 19 (1), 3-25.

Are We Encouraging Our Students to Think Mathematically?

Deborah Hughes Hallett

Harvard University

Introduction

In her 1990 paper, *Pedagogy and the Disciplines*, Ursula Wagener, of the University of Pennsylvania, describes a mathematics class:

> *A graduate student teacher in a freshman calculus class stands at the lectern and talks with enthusiasm about how to solve a problem: "Step one is to translate the problem into mathematical terms; step two is ... " Then she gives examples. Across the room, undergraduates memorize a set of steps. Plugging and chugging—teaching students how to put numbers into an equation and solve it—elbows out theory and understanding.*

The teacher in this classroom knows the subject matter. Her delivery and pacing are impeccable. Yet she teaches mathematics as a bag of tricks, rather than as an understanding of fundamental principles.

Is this an accurate description of the mathematics classes that we are teaching?

In order to help students learn to think mathematically, we first need to understand how they are thinking. Do students demand "plug and chug" from us? It is not so much that they demand it as they expect it. If we want to change their expectations, we must first understand their view of mathematics. Many students come to college with attitudes and expectations that may startle us.

Students' Expectations of First-Year Math Courses

One window into students' thinking processes is their comments on end-of-semester questionnaires. There are frequent comments about the instructor's clarity (or lack of it). Accessibility, sense of humor and accent are also important to students. They are also deeply concerned, and sometimes incensed, by issues of equity. Within a single course with common exams, some instructors are better than others, some give more handouts than others, some give review problems closer to the exam questions than others.

Another common complaint is that the instructor or the text didn't give the students enough help with the homework problems. For example:[1] "We spent hours doing some problems which [the instructor] didn't tell us how to do." Several would agree with the student who said the thing he liked least was that "...the examples provided did not help with homework." Another suggested, "I wish we could have gone over one example like the homework problems ... "

---

[1] Source: Course evaluations.

before we did the homework." Since many students didn't read math texts much in high school, but simply used them as a source of problems and examples, the kind of help they are asking for is a worked example, given in class or the text, which *closely* parallels the homework problem. To confirm this, the students in two courses at Harvard were surveyed and asked whether or not they agreed with the statement:

> *If you can't do a homework problem you should be able to find a worked example in the text that will show you how.*

Students in calculus gave it 4.1 out of 5 (where 5 indicates strong agreement); those in precalculus gave it a 4.7 out of 5. Further evidence of students' expectations came from the student who suggested that review problems should have the relevant section of the text listed after them in parentheses. When surveyed, his classmates agreed (4.2 out of 5 for calculus, 4.8 out of 5 for precalculus). More explicit still was the student who remarked that the best thing about the teaching in his section was that things were explained "in a cookbook fashion."

As instructors, we can learn a lot about how students think by listening to what they find difficult. The following two examples reminded me that although I feel it is important to know the meaning of one's computations, I do not always succeed in getting my point across.

- **Example: Wanting a 'Step-by-Step Approach.'** A student who had done badly on a Calculus II hour exam came to me complaining that he needed more step-by-step instructions on how to do the problems. Trying to narrow down the request, I asked what topics he already knew. One of them he was sure he knew, he said, was Euler's method. He had recently earned full credit on the following exam questions:

  > Consider the differential equation $dy/dx = x^2 + y$. **(a)** Use Euler's method with two steps to approximate the value of y when $x = 2$ on the solution curve that passes through (1,3). Explain clearly what you are doing on a sketch. Your sketch should show the coordinates of all the points you have found. **(b)** Are your approximate values of y an under- or over- estimate? Explain how you know.

  To check, I asked the student if he could draw a picture of the calculation he did for Euler's method. After a bit of thought, he said yes, he was sure he could draw such a picture. To check still further, I asked him what Euler's method was calculating. There was a *long* silence. Finally he said, rather hesitantly, that he thought it was the arc-length—the arc length along the polygonal curve.

Thus I would suggest that a more step-by-step method is not what was needed—but rather much greater attention on the part of both the instructor and the student to the meaning of the computations being performed. We should realize that exam problems which start, "Use such-and-such a method to do such-and-such," may not be testing all that we want to test. What if this exam question had been worded, "If $dy/dx = x^2 + y$ and $y(1) = 3$, estimate $y(2)$"? However, rewording this problem in this way would have caused some complaints, as students clearly agree (4.1 out of 5 in calculus and 4.6 out of 5 in precalculus) with the statement that:

A well-written problem makes clear what method to use to solve it.

- **Example: 'Vaguely Worded' Problems.** Another student doing badly in Calculus II came to me after an exam to find out what to do. The problem was, as he described it, that although he understood the basic ideas, he couldn't apply them because of the "vague" way in which the problems were worded. The example he chose to illustrate this was the following question:

  > Alice starts at the origin and walks along the graph of $y = x^2/2$ in the positive x-direction at a speed of 10 units/second. **(a)** Write down the integral which shows how far Alice has traveled when she reaches the point where $x = a$. **(b)** You want to find the x-coordinate of the point Alice reaches after traveling for 2 seconds. Find upper and lower estimates, differing by less than 0.2, for this coordinate. Explain your reasoning carefully.

The student had been unable to do this question because he hadn't realized that it was about arc length. He felt quite strongly that the wording of the question should have mentioned arc length specifically.

Again, the problem here seems to be how to teach students to do problems which do not explicitly ask for a certain computation, as well as how to get them to believe that such problems are reasonable.

How Do Students' Expectations Develop?

Most students come to college expecting new experiences. At a residential institution, we need to constantly remind ourselves that many students have never lived away from home before; many have never met students from different backgrounds, religions, or ethnic groups. It is easy to forget that many students have no idea what anthropology is and have never had the chance to take psychology or economics or visual arts. Most freshmen are eager for new experiences, but they are also insecure. Are they the admissions office's one mistake? Is it really worth their family making the financial sacrifices that are required to send them to college? One of the best ways for freshmen to reassure themselves that they are "doing OK" is to get decent grades—which often means *very* good grades, since they are often used to all A's. This makes them want to take some courses which are familiar and predictable so that they can be sure they will do well. This is particularly true in math and science where the grading and workload are inclined to be tougher than in other fields.

Undergraduates are busy people. It is easy for us to underestimate the pressure to join extracurricular activities and our students need to hold a job (or even two). Consequently, undergraduates prize courses and instructors which do not take too much of their time outside of class. A course in which they are expected to read about some topics on their own may have students struggling to get out of it, not because they can't do the work but because it takes too much time and they don't want to have to "teach the course to themselves." The feeling that everything "ought" to be covered in class is reflected in student reactions to mathematics. As an example, one student said that the best thing about the teaching in his section was the fact that

> *It was possible for me to learn the material without studying on my*
> *own too much if I paid attention in class.*

Thus most undergraduates, like most faculty, have more to do than they can reasonably fit into their schedules and tend to look for short-cuts. As an analogy for the way in which some students go about learning mathematics, imagine what it takes to become proficient at an office manager's job. Most undergraduates regard learning mathematics in much the same way a mathematician might regard learning such a job: as something you have to be shown how to do.

This view of mathematics is partly reasonable—we don't expect students to come up with the idea of a derivative by themselves, or to figure out the Fundamental Theorem of Calculus on their own. However, carried to extremes it can lead to absurd results. Consider, for example, the student (who had already taken BC Calculus) who was complaining rather loudly that he shouldn't be asked to suggest formulas for functions whose graphs he had been given. He said that he "did graphs" in the other order: If we gave him the formula, he'd draw the graph, but he wasn't doing this. When it was suggested that he might need to find formulas to fit lab data in his intended major (chemistry), he announced that he'd done experiments before and that one didn't *ever* need to do such a thing. It is important to realize that the vehemence that leads students to dig in their toes as completely as this one is born, at least partly, out of terror, not out of a real desire to be closed-minded.

A certain amount of predictability is entirely appropriate and necessary. Too much, however, means that it is easier for students to memorize how to do a long list of "types" of problems than to learn the basic ideas of mathematics. Many of our students are sufficiently diligent and sufficiently scared of not performing well that they will willingly master a very long set of problem-solving procedures, even by memorization. For example, while trying to figure out how to prepare for a calculus final, one student asked me if he should do the review problems over and over again until he could do them without looking at the solutions. In this way they differ greatly from the "creatively lazy" professional mathematician who would much rather figure out how it all works than memorize anything.

Most students coming to college know, at least in theory, that learning mathematics involves developing understanding. Many of them, often on their own, have developed an understanding of some topics. However, few of them have actually taken courses where an understanding was really *required*—in virtually every case, it was possible to do well just by learning to do all the types of problems shown in class. Consequently, asking students to do problems which have not been modeled for them in class or in the reading is inclined to strike them as unfair.

The Instructor's Role

There is ample evidence that the students' views of mathematics are often startlingly different than ours when they start college. What can be done about it? Is encouraging students to think mathematically part of our job, or should we just work with those students who are naturally inclined this way?

I suggest that we may be letting our students down if we do not try to broaden and deepen their thinking. It is, of course, much easier for us to teach a course in which we focus on template problems and algorithms. It's even possible to put the burden on the students by saying that they won't accept anything else. But is that really true? Perhaps, as George Rosenstein, Franklin and Marshall, suggests,[2] we have played a role in allowing these expectations to become established:

> I'm convinced there has been a conspiracy between math teachers
> and math students. The terms are that the teachers can do whatever
> they want in class, but will ask for only the well-practiced or routine
> on the exams. In return, the students will be cooperative and diligent
> at learning the manipulative or template material that is stressed on
> homework assignments and quizzes.

Our biggest challenge as teachers is to understand our students' thinking patterns well enough that we can affect them. Learning to think more independently is a difficult, often frightening, process for students. Thus, besides gaining an understanding of our students' thinking, we need to understand their feelings well enough to gain their trust.

How Should Our Students' Views of Mathematics Affect Our Teaching?

Before considering how to react to our students' view of mathematics, we need to consider what is meant by understanding. Here, too, there is a difference between professional mathematicians and first year undergraduates. In most students' minds, understanding is being able to visualize (or otherwise internally represent) concepts and the relationships between them. The notion which mathematicians might call, or include in, understanding—knowing the theoretical and logical connections between concepts—is not what most students mean by understanding. For most students, the visual form of understanding precedes the more theoretical version. Thus, to take students toward a rigorous thinking, we must first establish a solid intuitive, visual understanding.

Each department and each faculty member needs to decide exactly how to approach students' views of mathematics. Not to challenge these views at

---

[2]From Project CALC Newsletter, October 1991.

all is not doing our students justice, as well as not in the best interests of the profession. However, challenging them too much alienates students from the subject and teaches them little. The most useful guide to how much to challenge them is a robust understanding of students. To acquire this, each instructor is well advised to use every course they teach as an opportunity to learn about their students. Some techniques that may be useful:

1. Include problems on tests that ask students to sketch or explain. Grade them yourself, or look them over before handing them back.

2. Ask students to read some mathematics and summarize what they have read in a paragraph—and find time to read, or skim, the paragraphs.

3. Arrange an e-mail discussion group for a course. Look over the postings every few days.

4. A particularly good way to gain insight into what students are thinking is the "one-minute paper," advocated by Richard Light at Harvard. At the end of class ask students to take a piece of paper and write on it:

   - The most important thing learned that day.
   - The most confusing thing that day.
   - One question they have that remains unanswered.

   The pieces of paper are handed in as the students leave the room; the instructor reads through them before planning the next class. The answers to these questions can often be illuminating—and occasionally devastating—to the instructor. It becomes easier to see what the students are thinking, and addressing their misconceptions becomes more urgent. Acknowledging and acting on the responses can markedly improve communication with the class.

Conclusion

Successful teaching involves knowing what your students are thinking. However, it is often hard for instructors to "hear" students—especially as the students' thoughts are often not similar to their own. Consequently, any efforts to listen to students' thinking about mathematics are likely to improve one's teaching.

Big Business, Race, and Gender in Mathematics Reform

David Klein

California State University at Northridge

The mathematics reform movement may have positive attributes, but that is not what this appendix is about. This essay is divided into three sections, each taking a critical view of what has come to be called "mathematics reform." Rather than attempting an abstract definition of this term, I cite the principal documents and leaders of the reform movement on particular issues. The fault line separating the mathematics reform movement from its critics is nowhere more volatile and portentous than in California. The third and final section of this appendix is devoted to a short history of the conflict over mathematics reform in that state, with a focus on the controversial California mathematics standards. This set of standards has received widespread praise from prominent mathematicians and strong opposition from the mathematics reform community. As explained in the last section, this conflict helps to define, in *practical* terms, the mathematics reform movement.

The second section challenges assumptions about ethnicity and gender in the reform movement. Multiculturalism and mathematics for "all students" are recurring themes among reformers. Prominent reformers claim that learning styles are correlated with ethnicity and gender. But reform curricula, while purporting to reach out to students with different "learning styles", actually limit opportunities. Fundamental topics, including algebra and arithmetic are abridged or missing in reform curricula without apology.

Big Business and the mathematics reform movement have at least one thing in common. They both militate for more technology in the classroom. Calculators and computers are regular features in reform math curricula, and technology corporations routinely sponsor conferences for mathematics educators. The confluence of interests and the resulting momentum in favor of more technology is the subject of the first section.

Technology, Reform, and the Corporate Influence

The 1989 report "Everybody Counts" warned:

> In spite of the intimate intellectual link between mathematics and computing, school mathematics has responded hardly at all to curricular changes implied by the computer revolution. Curricula, texts, tests, and teaching habits—but not the students—are all products of the pre-computer age. Little could be worse for mathematics education than an environment in which schools hold students back from learning what they find natural. [17]

The imperative to integrate technology into the classroom goes far beyond mathematics courses. President Clinton calls for "a bridge to the twenty first century ... where computers are as much a part of the classroom as blackboards."

[18] Presently, four-fifths of U.S. schools are wired to the Internet and the rest are not far behind. Remonstrations by well-placed technology experts and educators, based on educational considerations, seem to warrant no delays [6], [18]. A 1996 report by the California Education Technology Task Force, a group dominated by executives in high tech industries, called on California to spend $10.9 billion on technology for schools before the end of the century. The task force claimed that "more than any other single measure, computers and network technologies, properly implemented, will bolster California's continuing efforts to right what's wrong with our public schools." [10]

A corporate perspective is also sweeping American universities with computer technology paving the way. The number of virtual universities and virtual courses is increasing exponentially. In 1997 there were 762 "cyberschools", up from 93 in 1993 and more than half of the nation's four year colleges and universities have courses available "off site" [4]. The second largest private university in the U.S., the University of Phoenix, offers on-line courses to 40,000 students from a faculty with no tenure. Other examples[1] and a recent history of technology and the corporatization of universities may be found in David Noble's interesting essays [16].

Will the computerization of schools improve education? The Los Angeles Times reports that "many critics worry that education policy is increasingly being driven by what companies have to sell rather than what schools need ...Computer companies want more technologically savvy consumers, for example, to increase the penetration of computers beyond the 40% of homes in which they are now found. And they argue that increased use of technology in schools will help fill a growing shortage of computer literate workers." [10]

Corporate foundations regularly fund mathematics reform projects, as for example, the "Exxon Symposium on Algebraic Thinking" for the Association of Mathematics Teacher Educators Conference, held in January 1998, with Texas Instruments hosting one of the dinners. Conversely mathematics reformers embrace a corporate vision of education, which includes the de-emphasis of basic skills and a greater reliance on technology. Consider, for example, the following promotional material for the K-6 curriculum MathLand from Creative Publications:

> Business leaders have expressed interest in changes in education as
> well—changes that go beyond what a traditional standardized test
> can measure. Recently, the US Departments of Labor and Education

---

[1]During the mid-1990's the California State University administration initiated an unprecedented partnership with technology giants Microsoft, GTE, Fujitsu and Hughes Electronics. The joint venture, called the California Educational Technology Initiative, or CETI, will, if implemented, wire up the 23 campuses of the CSU with state-of-the-art telephone and computer networks, as well as invest billions of dollars in education-related electronics. By the Spring of 1998, a dozen CSU campus faculty senates passed resolutions asking for delays and criticizing the merger. The California State Student Association passed its own resolution denouncing CETI and opposing any "privatization of the California State University as a whole." Microsoft and Hughes subsequently pulled out, but CSU Chancellor Reed continues to seek new corporate partners. The implications of such a partnership are not fully worked out, but incentives for the faculty to market computer products to students, and the creation and marketing of courseware have been seriously considered.

formed the Secretary's Commission on Achieving Necessary Skills (SCANS) to study the kinds of competencies and skills that workers must have to succeed in today's workplace. According to the SCANS report *What Work Requires of Schools: A SCANS Report for America 2000*, business leaders see computation as an important skill, but it is only one of 13 skills desired by Fortune 500 companies. These skills are (in order of importance): teamwork, problem solving, interpersonal skills, oral communication, listening, personal development, creative thinking, leadership, motivation, writing, organization skills, computation, and reading. [15]

The California Mathematics Council (CMC), an affiliate of the National Council of Teachers of Mathematics, boasts 12,000 members. In an open letter to the California Board of Education dated April 17, 1996, the CMC included the same ordered list of basic skills with reading and computation given last. Citing unspecified "educational research" and "neuro-biological brain research", the CMC letter endorsed the direction of the 1992 reform-oriented California Mathematics Framework and added:

> Equally impressive is that these changes in the way we teach mathematics are supported by the business community. *What Work Requires of Schools: A SCANS Report for America 2000* concludes that students must develop a new set of competencies and new foundation skills. It stresses that skills must be learned in context, that there is no need to learn basic skills before problem solving, and that we must reorient learning away from mere mastery of information toward encouraging students to solve problems.

Learning in order to know must never be separated from learning to do. Knowledge and its uses belong together (*A SCANS Report*) [2]

The NCTM *Curriculum and Evaluation Standards* also recommends that "appropriate calculators should be available to all students at all times" and "every student should have access to a computer for individual and group work." Reform texts place little restriction on technology. The second edition of the Harvard Calculus text instructs that students "are expected to use their own judgment to determine where technology is useful." [7] The 1992 California Mathematics Framework recommends that calculators be available at all times to all students, including Kindergarten students, and asks, "How many adults, whether store clerks or bookkeepers, still do long division (or even long multiplication) with paper and pencil?"

None of the above is intended to suggest that technology should not be used in mathematics classes. Nor do I suggest any kind of conspiracy theory. I have incorporated the (limited) use of computers in some of my own classes at California State University, Northridge, and I agree with almost all of Professor Krantz' balanced discussion in Section 1.10. My only reservation is Professor Krantz' suggestion that an entire "lower-division mathematics curriculum [should] depend on `Maple` (or `Mathematica`, or another substitute) and that [students] need to master it right away." This seems to me to be premature. It might be appropriate at some point, but a compelling curriculum with this feature should be presented, vigorously reviewed, and thoroughly tested on real students first.

The use of technology in mathematics education should be considered against the backdrop of extremely powerful business interests which seek to create new consumers of technology. Incorporating ever more technology into the classroom may or may not be consistent with good educational practices. Large-scale implementations of technology in the classroom receive tremendous momentum from funding agencies—at times, far beyond what the results merit. With the huge sums of money involved in computerizing education, the educational merits of technology are rarely discussed. For politicians and entrepreneurs, no justification is necessary, but educators should demand clear evidence of the beneficial effects of technology before it is incorporated in classrooms.

The calculator is one of the staples of the reform mathematics movement from Kindergarten through calculus and beyond. Mathematics instructors, including calculus teachers, regularly allow students to use calculators on examinations and contort their tests to avoid giving points for mere button pressing skills. I agree with Professor Krantz' assertion in Section 1.10 that "if a student spends an hour with a pencil—graphing functions just as you and I learned— then there are certain tangible and verifiable skills that will be gained in the process." I don't think it is unreasonable to require students to demonstrate these and other skills on examinations without calculator assistance.

At the elementary school level, arithmetic is a victim of technology in the reform curricula. Long division in particular is frequently a target for elimination. For example, long division with more than single digit divisors was consciously eliminated from the proposed California math standards by the Academic Standards Commission, and the California Mathematics Framework makes it clear that "clerks or bookkeepers ... do [not do] long division ... with paper and pencil." In addition to sharpening estimation skills, mastery of the division algorithm is important for understanding the decimal characterization of rational numbers, a middle school topic, as well as quotients of polynomials and power series in later courses.

In a society that worships technology, it is all too easy to surrender the integrity of sound traditional curricula to machines, their corporate vendors, and reform-evangelists.

Gender, Race, and Ethnicity in the Reform Movement

One of the themes of the mathematics reform movement is that women and members of ethnic minority groups learn mathematics differently than white males. The thesis that learning styles are correlated with ethnicity and gender is widely accepted in education circles and its validity is not assumed to be restricted to mathematics. One example of this ideology occurred when the Oakland School Board resolved that Ebonics is genetically based [19]. Mainstream views from the academy are similar. In a well-referenced study on how African Americans learn mathematics, published in the Journal for Research in

Mathematics Education, one finds [14]:

> Studies of learning preferences suggest that the African American students' approaches to learning may be characterized by factors of social and affective emphasis, harmony with their communities, holistic perspectives, field dependence, expressive creativity, and nonverbal communication ... Research indicates that African American students are flexible and open-minded rather than structured in their perceptions of ideas ... The underlying assumption is that the influence of African heritage and culture results in preferences for student interaction with the environment and that this influence affects cognition and attitude ...

The Journal of American Indian Education devoted an entire special issue to the subject of brain hemispheric dominance and other topics involving Native American learning styles. Included is a reprint of Dr. A. C. Ross's paper, "Brain hemispheric functions and the Native American," that asserts Native Americans are "right-brained". Ross explains that the "functions of the left brain are characterized by sequence and order while the functions of the right brain are holistic and diffused." Elaborating, he maintains that "left brain thinking is the essence of academic success as it is presently measured. Right brain thinking is the essence of creativity." Citing earlier research, Ross concludes that "traditional Indian education was done by precept and example (learning by discovery) ... creativity occurs in the learning process when a person is allowed to learn by discovery. Evidently, traditional Indian education is a right hemispheric process." [9] The final article in the same journal takes issue with this point of view and laments that "a veritable right-brain industry has developed" and warns of the dangers to Indian education by characterizing this entire ethnic group as right brained.

The view that women and minority group members learn differently from white males is far from marginal within the mathematics reform movement. A radio interview of NCTM President Jack Price, independent textbook publisher John Saxon, and Co-Founder of Mathematically Correct, Mike McKeown, occurred on April 24, 1996. The KSDO radio show on Mathematics Education, hosted by Roger Hedgecock, was held in conjunction with the annual meeting of the National Council of Teachers of Mathematics, in San Diego that year. During the interview, President Price asserted:

> What we have now is nostalgia math. It is the mathematics that we have always had, that is good for the most part for the relatively high socio-economic anglo male, and that we have a great deal of research that has been done showing that women, for example, and minority groups do not learn the same way. They have the capability, certainly, of learning, but they don't, the teaching strategies that you use with them are different from those that we have been able to use in the past when young people, we weren't expected to graduate a lot of people, and most of those who did graduate and go on to college were the anglo males. [13]

The reform movement presupposes that broad classes of non-white males learn "holistically", that mathematics should be integrated with examples and connected as widely as possible to other human endeavors. Algebra and arithmetic are particularly short changed as "mindless symbol manipulation" and "drill and kill." To cite one typical example of this, a mathematics educator wrote on the Association of Mathematics Teacher Educators listserve, "I know this may come as a shock to some mathematics professors out there, but few students find manipulating x and y engaging."

It is clear that proponents of reform are acting out of a sincere desire to improve mathematics education for all students. But the mathematics community should be suspicious of trends which draw legitimacy from racial or gender theories of learning.

No one disputes that culture plays a role in academic achievement. The Los Angeles Times published a special report entitled, "Language, Culture: How Students Cope" as part of a three day series of special reports on education [11]. The Times report explicitly discounts any link between race and ability, but acknowledges that "ethnic differences [in academic accomplishments] remain, even after accounting for income, parent education or language a student speaks at home." High achievement by Asian American students is a result of hard work and a strong emphasis on the importance of education, and this contrasts sharply with the "complacency that hampers so many of California's white students, who have shown a sharper drop on reading scores this decade than either blacks or Latinos." "The burden of acting white" is a theory that African American students "resist schooling to protect their self-image and distinguish themselves from a majority culture that too often devalues their abilities." Many Latinos, for cultural and economic reasons, may see pursuing an education as selfish, since getting a job instead would contribute directly and immediately to family members [11].

Investigating cultural reasons for differences in academic achievement is quite different from proposing that members of different ethnicities and genders actually learn mathematics in different ways. The latter point of view is especially serious when it leads to new, watered-down mathematics curricula.

There should be no doubt that minority students can thrive in traditional programs. Take the case of Bennett-Kew Elementary School in Inglewood, California. According to Principal Nancy Ichinaga, 51% of the students are African-American and 48% are Hispanic (mostly immigrant with Limited English Proficiency). Approximately 70% qualify for subsidized lunches. Below are its 1997 California Achievement Test results, with Normal Curve Equivalent scores (similar to percentiles):

Grade	1	2	3	4	5
Math	62	79	81	75	68

Bennett-Kew believes in high, explicit standards for all students. The mathematics standards are not just year-by-year, but month-by-month. There is regular diagnostic testing of student progress and immediate remediation. The school is committed to direct instruction and does not use newer books. While discussing the mathematics reform movement with me, Principal Ichinaga remarked, "Reform is for the birds."

The traditional approach to teaching calculus used by the legendary teacher Jaime Escalante is another example of minority students thriving in a traditional mathematics program. In 1974, Escalante took a job teaching basic mathematics at Garfield High School which was in danger of losing its accreditation because discipline and test scores were so bad. Five years later, insisting that disadvantaged and minority students could tackle the most difficult subjects, he started a small calculus class. The effect was to raise the curriculum for the entire school. In 1982, 18 of his students passed the Advanced Placement calculus exam. This was the subject of the movie, *Stand and Deliver*. Working with his fellow calculus teacher Ben Jimenez, and Garfield Principal Henry Gradillas, Escalante sent ever increasing numbers of students to leading universities with AP calculus credit.

By 1987, Garfield High School had more test takers than all but four high schools in the United States. The number of test takers reached its peak of 143 students in 1991, the year Escalante left Garfield. The passage rate was 61%. The numbers have declined ever since. By 1996 there were only 37 test takers with a passage rate of 19%. It is interesting that former Principal Gradillas' career declined after the spectacular successes of his high school. After finishing his doctorate in 1987, he "expected to be given an important administrative job that would help spread the school's philosophy to other parts of Los Angeles. Instead he was told to supervise asbestos inspections of school buildings. District officials denied they were punishing him, but one said privately that Gradillas was refused better assignments because he was considered 'too confrontational'." [21]. Rather than studying his effective methods, Escalante is shunned by the mathematics reform community. The disapproval is mutual. According to Escalante, "whoever wrote [the NCTM math standards] must be a physical education teacher." [3]

> The calculus reform movement is inextricably linked to the K–12 mathematics reform movement. Consider, for example, the following statement from the preface of the first edition to the Harvard Calculus text [7]:

> We have found this curriculum to be thought-provoking for well prepared students while still accessible to students with weak algebra backgrounds. Providing numerical and graphical approaches as well as the algebraic gives the students several ways of mastering the material. This approach encourages students to persist, thereby lowering failure rates.

Lower failure rates at the cost of eviscerating the algebra component of calculus is harmful to students of all ethnicities and both genders. Algebra and arithmetic are consistently de-emphasized in reform curricula in exchange for the more "holistic" calculator assisted "guess and check" routine. The entire reform program mortgages future opportunities to attend to the immediacy of high failure rates. The de-emphasis of algebra in reform calculus justifies and caters to the K–12 reform mathematics program.

Calculus proofs and even definitions require students to be competent in algebra. Calculus reform texts tend to relegate both of these to appendices, sparing students the necessity even to turn a few pages in order to avoid them. Instructors who wish to include definitions, such as the definition of a limit and/or a few proofs, must overcome additional psychological resistance because of the location of these topics in the textbooks. When a proof or definition is placed in an appendix, it sends the message to the student that the topic is not important and may be safely skipped.

The emergence of these trends at a time when greater numbers of previously under-represented students are attending universities should cause some reflection within the mathematics community. Are we expecting less of these students? If so, is it because they learn mathematics differently from students of an earlier era, or is it because their mathematical preparations are deficient? I think it is the latter, and I believe that the mathematics community would do well to purge itself of any hidden assumptions that non-Asian minority students learn mathematics differently from anybody else. The focus should be on raising the level of mathematics education in K–12, not on how best to lower it in the universities.

The Politics of Mathematics Reform in California

Nowhere has the conflict over mathematics education reform been more contentious than in California. California led the United States in institutionalizing K–12 mathematics reform. The 1992 California Mathematics Framework is based on the 1989 NCTM Standards and has served as a guide for politically powerful reformers, like the California Superintendent of Instruction, Delaine Eastin (elected in 1994), as well as countless specialists in the state's Colleges of Education who have used it as course material for K–12 student teachers. But California's commitment to the principles of mathematics reform predates the NCTM Standards. For example, one finds in the 1985 "California Model Curriculum Standards, Grades Nine Through Twelve":

> The mathematics program must present to students problems that utilize acquired skills and require the use of problem-solving strategies. Examples of strategies that students should employ are: estimate, look for a pattern, write an equation, guess and test, work backward, draw a picture or diagram, make a list or table, use models, act out the problem, and solve a related but simpler problem. The use of calculators and computers should also be encouraged as an essential part of the problem-solving process. Students should be encouraged to devise their own plans and explore alternate approaches to problems.

The educational philosophies behind the mathematics reform movement are canonical in America's colleges of education and have been for most of this century [8]. The broad principles of reform have been institutionalized in California state documents for well over a decade and have taken root in the schools. Reform curricula based on these principles are ubiquitous in California's elementary schools. The controversial curriculum, "MathLand", for example, has been adopted by 60% of the state's public elementary schools, according to its publishers [20], and there are many other similar curricula widely in use. Secondary mathematics curricula such as Interactive Mathematics Program and College Preparatory Mathematics originated in California and are widely used throughout the state at the time of this writing. The alignment of these and other self-described reform curricula with the NCTM Standards seems to be uncontested. Indeed, much of the development and implementation of these curricula has been funded by the National Science Foundation and other powerful, reform dominated institutions. In particular, MathLand, perhaps the worst of all reform curricula, has been promoted through the NSF funding.

California is experiencing a backlash at the grass-roots level against the general education reform movement (including Whole Language Learning and "Integrated Science"), and mathematics reform in particular. Reacting to the de-emphasis of arithmetic and algebra in the reform curricula, and the over-reliance on calculators, parents' education organizations have emerged all over the state, several with their own web sites containing material starkly critical of "reform math" or "fuzzy math". I am associated with the largest and best known of these groups, "Mathematically Correct".

Of particular concern to parents and teachers critical of the reform movement is the lack of accountability and measurable standards of achievement in the schools. "Authentic assessment" in place of examinations with consequences, and little if any importance placed on student discipline and responsibility in the reform literature, help to make reform math an object of ridicule among vocal parents' groups. It is noteworthy that all parties acknowledge the importance of better teacher training.

The conflict between the mathematics reform movement, on the one hand, and parents' organizations combined with a significant portion of the mathematics community, on the other hand, reached a turning point in December, 1997. At that time the California Board of Education rejected the reform-oriented draft standards from one of its advisory committees—the Academic Standards Commission—and, with the help of Stanford mathematics professors Gunnar Carlsson, Ralph Cohen, Steve Kerckhoff, and Jim Milgram, developed and adopted the California Mathematics Academic Content Standards [1].

Unlike the Academic Standards Commission proposal, these new standards made no pronouncements about teaching methods, only grade- level benchmarks. The reaction from the California mathematics reform community against this lack of coercion was swift and harsh. Their response was to claim that the official math standards, written by the Stanford mathematicians, lowered the bar. Turning reality on its head, State Superintendent Delaine Eastin charged, "[The State Board of Education Standards] is 'dumbed-down' and is unlikely to elicit higher order thinking ..." Judy Codding, a member of the Academic Standards Commission and the powerful National Center on Education and the Economy put it bluntly when she said, "I will fight to see that [the] California Math Standards are not implemented in the classrooms" [22].

Other Reformers with national stature echoed the outrage. Luther Williams, the National Science Foundation's Assistant Director for Education and Human Resources, wrote a retaliatory letter to the California Board of Education widely interpreted as threatening to cut off funding of NSF projects in California. The lead story in the February 1998 News Bulletin of the NCTM, *New California Standards Disappoint Many*, began with the sentence, "Mathematics education in California suffered a serious blow in December." The article quotes a letter from NCTM President Gail Burrill to the president of the California Board of Education that included the statements: "Today's children cannot be prepared for tomorrow's increasingly technological world with yesterday's content . . . The vision of important school mathematics should not be one that bears no relation to reality, ignores technology, focuses on a limited set of procedures, . . . California's children deserve more." Presumably the accusation that technology is ignored refers in part to a policy decision of the California Board of Education that state-wide exams based on the new math standards will not include the use of calculators—a serious blow to the corporate/reform ideology.

Joining the reform math community, the state-wide chairs of the Academic Senates of the UC, CSU, and California Community College systems, none of whom were mathematicians, issued a joint statement condemning the adoption of California's math standards and even suggested that "the consensus position of the mathematical community" was in opposition to the new standards, and generally in support of the rejected, reform- inspired draft standards written by the Academic Standards Commission.

In opposition to the reform community and in support of California's new math standards, more than 100 California college and university mathematicians endorsed an open letter addressed to the Chancellor of the 23-campus California State University system. The open letter disputed the existence of such a consensus and urged the Chancellor to "recognize the important and positive role California's recently adopted mathematics standards can play in the education of future teachers of mathematics in the state of California." Among the endorsing mathematicians were several department chairs and many leading mathematicians [12].

Further contradicting the reformers' claims against California's math standards, Ralph Raimi and Lawrence Braden, on behalf of the Fordham Foundation, conducted an independent review of the mathematics standards from 46 states and the District of Columbia, as well as Japan. California's new board-approved mathematics standards received the highest score, outranking even those of Japan [5].

The sharp conflict over the California math standards defined, in practical terms, the mathematics reform movement. Reformers denounced the state's standards in public forums and the press, while traditionalists and critics of reform defended the standards. Based on purely mathematical considerations, the board-approved California standards are easily seen to be superior to the rejected, reform-oriented version offered by the Academic Standards Commission. A careful and well-written comparison these two sets of standards by Hung-Hsi Wu is available on the Mathematically Correct web site [22].

The extent to which the California math standards will be taken seriously by school districts is difficult to predict. The superintendent of the Los Angeles Unified School District, the second largest school district in the U.S., admonished LAUSD personnel to take no action to implement the new standards, arguing instead that the already existing LAUSD standards were superior. A refutation and insightful comparison of the LAUSD math standards with the California standards was developed by Jim Milgram, Mathematically Correct Co-Founder Paul Clopton, and others. It is also available on the "Mathematically Correct" web site [13]. The LAUSD K–12 math standards are vague and repetitive, trigonometry is completely missing, and third graders are encouraged to use calculators.

Opposition to California's mathematics standards from reform leaders continues as of this writing. Former NCTM president Jack Price wrote in a letter published by the Los Angeles Times on May 10, 1998:

> . . . if the state board had adopted world-class mathematics standards for the 21st century instead of the 19th century, there would have been a great deal of support from the 'education' community.

This sententious observation encapsulates the topics discussed in this essay. For the reformers, "world-class mathematics standards for the 21st century" eluded the Stanford mathematicians who wrote California's 1998 math standards. Missing are the greater emphasis on technology—an end in itself—and pedagogical directives harmonious with the reified "cognitive styles" of the racially diverse populations of the 21st century. The "19th century" arithmetic, algebra, geometry, and trigonometry highlighted in California's 1998 standards will have diminished value in the postmodern epoch of technological wonderments envisioned by math reformers.

Perhaps the academic community should consider whether the discipline of mathematics education—much more so than mathematics—needs fundamental alterations for the 21st century.

References

[1] Available at: http://www.cde.ca.gov/board/k12math_standards.html

[2] California Mathematics Council Open Letter,
http://wworks.com/ pieinc/scan-cmc.htm

[3] Charles Sykes, *Dumbing Down Our Kids: Why American Children Feel Good about Themselves but Can't Read, Write, or Add*, St. Martin's Press, 1995 p. 122

[4] *Forbes*, June 16, 1997, I got my degree through E-mail

[5] Fordham Report: Volume 2, Number 3 March 1998 *State Mathematics Standards* by Ralph A. Raimi and Lawrence S. Braden,
http://www.edexcellence.net/standards/math.html

[6] David Gelertner, Should Schools Be Wired To The Internet? , No–Learn First, Surf Later, *Time*, May 25, 1998

[7] D. Hughes Hallett et al, *Calculus*, John Wiley and Sons, New York, 1992, 1998

[8] E.D. Hirsch Jr., *The Schools We Need; Why We Don't Have Them*, Doubleday, New York (1996)

[9] *Journal of American Indian Education*, Special Issue, August 1989

[10] *Los Angeles Times*, June 9, 1997, High Tech Sales Goals Fuel Reach into Schools

[11] *Los Angeles Times*, May 18, 1998, all of Section R

[12] Open Letter to CSU Chancellor,
http://www.mathematicallycorrect.com/reed.htm

[13] Mathematically Correct, http://www.mathematicallycorrect.com/

[14] Carol Malloy and Gail M. Jones, An Investigation of African American Students' Mathematical Problem Solving, *Journal for Research in Mathematics Education*, 29, no. 2, March 1998, pages 143-163

[15] MathLand,
http://www.mathland.com/assessInMath.html#assess_LAData

[16] David Noble, Digital Diploma Mills: The Automation of Higher Education, *Monthly Review*, Feb. 1998, Selling Academe to the Technology Industry, *Thought and Action: The NEA Higher Education Journal*, XIV, no. 1, Spring 1998

[17] National Research Council, *Everybody Counts: A Report to the Nation on the Future of Mathematics Education*, National Academy Press, Washington, D.C., 1989

[18] Todd Oppenheimer, The Computer Delusion, *The Atlantic Monthly*, July 1997,
http://www.theatlantic.com/issues/97jul/computer.htm

[19] *San Francisco Chronicle*, December 26, 1996, Ebonics Tests Linguistic Definition

[20] *Time*, August 25, 1997, Suddenly, Math Becomes Fun And Games. But Are The Kids Really Learning Anything?

[21] *The Washington Post*, May 21, 1997, A Math Teacher's Lessons in Division

[22] Hung-Hsi Wu, *Some observations on the 1997 battle of the two Standards in the California Math War*,
http://www.mathematicallycorrect.com/hwu.htm

Will This Be on the Exam?

William McCallum

University of Arizona

1. The Proof of the Pudding ...

Many of us in academe remember with pleasure our experience, as students, of the final exam period: the few days before each exam, studying the textbook, memorizing facts and formulas, grasping anew the basic concepts, and constructing for the first time a complete picture of the course. We went into each exam with a sense of anticipation, and left with the pleasure of a job well done. As a teacher, I enjoy constructing final exams, balancing easy questions with hard, rote skills with conceptual understanding, theory with applications; providing some lucky student with the experiences I had.

The tests and exams which determine a student's final grade form the definitive statement, from the student's point of view, of what is in a course.[1] Although it is important for instructors to think about standards, skills, conceptual understanding, and applications, it is the exams that state the instructor's true expectations, and it is the student's performance on those exams that indicates how well he or she has met those expectations. There is often a serious contradiction between the envisioned syllabus, the subject of discussions in textbook committees, and the real syllabus, represented by the final exam. The question that forms the title of this article, so often asked by our students, would be a good question for us to ask ourselves whenever we are discussing the curriculum. Our discussions would be more cogent if we started at the end of the course, so to speak, with a frank admission of what we really expect our students to be able to do, rather than at the beginning, with the syllabus or the textbook's table of contents. It doesn't much matter what you teach or how you teach it if the students can study for your final by memorizing soon-to-be-forgotten procedures.

2. A Test for Tests

A couple of years ago I obtained some sample final exams from the mathematics department of a prestigious university. One was a calculus exam, which, out of curiosity, I gave to `Mathematica`. Overall, it did very well, although on some questions it would not have earned full marks, as when it correctly stated that an infinite series diverged, but failed to give the test that confirmed this. Perhaps it did not show as much work as a grader would have wished, but it would be easy to customize it to do so (indeed, there are calculators that will 'show the steps' in symbolic differentiation and integration). Most strikingly, there was not a single question that did not have a corresponding `Mathematica` command. Barely one asked for a demonstration of conceptual understanding or an ability to select the appropriate tool (beyond knowing whether to type

---

[1] Although I will frame this discussion in terms of exams, my comments apply to any other form of final assessment.

`Integrate[]`, `Differentiate[]`, or any others of a small range of commands). Furthermore, barely any explanation of the answers was requested (although perhaps the need for explanation was tacitly understood).

What conclusions can one draw from this? I would like to avoid drawing conclusions on the issue that has been most bitterly contested, the issue of how technology should be used in teaching and examining students, and what technical skills remain vital in the age of technology. Rather, I would like to point to a conclusion that is indisputable, no matter what one's attitude to technology. The fact that a computer algebra system was able to perform so well on this exam is an indication of its low intellectual content. Imagine an English final exam, at the college level, that consisted of nothing but spelling and the elementary grammar built into current grammar-checkers.

It might be argued that there is nothing wrong with such exams in mathematics, or with the courses that go with them. There is indeed a tradition of techniques courses, which aim explicitly to teach nothing more than mathematical techniques for engineers and scientists. I have heard it argued that freshman calculus should properly be regarded as such a course. Whether or not such courses still make sense in the age of technology is a matter of often vociferous debate, and I don't want to get into that debate right now. Rather, let me, for the rest of this discussion, limit myself to courses which are intended to be more than techniques courses.

For such courses, I'd like to propose a test for evaluating exams, inspired by my `Mathematica` experiment. When you write an exam, ask yourself:

Can all of the questions be answered satisfactorily using purely mechanical procedures?

By mechanical procedures, I mean not only computer algebra systems, but problem templates, pencil and paper algorithms, anything that a computer could be programmed to do. Common sense applies: I exclude Rube Goldberg contraptions narrowly designed for one specific question. If all of the exam, or a large part of it, can be handled mechanically, with output that would be acceptable to you as a grader, then it doesn't pass my test.

I am not arguing that no exams should be limited to mechanical skills, but that there should be some exam that isn't so limited, some point where the course, as represented in the questions you grade your students on, should rise to a higher intellectual plane. Furthermore, it is quite possible to reach this higher plane and still allow computer algebra systems on the exam. Such exams, however, would at some point request mathematical reasoning and verbal explanations, which cannot be supplied by the computer.

Finally, I want to draw attention to the stipulation in my criterion that questions be answered satisfactorily. In applying the criterion, it is important to consider not only the text of the questions on your exam, but your grading standards. The first example in the next section illustrates this point.

3. Good Questions and Good Answers

What are the consequences of my proposed criterion? Here are some general principles of question design and grading standards that can drawn from it. Again, let me emphasize that I do not mean to exclude mechanical questions altogether; however, in what follows, I'll give examples of questions that, if included on an exam, would raise it above the purely mechanical level. I have used all these questions on final or midterm exams in courses that I have taught, with the exception of the first one, which I use in a handout to students on how to write mathematics, and the last one, which I made up for this article. To better illustrate my principles, I have chosen easy questions. At least, they should be easy; however, students do not find them so. If you find this hard to believe, I encourage you to try them yourselves, under the following conditions: **a)** Give them as written, without any added hints or suggestions. **b)** Do not prepare or warn the students in any way.

Require answers to be complete sentences. One thing that computers can't do very well (yet) is think about why they did what they did, and write about it in convincing English prose. On the other hand, this is something we want our students to be able to do. Consider the following extremely easy question.

- *Question.* Find the equation of the line through the points $(-1,0)$ and $(2,6)$.

Here are two answers, using the same method, but written very differently. The method might not be one you prefer to teach; however, it is not the method but the difference in writing which illustrates my point.

- *Answer 1.*

$$m = \frac{6-0}{2-(-1)} = 2$$
$$y = 2x + b$$
$$0 = 2(-1) + b$$
$$b = 2$$
$$y = 2x + 2$$

- *Answer 2.* The slope of the line is

$$m = \frac{6-0}{2-(-1)} = 2,$$

so the equation of the line is $y = 2x + b$, for some constant b. Since $(-1,0)$ is on the line, $0 = 2(-1) + b$, so $b = 2$, and therefore the equation is $y = 2x + 2$.

Answer 1 demonstrates the ability to follow an algorithm; Answer 2 demonstrates that ability and, in addition, an understanding of why the algorithm works, and an ability to write clear mathematics. By many grading standards that I have seen, Answer 1 would be considered perfectly correct. I am proposing that Answer 1 be given very little credit indeed.

Ask questions that require students to decide for themselves what techniques to use and how to fit them together. Here is a question that may seem very easy, but that gave students in our multivariable calculus course a lot of trouble on the final exam:

- *Question:* Consider the plane $2x + y - 5z = 7$ and the line with parametric equation $\vec{r} = \vec{r}_0 + t\vec{u}$. (a) Give a value of $\vec{u}$ which makes the line perpendicular to the plane. (b) Give a value of $\vec{u}$ which makes the line parallel to the plane. (c) Give values for $\vec{r}_0$ and $\vec{u}$ which make the line lie in the plane.

Consider part (a) of this question. I suspect that most of the students in this course knew how to find the normal vector to a plane, given its equation, and how to find a parametric equation for a line parallel to a given vector; if they had been given questions which explicitly asked them to perform these tasks, they would have had less trouble. Looking at their answers to part (a), however, I realized that their difficulty was that it was up to them to choose these techniques and put them together. The problem was exacerbated by the fact that the two techniques came from different parts of the course. Many students simply did not know where to start. Their problem was not a lack of facility in using tools, but a lack of judgment about which tools to use, and when to use them.

Test conceptual understanding. I have been surprised in discussions with my colleagues to discover that this principle is not unanimously endorsed. Indeed, it is often used as a defense for teaching calculus as a pure techniques course that nothing more can be expected from students. Another argument that I have heard is that conceptual understanding flows eventually from a rigorous grounding in the 'basics', so that it is not necessary to test it separately. This is not borne out by my experience with the following question.

- *Question.* Suppose that $f(T)$ is the daily cost to heat my house, in dollars, when the outside temperature is T degrees Fahrenheit. (a) What are the units of $f'(103)$? What does $f'(103) = 0.87$ mean? What are the practical consequences of this fact? (b) If $f(103) = 10.54$ and $f'(103) = 0.87$, approximately what is the cost to heat my house when the outside temperature is $100°$?

The correlation between ability to answer this question satisfactorily and technical ability is weak at best: I have seen many students with excellent algebraic skills who are unable to answer the question at all, and many who are quite weak in algebra who give very good answers. It seems wise to me to test both dimensions, rather than just one.

Avoid Excessive Use of Templates There's not much point in giving an example here: Any question can serve as a template, with sufficient repetition of the type. The traditional calculus course provides many examples, but the different calculus courses that have been developed over the last ten years are just as susceptible to the problem.

The problem with template problems is that students can memorize how to solve them. A template problem on a final exam thus becomes a test of the student's ability to memorize. There are, of course, times when that is exactly what one wants to test (rules of differentiation and integration). But much of the time it leads to absurd results. For example, a standard calculus question is to ask the student to compute $\lim_{x->a} f(x)$, where $f(x)$ comes from a fairly well-defined and limited class of functions (e.g., rational functions). Students will memorize how to answer such questions using some version, perhaps fancier, of the following rules: Try setting $x = a$; if you get $0/0$, cancel powers of $x - a$ and try again. This is good algebra practice, but a poor test of whether students understand the concept of limit. The first question in the next paragraph provides, I think, a better test of that concept.

Ask Students to Reason From Graphical and Numerical Data, in Addition to Reasoning Algebraically Asking students to work with different ways of representing the same object encourages them to come to grips with the underlying concepts and not rely on memorization.

- *Question.* There is a function called the error function which is used by statisticians and denoted by erf(x).

  (a) Given that
  $$\mathrm{erf}(0) = 0$$
  and
  $$
  \begin{aligned}
  \mathrm{erf}(1) &= 0.299793972 \\
  \mathrm{erf}(0.1) &= 0.03976165 \\
  \mathrm{erf}(0.01) &= 0.00398929,
  \end{aligned}
  $$

  estimate erf$'(0)$, the derivative of erf at $x = 0$. Only give as many decimal places as you feel reasonably sure of, and explain why you gave that many decimal places.

  (b) Given the additional information that
  $$\mathrm{erf}(.001) = 0.000398942,$$

  would you change the answer you gave in (a)? Explain.

This question is subject to the following criticism, which is often raised against questions based on numerical data: From a strictly logical point of view, there is nothing that one can say about erf$'(0)$ using the given information. Nonetheless, it is my contention that a student has demonstrated a good intuitive understanding of the concept of limit if he or she answers part (a) by giving some sensible approximation such as erf$'(0) \approx .40$, then improves it to erf$'(0) \approx .3989$ for part (b), and explains his or her choices by pointing out when the digits in the first four decimal places of the difference quotient appear to stabilize.

Here are a couple of questions based on graphical reasoning.

- *Question.* Below is the graph of the derivative of a function f; i.e., it is the graph of f'.

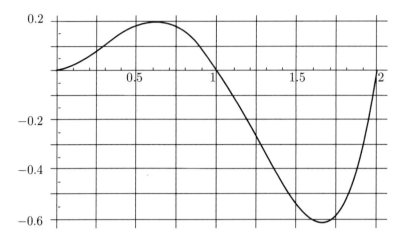

You are told that $f(0) = 1$.

(a) On what intervals is f increasing? Explain your answer.

(b) On what intervals is the graph of f concave up? Explain your answer.

(c) Is there any value $x = a$ other than $x = 0$ in the interval $0 \le x \le 2$ where $f(a) = 1$? If not, explain why not, and if so, give the approximate value of a.

You might ask: What is the difference between this question and one where the students are given a formula for f' and then asked the same questions about it? I think that both are fairly good questions. However, the algebraic version of the question is subject to the following problem: Many students have learned a recipe for determining where a function is increasing, decreasing, concave up or concave down. It involves setting formulas equal to zero, solving equations, drawing sign diagrams, taking derivatives, and so on. It is possible to be able to perform this procedure and still not understand that a function is increasing where its derivative is positive. That is, it is possible for a student to be able to answer the algebraic form of the question and be completely helpless when confronted with the graphical version.

Of course, there could also be students who can answer the graphical version, but are unable to answer the algebraic one. That is why I said that graphical and numerical reasoning should be introduced in addition to, not instead of, algebraic reasoning. If you keep approaching things from different points of view, you have a better chance that students will attempt to understand the underlying concept rather than rely on an ever increasing store of memorized procedures.

It is also important to ask questions that require students to translate between different points of view. Here is a simple example.

- *Question.* Below are graphs of $\sin(ax)$ and $\sin(bx)$, where $a > b > 0$. The scale on the axes is the same in both graphs. Which is the graph of $\sin(ax)$, (I) or (II)? Justify your answer.

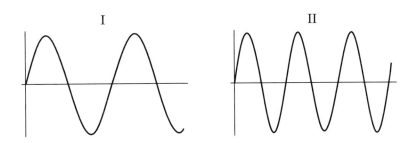

To answer this question, the student is required to relate the visible difference in periods to the algebraic fact that $a > b > 0$. This requires solving the equations $\sin(ax) = 0$ and $\sin(bx) = 0$, and then giving a geometric interpretation to the algebraic form of the solutions. (One can imagine arguments involving experimentation with a graphing calculator which do not require any algebra; however, such arguments would be weaker and, presumably, worth less in the grading.)

4. ... Is in the Eating

There are many ingredients in addition to the exams that go to make up a curriculum: the textbook, the method of teaching, and the use of technology, for example. All are important and are rightly the subject of discussion. However, they all represent the good intentions of the teacher, the hopes and plans. Our discussions about the curriculum are so volatile, I suspect, because they are grounded in fantasy and wishful thinking, rather than a realistic picture of what is actually going on inside our students' heads. Good intentions are often unrealized; hopes dashed; plans modified. It is the exam in December or May that reveals what is really in the course, not the book we chose in June or the syllabus we distributed in August.

Whether you are a Sage on the Stage or a Guide on the Side; whether or not you teach the Mean Value Theorem; whether you teach in a classroom full of computers running `Maple`, or ban calculators from your course altogether; it is your exams, and your students' answers, that tell what you have achieved as a teacher. Moreover, it is your exams that tell the students what they really must study. Any course that has been stable for a few years has a reputation, communicated through copies of previous exams and through the experiences of former students. Students pay attention to this reputation; they look at previous exams; they listen to the folklore surrounding the course. Much more than anything said by the Sage or the Guide, it is the course's reputation that determines what students study in the days before the final, and thus determines what knowledge and abilities they ultimately retain.

Teaching or Appearing to Teach: What's the Difference?

Kenneth C. Millett

University of California, Santa Barbara

Ask most professors, department chairs, or deans about the quality of teaching at their institution and they will tell you that their faculty are excellent teachers. How do they know? Students tell them so is the most likely answer. Anonymous student evaluations at the end of the course commonly provide the required evidence. And, they will say, very few students complain about teaching. I wonder, however, if there is real evidence that teaching and learning are actually occurring? Without concrete evidence that students are learning, how can there can there be credible evidence that teaching has occurred, excellent or otherwise? To put it more succinctly, you can't have good teaching without good learning. So, the question is "What has actually been learned?" By ignoring this fundamental question and relying solely on student opinion for the definition of teaching quality, administrators, professors, and students become co-conspirators in perpetrating a fraud: This is the tacit agreement, quoted anonymously at the start of Chapter 1 of Krantz's book, between instructor and student to the effect that instructor will pretend to be teaching, the student will pretend to be learning, and both will state that the other is doing a good job. One has an illusion of good teaching without the substance, a very real occurrence when the appearance of acceptable teaching performance is a condition of employment.

There are many ingenious ways in which this fraud is perpetrated. For example, an instructor could offer an "exam review sessions" and "practice exams" that provide students with model questions from which exam questions are superficial modifications. Another approach is to focus on the memorization of basic vocabulary and the acquisition of basic algorithms that can be applied to a small set of stereotype problems. The result? Students are happy, often believing that they have learned something because they did well on the test. They are happy with the instructor who has provided them with what they needed in order to be successful. Therefore, strong teaching evaluations are given. But what has the student actually learned? Quite often, it seems, students have been successfully trained to respond to specific stimuli with specific predictable responses. "Automaticity" and accuracy are the goals, we are told by the "successful teacher". Is it an appropriate college or university goal to create human automata? Especially in this computer era, we must set higher expectations for our students than to become the new intellectual revolution assembly line robots. Alas, many elements of the educational system appear to be designed for this purpose. And students, parents, teachers and administrators, at all levels, believe that this is the goal of education. To be a truly successful teacher, one must confront the issue of what is learned by the student. In the end, this is what really counts.

Possessing a deep understanding of course material, acting in a manner that is respectful of student and oneself, speaking clearly, writing clearly, meeting office hours, etc. are all things that support teaching. But, in fact, one can quite successfully follow the prescriptions of the "how to teach" book without any real teaching ever occurring. One way to address what is missing, I believe, is to pay serious attention to what learning is sought in the course. One should first have a clear understanding of the goals of the course. "learning calculus" is not sufficient. Useful goals are clear, concise statements that are understood by instructors and by students. Most importantly, they must be expressed in a way that both can know whether these goals have been achieved or not. Second, both instructor and student should understand how to achieve these goals. This will avoid the common pitfall of purposeless work that absorbs energy, creates the illusion of learning, but is ultimately ephemeral in intellectual substance. Third, instructor and student must understand the manner in which the achievement of course goals will be measured. Here lies one of the more inviting traps in the world of education. It is quite common, at all levels, to equate the ability to "do mathematics" with "passing the test." I recall a student who finally understood the issue. After a long discussion of a test problem, she said, 'I get it, you want me to understand calculus, and I just want to get an 'A' in the course." Neither are professors immune. Ask a colleague about how a course went and you will usually hear, "Oh, about half the students passed."

A key to addressing this dimension of teaching is to have strong, clear goals. Select materials and strategies that assist the achievement of these goals. Regularly analyze whether or not these strategies are doing so. Use these goals, and only these goals, to measure student achievement in the course. Finally, evaluate your own performance as well as your students' in terms of these goals. Teachers are learners too: continuously learning ways to increase their effectiveness. These are the principal dimensions of teaching that I will discuss in this essay. I will close, however, with a short report on a discussion of the book that took place over two months at Santa Barbara. Participants in the discussion were graduate and undergraduate students who are the staff of UCSB's NSF-supported California Alliance for Minority Participation (CAMP) Achievement Program. They organize and facilitate the work of study groups or provide support of students interested in science, engineering, mathematics, technology based careers. I wish to share their reactions derived from our reading and discussion of the book by way of broadening our perspective on the recommendations found there.

Goals

Why focus attention on goals? They are necessary to define the nature and extent of the success. While they are sometimes informal or part of the institutional cultural, they are most useful when they are articulated in a form understandable by new faculty members and students. Formulating these goals is one of the most important responsibilities of a faculty. It seems, this is a responsibility that is frequently avoided or trivialized. Why? It is work that takes time away from students, research, and other scholarly activities. And it is very difficult. During a recent meeting a colleague said, "If we have to agree on the goals for this course, we will never get anywhere. We have as much chance as proving that God exists!" We never did agree on goals. Some argued that they were unnecessary, others said we already knew what they were, and others, disagreeing, that they were essential in order to evaluate proposed texts and syllabi. There was never a resolution of this issue. Nevertheless, a text and syllabus were adopted. In such circumstances it is little surprise that, for example, the course failure rate varies between 10 percent and 70 percent according to who is the instructor. As far as I can determine, these events reflect profoundly different courses despite sharing the same name, text, and syllabus. Individual instructors formulate their own goals, as they must, in such circumstances.

So, what are good useful course goals? Ideally, each course offered by your department has a few clear academic goals but often this in not the case. Goals are not the same as list of topics, chapters, or sections to be covered. Goals must make sense to you and your students. All of your must be able to know whether or not they have been attained? To give a first example, let me return to the earlier meeting discussion of calculus texts and syllabi. A colleague said, "I know one of Millett's goals, he wants his students to be able to solve problems they have never seen before." And it's true, that is one goal I have for many courses. It is not a goal shared by my colleagues, neither explicitly nor implicitly. But I think that it is a good an example of a goal. Easy to understand and advancing a key component of a university education, especially in mathematics. Other beginning calculus course goals are: "Understand the relationship between math and the 'real world'." "Assume personal responsibility for learning." "Strengthen critical thinking and problem solving skills." "Acquire a concrete understanding of fundamental concepts." "Develop basis calculus skills." "Increase the number of students making steady progress in math and science programs." I have twenty four such goals, many of which naturally expand into a more detail description. Those associated to the "basic calculus" skills will seem familiar to those who have used the Harvard Consortium curricula. I share these goals with my students and point to them, where appropriate, in explaining assignments or course structure. In a second year introduction to advanced mathematics course, goals included: "Be able to read and explain mathematics." "Be able to distinguish between a complete and correct proof and a faulty argument." "Be able to write and explain simple proofs." And, finally, in a linear algebra course, a very challenging goal was: "Be able to extend concepts, methods, and results from finite dimensional vector spaces to Banach or Hilbert spaces." These are the kind of goals I use in contrast to "Calculate the derivatives and integrals of polynomial functions," or "Create a truth table for simple propositions." or "Be able to state, prove the Hahn-Banach theorem." At all levels, students often believe that they need only memorize an algorithm or body of material in order to be successful. They do this in much the same way that they would memorize a poem or learn to juggle three balls. While both may be good things to learn, they are not good models learning mathematics in a form that will be useful beyond a grade in a specific course. And becoming able to do so is, after all, what we are all about. Our goals should enable this to occur.

In addition to the challenge of developing a faculty consensus on goals, there are other dimensions that should be considered. Are the goals understandable by students as well as instructors? Are the goals appropriate for the students enrolling in your course? Students will need to have had the appropriate preparation to have a fair opportunity to achieve the goals. Are there appropriate and sufficient resources to goals? For example, is the size of the class appropriate? Does the textbook promote or inhibit student success in achieving these goals? As I mentioned earlier, in my calculus courses one of the goals is to develop an ability to solve unfamiliar problems. To quote a former student, "To learn what to do when you don't know what to do." Textbooks organized around a collection of standard problem types and training in the use of standard algorithms have proven to be an insurmountable barrier to achieving this goal. Since many of the traditional texts have this structure, they would not qualify for adoption in a course that I would teach. The reason is that most of my students come directly from high school experiences that have convinced them that this is what math is about. It is extremely difficult to be open up their thinking to alternative views on this matter. Thus, if this is one of the goals you will not be able to use such texts and, if you are required to use such a text, this would be an unachievable and, therefore, inappropriate course goal.

Another example in which the specific goals are critical would be determining the array of teaching strategies to be employed in the course. The abilities and experience of teaching assistants or homework readers also have significance for the choice of goals and the support or training needed to enable them to help in the achievement of the goals. In large public universities many graduate students have not been educated in this country. They, and many of those who have, find it very difficult to stimulate, organize, or guide small group student discussions of assigned problems. In some cases, I have failed to get them to even attempt to do so. They have become obstacles to achieving the goals and, in some cases, have undermined my best efforts with certain groups of students. The sad result is that rather than helping these students they have caused some to achieve a level of performance needed to continue in mathematics. These students have had to seek an instructor who goals were compatible with their own thinking and retake the course. For example, if one of my goals is to increase the quality and quantity of mathematical communication and the strategy of small group discussions is quite effective in providing opportunity to practice explaining calculations or solutions of problems. As I do not give full credit for work without full explanations, students who are unable or unwilling to provide them will not be fully successful in the course.

Strategies

Goals without strategies to attain them are useless. Strategies depend not only on the course goals but, as do the goals, depend upon the circumstances, the preparation and attitude of the assistants and the students in the class. Many experienced instructors have a large and varied array of instructional approaches available to them. Novices may be constrained to the most familiar, those they have observed as students and which they feel most able to imitate. What ever the choice or choices, the selection is determined in relationship to the goals. If the goals are superficial, then superficial strategies are all that are required. If, however, you have chosen a more challenging array of goals, then you may need to improvise, modify, create, experiment, etc. over a period of time. For everyone, an ongoing evaluation of the observed results needs to take place routinely in order to determine to whether learning is actually occurring and course goals are being accomplished. The real teaching professional continuously elaborates upon previous approaches, develops new ones, and measures progress in order to make "mid-course" modifications.

In public universities we are often called upon to teach classes of quite different sizes and purposes. The strategies will change even though a goal may remain the same. For example, the development of stronger mathematical communication abilities and the acquisition of self-confidence and personal responsibility for learning are among my beginning calculus course goals. As a consequence, even in classes of 150 students, I frequently use a "small group activity" strategy that requires students to interact mathematically with classmates and asks them to share, even in the large class setting, their results and questions with the whole class. Having a seventy-five minute class requires that I monitor the extent of engagement of the students with the material I am trying to communicate. They often "fade" after about fifteen to twenty minutes. At this point I might choose to introduce a new topic or idea by giving them a "surprise group quiz". Of course my purpose is NOT to test them on material I have not yet discussed, but they should have read, but to get them thinking about the problem or topic, to change the energy level, to redirect and re-engage their thinking, and to prepare them for what I would like to talk about next. This, I find, can work surprisingly well. I give them about five to ten minutes, ask them to share their conclusions or solutions. Often they use the board or overhead. I ask for alternative solutions, for different answers, In general, I work hard to express appreciation for every student's contribution. When there are competing answers or solutions, I ask that they all be explored in a supportive and civil manner that does not compromise on our goal seeking a correct and understandable answer. This can work VERY well and usually provides me with all the material I need to introduce and discuss the topic I had in mind. But you need to be "on your toes" to fully take advantage of this approach. I once asked a class to work on a problem leading to multiple integrals. Rather than proceeding as I had expected, one group of students took an approach leading to the Lebesgue integral while others took the expected path. The resulting discussion required about an hour, brought out some misunderstandings about functions of two variables, lead to a comparison of numerical estimations, and to a comparison between the two approaches in terms of an ability to do symbolic calculations. I was exhausted by the end of the class. Years later students who participated in the discussion would bring it up in conversations about their experience in my class. This only happened once in recent times, but it is an excellent illustration of the sort of thing I am seeking from my students and how one can go about trying to make it happen.

In another experiment, I have been trying a strategy to increase the intensity and effectiveness of interactions with students and to improve the level of student performance on homework. I have converted my office hours into a three hour session in the "math lab" during which each student is required to personally present her or his homework each week. From their papers I read the name of each student and greet them. This helps me to know them better and gives the class of 150 students a smaller feel. I then scan a selected problem or two and give the student my quick assessment of the quality of the submitted work. If it is up to "standard" I tell them so and wish them a good day. If it is not, I tell the student where there appear to be deficiencies and offer them a chance to do an immediate revision, with the assistance of the graduate student staff working in the lab. In smaller classes, I have asked the students to come to my office rather than the lab. I have them sign up for a specific Friday interview time, about 6-7 minutes each for a class of 45 students, either with me or a graduate assistant. We receive their homework, scan a problem, ask them a follow-up question or two, and, if time allows, discuss questions that they might have about the course material. If the homework is not up to the course standard I offer them the option of continuing to work on it over the weekend to attempt to improve the work. In some cases I have refused to accept the work because it was so poorly done and have insisted that the student make a brief appointment with me the next week to review their progress. In short, I would only accept work that appeared to be of passing quality.

I mention these examples to expand the range of strategies that might be considered and because they represent strategies that are new to me, even after teaching for about thirty-five years. While they may be familiar or time worn to some, I find myself continuing to evaluate the degree to which they help students enrolled in my classes reach a higher level of accomplishment. This must be directly tied to the course goals. It is worth the effort? There are at least two differing views or priorities that seem to determine the answer. For those willing to devote roughly 30 hours per week to teaching, as I have tried to do, the increased level and rate of success and greater interaction with students supports an affirmative response. For those needing to minimize the number of hours required to be evaluated as a good teacher, strategies of this sort are not the best. I work on a base of 60 hours per week with at least 30 devoted to research, scholarly and administrative work and the remainder devoted to teaching. In a research university there is an intersection of the two due to the time spent supervising graduate research.

Success or Failure?

The bottom line is you haven't been teaching if your students haven't been learning! So both you and your students need to know whether the course goals have been achieved or not and to what degree. You will need to be able to determine grades or award credit for the course. But also, you must consider future changes in goals, strategies, or methods of assessment based on your experience in each course. What sorts of things should be considered in the first step, the evaluation of student work?

There are many approaches to grading, ranging from the implementation of a uniform standard examination system found in multi-section courses at some large public universities through complete individual instructor autonomy. In some cases a "grading curve" is imposed on the instructor while in other institutions a "standards based" approach is employed. I am, and have been for a most of my career, a believer in the standards approach. This came about many years ago when I discovered great differences in performance between an 8 AM calculus class and an 11 AM class that I was teaching. As these differences were consistent over a period of several terms, I have adopted a more objective approach to awarding grades rather than using a "curve" for each of the classes. I needed to develop, articulate, and teach according to a set of appropriate course goals. And I needed to evaluate student work based upon these goals. With the advent of common final exams and an ability to compare performance and grades across sections, I have had to raise my average grade a bit in order to insure equity across instructors. While many do not subscribe to this quantitative approach, I strongly recommend it as a consistent and fair method. Instructors will need to look closely at this question before beginning the course in order to insure that students are informed on exactly how grades will be determined.

How does one attempt to measure student achievement? I mentioned above my experiments with weekly short interviews. We gave credit (the amount was the equivalent of a graded homework problem) if a student can explain a problem and/or its solution at a level appropriate to the course. If relevant work was not submitted for the chosen problem, the grade was 0 according to a scale that I will discuss later. The underlying principal I have adopted with respect to the interview questions and, indeed, any work asked of students is: "If it's important to the course goals, we will ask you and, if not, we won't." For example, students are asked to explain their solutions to assigned homework problems. They can also be asked to discuss questions that have not been assigned or material from the reading assignment in the text.

With respect to examinations, I believe that there should be "no surprises." For my students, not to see an problem for the first time on an exam would be surprising since one goal is developing an ability to solve unfamiliar problems. If it is important in the course, it should be part of the evaluation. If not, it has no place there. No "trick questions" is another maxim I try to follow. Perhaps the most radical thing I do connected with the adoption of goals is to evaluate student performance on against a five point scale as follows. Each unit of work, interview, homework problem, or exam problem (sometimes portion of a multi part problem) is awarded credit according to the scale: 0 points: no progress or relevant information; 1 point: some visible progress that could lead to a solution or correct response; 2 points: significant progress, many major elements present, a partial explanation or proof; 3 points: essentially complete and correct solution but with only minor gaps, errors, or lack of explanation; 4 points: fully correct and complete solution including explanation or proof as appropriate. This evaluation scheme has been difficult for graduate assistants, undergraduate readers, and my students to adopt and to understand. In addition, it has been difficult to use in making comparisons with the credit awarded by other instructors (but their lack of specific measurable goals makes this difficult under any circumstances). Alas, many students confuse the numbers with grade points and while there is a connection it has not been easy for them to internalize after years of experience with another system. I actually believe this has been helpful in focusing attention on what they have actually accomplished. And there is no pleading for "partial credit". Even if they have presented pages of material, if there is nothing that could lead to a solution the mark is 0. The discussions that I have had with both students, graders, and assistants have helped us all keep a focus on the goal fully complete and correct solutions, with full explanations and proofs in contrast to an accumulation of points based on "partial credit" as a strategy for success. Since I allow only integral grades decisions have to be made as to whether on not work meets the standard for a 4. These discussions can lead to the creation of a rubric and benchmark solutions and can be the focus of a rich "professional development" opportunity as you interact with graduate students or undergraduate student readers. What does a "4" really look like? How can two solutions that look very different both merit the same mark? Isn't one "better" than the other? The development of model problems and exemplars to illustrate what each of the marks means can also be a helpful activity. By the way, current translation to letter grades makes 3.3 and above an A (one student received a 3.9 in one of my courses this term) and below 2.0 is not passing (for my institution, this means a "C$-$" or lower).

What has been the results of these experiments? Here is one point where the goals are critical. I could and have responded to such questions from colleagues and others by recounting anecdotes. For some purposes, this might be sufficient. But, at least for myself, I need an approach that will help me decide on continuing, modifying, or abandoning a particular strategy in favor of more productive ones. I need to weigh the costs against the benefits, for my students, my assistants, and myself. Anecdotes are not adequate for this task. Neither are the "feelings" that we often cite, even those based on thirty years of teaching experience. An approach that is less subjective and influenced by a desire for confirmation of prejudices, one that is as objective and analytical approach as one can easily construct is required. One must recognize and respect the limitations of drawing conclusions, especially with respect to ones own teaching, from the course data. But I know of no other choice. I try to bring the same quality of thinking to my teaching as I bring to my research, and the same skepticism and demand for "proof". At least proof to the extent that it is feasible in this context.

One very useful task that I have done in the past is an item by item analysis of the final exam work do determine whether or not the problems appeared to capture the information that was being sought. This has proved very useful while teaching first year calculus and has lead to some important changes in the types of problems that I have asked as well as leading to shifts in emphasis during the course. What then have I observed in my introduction to proof course? An analysis of final exams and homework has high lighted the persistent problem of students to attempting to memorize certain proofs, either as found in the text or gotten from class discussion sessions or lectures. When grading 45 of them, one recognizes them their similarity, identical notation across students and a loss of precision or focus. In addition, one finds fragments of arguments and a lack of completion or resolution. This sense has been confirmed by students in our interviews. This is a common problem. One faced by all instructors in this course and is a frequent topic electronic discussion groups concerned with teaching mathematics at a university level. I have tried some of the approaches suggested there, such as asking students to explain a question or problem prior to attempting a solution. This has been quite helpful. Under stress, however, my students seem to fall back on the memorization habits learned at a younger age. They wish to depend on memorization and pattern problems or algorithms, a successful method in high school and earlier. Mostly, these students do not yet have confidence in their own intelligence or ability to think. While I wish I could be certain that I have made progress with my students, and I know I have with a few, I am not satisfied with the result.

Parenthetically, this specific problem is a major reason for my interest in issues of K–12 mathematics education. It is also the reason I am quite dismayed by many of the recent proposals to implement a mathematics curricula promotes the acquisition of facts, standard algorithms, automaticity, and abandons efforts to develop a stronger capacity to think and communicate mathematically, to solve unfamiliar problems, to successfully use mathematics in a wide range of circumstances. Mathematics often appears to mean pages of exercises, multiple choice tests and has nothing at all to do with what should be going on "between the ears". Selecting students who are successful at this sort of anti- intellectual exercise for admission to a university where, if we stand for something at all, it is valuing curiosity, creativity, individual thinking, and puzzling things out makes no sense. It seems to be a process designed to optimize the frustration of professors and their students. Addressing this is the real remedial educational required at the university.

What is the bottom line? Less than 10% of the students received failing grades in my class compared to the historical average of 32%. A review of class records shows that the failures were precisely the students who did not present themselves for weekly interviews. Although some students complained about the level of effort required of them, many appreciated the opportunity to receive weekly information on their performance. Approximately ten students dropped the course after the first week or two and others may have avoided enrolling due the intensity of effort required. This an important factor that may have lead to a lower failure rate as I have observed the same behavior in calculus classes. There were 26% A's compared with the historical average of 12%. In order to insure that my standards were similar to those employed by other instructors I reviewed the graded exams from two other courses and talked with several more instructors. In some cases I used exam problems structurally parallel to theirs. I even compared grading standards to try hold to the same or better expectation for performance. These grades are sufficiently different from the historical experience that, if I should ever be asked to teach the course again, I will want to look again closely at the standards that I employed for this course.

I believe that these strategies have impacted several important problems. First, the number of students enrolled in the course who are truly engaged in a serious attempt to pass the class been significantly increased and the failure rate reduced. I believe the pressure of meeting once a week and being confronted, in as sensitive and as direct a fashion as possible, is very much appreciated by most of the students, despite their nervousness. Second, I and my assistants have a much richer and more immediate feedback on how well the students are doing and can make timely adjustments to the course to respond the needs or to take advantage of opportunities.

Student Perspectives

At the University of California, Santa Barbara, I am the Regional Director of the California Alliance for Minority Participation (CAMP) and also coordinate the associated Achievement Program Academic Workshops. These workshops are affiliated with critical barrier courses taken by students pursuing degrees in science, engineering, or mathematics. They are organized and lead by outstanding graduate and undergraduate student staff recruited and trained for this work. During the Spring Quarter these staff members[1] participated in weekly staff meetings as part of their own leadership development and professional growth. During these meetings we discussed the second draft of *How to Teach Mathematics* by Steven Krantz as a means to stimulate thinking about how we could improve the effectiveness of our workshops. In this concluding section, I will share some of the positions developed during these discussions.

First, they were impressed and encouraged by the discussion of changes in opinion of the author and reports of other professors concerned with improving the quality of college and university teaching. The fact that professors might continue to work at becoming "better" teachers was new to many of the participants. The conversations frequently returned to the topic of respect. It arose in several forms: a professor's respect for him or herself, for colleagues and for students, and a student's respect for the professor, for fellow students, and for themselves. For example, as future teachers and current leaders of workshops, there was an appreciation of the issue of self-respect and its significance in terms of their own preparation and development of the skills needed to be successful leaders. The importance of instructors treating students with civility, courtesy, and respect arose in almost every discussion. They felt that often professors did not respect their students. Among the other ways in which this lack of respect was manifested included not being prepared for the class or not having mastered the material, not arriving on time, not ending the class on time, and by not insuring a productive class environment by allowing students to wander in and out of class, to talk in class, or to otherwise disrupt the class. The fact that this issue arose so frequently in the book reflects its importance to Krantz and to them.

While Krantz tells the reader that his book is not a recipe, there are many structural ways in which the book appears to be just that. The suggestions, it was felt, were helpful for all instructors but especially for those not familiar with the K-16 educational culture in the United States. At UCSB we have not hired a mathematician who has received an undergraduate education in this

---

[1]Spring Quarter 1998 Achievement Program staff are graduate students Doli Bambhania, Kathi Crow, Ana Garza, Nancy Heinschel, Michael Saclolo, Becca Thomases, and Jeremiah Thompson and undergraduate students Gladis Aispuro, Maria G. Arteaga, Carla Billings, Roxana Cervantes, Hozby Galindo, Analilia Garcia, Hector Garcia, Nicolas Hernandez, Elizabeth Hutchins, Mason Inman, Katrina Jimenez, Christina Luna, Patrick Murphy, Erica Ocampo, Manuel Salcido, SuGen Shin, Shannon Shoup, Edgar Torres, and Ahmad Yamato.

country in more than a decade. While sympathetic to the challenges of teaching in a foreign language and culture, the students were concerned that instructors provided them with high quality instruction. Concerns with language and with respect for women and persons other national origins where the principal ones mentioned.

The question of preparation provoked a rich discussion. Graduate students, especially, felt that the amount of time that was required for new teachers was undervalued. Furthermore, one can not really "over-prepare". Rather, one could prepare badly. One analogy is that of the actress or actor who has performed the same role in a play hundreds of times. The lines are not the issue. The challenges are to have a deep understanding of the role, to make character "come alive" time after time, to directly engage each audience, and to "fill the room with your presence," even while fighting a flu bug or distracted by personal problems. For professors, making learning living intellectually engaging experience, sustaining interest and progress over the period of weeks and months at a high level, and stimulating a search for new and deeper meaning are all critical elements of the teaching craft.

The uses of various pedagogical methods provoked an energetic discussion as well. For example, while some instructors use transparencies or computer displays effectively while others fall into the trap of merely displaying a series of images too rapidly and without sufficient impact. Their use can be a problem for students. Don't "read the book to the class." Students should be reading it themselves. Repeating material in the book undermines the value of the class meeting. Much more should be expected from the instructor by way of establishing priorities and making choices of material! Lectures by charismatic speakers and small seminar discussions which are "content-free" are a problem. Some instructors do not appear to be genuinely concerned with whether or not their students actually learn anything of substance. "I won't disrupt your life if you don't disturb mine," appears to be the guiding principle of some. Actual engagement or interaction between professor and student is to be avoided. In contrast, students argue that the focus should be on what is good for the students, not what is good for the professor. Maximize learning, not minimize disruption.

Don't "read the book to the class." Students should be reading it themselves. Repeating material in the book undermines the value of the class meeting. Much more should be expected from the instructor by way of establishing priorities and making choices of material!

Conclusion

I have very much appreciated this invitation to spend some time reading and reflecting on the second edition of this book. I believe that it will be a useful resource for persons wishing to improve the teaching of mathematics at the college and university levels. My student colleagues certainly reacted very favorably to the material in this edition and the discussions helped us all better understand some important elements. The CAMP staff discussions had a tendency to revisit certain key issues following the course of the reading. At one point, the recognition of recurring themes became the focus. For example, the concern for respect recurs often. Because it is so important to students and is seen to a serious problem that inhibits their learning, the CAMP staff were quite encouraged by the treatment it receives.

In one of the concluding discussions, I told of my perspective on the book and concerns I wished to address in this essay. These did not receive the same interest as did the topics in the book. Those issues are much closer to the immediate experiences of the students and are more familiar to them. As such, they represent an good collection starting places. Indeed, a more appropriate title might have indicated that this is really only an introduction to teaching mathematics. Excellent for novices, it does not address adequately some of the issues that I have tried to introduce in this essay. There is, I believe, very much more to being a successful mathematics teacher than is presented in Krantz's personal perspective. It is too easy to ignore the fact that unless our students are learning, we are not actually teaching. To ignore the implications of this undermines the fundamental and historical role of mathematics departments in most universities. How do we know our students are learning? What is it, exactly, that they are to be learning? What are the goals of our courses and of our teaching? Are we succeeding or failing?

As challenging as it might seem, we need to be able to answer these questions for ourselves, for our students, and for out institutions. It needs to be a major concern for persons starting out teaching careers. And it needs to be a continuing concern, especially for those of us who have been teaching for decades May our teaching efforts be described more favorably than with Macbeth's words, "Life's but a walking shadow, a poor player that struts and frets upon the stage and then is heard no more: it is a tale told by an idiot, full of sound and fury, signifying nothing."[2]

---

[2]William Shakespeare, *Macbeth*, Act V, Scene 5

Why (and How) I Teach without Long Lectures

J. J. Uhl

University of Illinois at Urbana-Champaign

1. Why I Gave Up Long Lectures

I could have benefited greatly from Steven Krantz's tips in 1962 when I taught my first class. In fact, I can see that over the years my lecturing style and techniques evolved to be remarkably similar to those Steven Krantz (SK) suggests. I was a very popular lecturer and recently won an MAA sectional award for distinguished teaching based in no small part on the lecture courses I gave at Illinois between 1968 and 1988. But for the last ten years, I have completely abandoned the long lecture method.

My last lecture effort was calculus in 1988. I thought I did a bang-up job, but the students did not respond with work anywhere near the level I was used to and have become used to after I gave up on introductory lectures—despite the fact that I had been giving the lectures largely in harmony with SK's recommendations.

Simply put, today's students do not get much out of long lectures, no matter how well they are constructed. The material comes too fast and does not sink in well. The students of the past responded by becoming quiet scribes. Today's students demand more action and accountability. That's why many students cut class and even when they come they often ask hostile questions such as "What's this stuff good for?" They do not read their texts. Some students even disrupt lectures. And as SK notes, many professors ask the questions

- Why won't my students talk to me?

- Why is class attendance so poor?

- Why won't students do their homework?

- Why do they perform so poorly on exams?

And then they shrug it off saying to themselves: "If only I had taught at Harvard things would be different. I would have bright and eager students." or "Students these days are impossible."

It is the lecture method of teaching that is impossible—the method of teaching via long lectures is crumbling under its own weight. This is true not just in mathematics. Across the University of Illinois, there is a major controversy about whether professional note takers may take notes and sell them to students who would rather not attend lectures. One of the first to note that the lecture system needed to be replaced was Ralph Boas in 1980: "As a means of instruction, lectures ought to have become obsolete when the printing press was invented. We had a second chance when the Xerox machine was invented, but we muffed it." Many math instructors are trying to teach today's students using only yesterdays tools and approaches. And neither the instructors nor the students pleased with the results.

Introductory lectures are not (and probably never have been) a particularly effective vehicle for introducing students to new material. A few strategically timed and strategically placed short follow-up lectures (sound bites) can be very effective. But the problem with introductory lectures is that they are full of words that have not yet taken on meaning and full of answers to questions not yet asked by the students. A further problem is that many lecturers fall into the trap of believing that their job is to think for the students. This effectively shunts the students to the sidelines—making them into mere scribes who verify in the homework and tests the math truths promulgated by the lecturer. As Bill Thurston put it: "We go through the motions of saying for the record what the students 'ought' to learn while students grapple with the more fundamental issues of learning our language and guessing at our mental models. Books compensate by giving samples of how to solve every type of homework problem. Professors compensate by giving homework and tests that are much easier than the material 'covered' in the course, and then grading the homework and tests on a scale that requires little understanding. We assume the problem is with students rather than communication: that the students either don't have what it takes, or else just don't care. Outsiders are amazed at this phenomenon, but within the mathematical community, we dismiss it with shrugs."

In summary, I do not disagree with SK's approach to lectures, as he gives some great advice, which I used to follow as well. However, I do question the necessity, importance, and educational quality of lectures as a method for students to learn mathematics.

2. What I Replaced Lectures With

Another piece of wisdom from Ralph Boas : "Suppose you want to teach the 'cat' concept to a very young child. Do you explain that a cat is a relatively small, primarily carnivorous mammal with retractile claws, a distinctive sonic output, etc.: I'll bet not. You probably show the the kid a lot of different cats saying 'kitty' each time until it gets the idea. To put it more generally, generalizations are best made by abstraction from experience."

Today my calculus, differential equations and linear algebra students get the experience they need through `Mathematica`-based courseware written by Bill Davis, Horacio Porta and me. The basic ideas are laid out in interactive `Mathematica` Notebooks in which new issues arise visually through interactive computer graphics. With this courseware, limitless examples are possible almost instantly. If the student doesn't get the point right away, then the student can rerun with a new example of the student's own choosing. They can use the courseware to touch and see the math "kitty" as many times as they want to. They see for themselves what the issues are before the words go on and generalizations are made. One of our favorite techniques is to give a revealing plot and ask the students to write up a description of what they are seeing and to explain why they see it. In these courses, conceptual questions are the rule and students answer them. Contrast this with the typical student problems assigned in traditionally taught mathematics courses.

Here is the story behind the evolution of our courseware and the way it is used: In 1988-90, when Horacio Porta, Bill Davis and I were developing the original version of the computer-based course Calculus&`Mathematica`, Porta and I offered regular introductory lectures at Illinois. We noticed poor attendance and asked the students why. The students uniformly replied: "We don't need them. We can get what we need from the computer courseware when we need it. What we do want is a followup discussion from time to time." We followed their advice and have never seen the need to go back. Our students taught us how to teach. Over the years, almost all teaching of Calculus&`Mathematica` (and sister courses DiffEq&`Mathematica` and Matrices, Geometry&`Mathematica`) has evolved to this model (sometimes known as Studio learning, a term coined by Joe Ecker for his `Maple`-based calculus course): All the student problems are freshly written with the idea of engaging the student's interest. Assignments are made on Thursday. Students work on each assignment for one week. One day before the assignments is due, a classroom session is held to discuss what the week's work was all about. Students come armed with questions and if they don't fill up the whole hour, then the instructor gives a several pointed mini-lecture addressing points the students should have picked up during the last week. All other class meetings are in the computer lab with the instructor answering student questions as they arise—at the ultimate teachable moment. This lab interaction between teacher and student (which is sometimes done via e-mail) is very important. No longer are the students the professor's audience; students are the professor's apprentices.

The students' weekly assignments count for at least half their semester grades. Because there are no other lectures, the whole course consists of student work. In this model , it is what the students do that is important rather than what the professor says and how professor says it. Still the influence of the instructor is pervasive and the course ends up satisfying Gary Jensen's and Meyer Jerison's criteria: Setting pace, teaching students to read, and fully engaging the student in the learning process.

This learning model cannot be accomplished with traditional printed texts and traditional lectures and lends itself rather well (but not perfectly) to Internet distance education. NetMath centers offering via the Internet calculus, differential equations and linear algebra courses for university credit supported by live mentors have formed at several universities and colleges. Here is a reaction from a high school teacher who sponsors NetMath Calculus&`Mathematica` in Alaska: "Jessica has really enjoyed the course, and her father, a veteran of traditional calculus courses, is very impressed with the understanding of the mathematics that this method imparts. He has done all the problems and loved it. There have been some loud arguments—most of which she has won."

3. Content Issues

The trouble with the lecture system is compounded by the fact that our undergraduate courses, for the most part, have been frozen in the past and have become unable to adjust to modern demands. Undergraduate mathematics courses today are nearly indistinguishable from the undergraduate courses I took in 1960. Peter Lax put it this way in 1988: "The syllabus has remained stationary, and modern points of view, especially those having to do with the roles of applications and computing are poorly represented ..." When I look over mathematics undergraduate courses during this century, I see a smooth evolution of new ideas and better mathematics through the period 1900–1960. Topics of limited interest such as haversines, common logarithms, Hoerner's method, latus recta, involutes, evolutes, Descartes's rule of signs all had their time in the sun but were de-emphasized in favor of more important topics. And then the content became frozen. There is a whole list of 20th Century topics that have been by and large rejected in today's mathematics classroom. A short list: The error function, singular value decomposition for matrices, unit step functions and their "derivatives", the Dirac delta functions in differential equations, using the computer to plot numerically solutions of differential equations, Fast Fourier Transforms, wavelets. There is plenty of what Peter Lax calls "inert material" in most of our current mathematics courses. It's time to get rid of it and open the door to some fresh, important material.

My bet is that the underlying cause of this is our current fanaticism about having one-size-fits-all uniform texts chosen by central committees who often lack the expertise to make significant changes. They just go on tinkering with what was done the year before. It seems the central committees do not trust the initiatives of individual faculty members, so they shackle them with obsolete material. Publishers respond in kind. And the publishers stay away from texts for modern courses because new, modern, original texts are unlikely to sell well. This is the reason that most well-selling traditional calculus texts are clones of George Thomas's calculus course of the 1950s.

The trend is for engineering, biology and science departments to begin teaching the mathematics their students need. Mechanical engineering departments are teaching lots of advanced calculus and differential equations. Electrical engineering departments are teaching lots of probability and complex variables. According to a source inside the Stanford University Computer Science Department, they have decided:

a. They'd like their students to have more math.

b. But not the kind of math that's coming from the math department.

c. It's never going to come from the math department.

d. They'll start doing it themselves.

No wonder Sol Garfunkel and Gail S. Young wrote: "Our profession is in desperate trouble—immediate and present danger. The absolute numbers and the trends are clear. If something is not done soon, we will see mathematics department faculties decimated and an already dismal job market completely collapse. Simply put, we are losing our students." Are mathematics professors and departments in extreme denial? I wish SK had dealt with these serious issues.

4. Specific Remarks about SK's Revision

SK: "We do not want our students to learn to push buttons. We want them to think critically and analytically. It has been argued that `Mathematica` and similar software to help students interact dynamically and visually with the graphics: Vary the value of a in the equation $y = ax^2 + bx + c$ and watch how the graph changes. That is not what I want my students to learn. I want them to understand that, for large values of x, the coefficient of a is the most important of the three coefficients. And changing its value affects the first and second derivatives in a certain way. And, in turn, these changes affect the qualitative behavior of the graph in a predictable fashion. AFTER [SK's emphasis] these precepts are mastered, the student may have some fun verifying them with computer graphics. BUT NOT BEFORE [my emphasis]."

Reaction: Why not before? Is this a moral issue? Certainly this is not an educational issue. Students (and research mathematicians) learn lots from examples. Here is my version: Have the students play with graphics, first varying a and coming up with a conjecture of what the influence of a is. Then ask them to explain why their response is correct. Then ask them to explain how changing the value of a affects the first and second derivatives in a certain way and how this is reflected in corresponding plots. In this way the students engage completely in Saunders MacLane's sequence for the understanding of mathematics: "intuition-trial-error-speculation-conjecture-proof." I don't care how many buttons students press; if it helps them to think critically and analytically, I'm all for it.

SK: In a programmed learning environment, whether the interface is a PC or with `Mathematica` Notebooks or with a MAC, the students cannot ask questions.

Reaction: Disagree. Students at computers can and do ask questions—lots of them.

SK: "What sense does it make to have a mathematics classroom, with a computer before each student, and the instructor delivering command to the students? ...People need to perform laboratory activities in their own time at their own pace."

Reaction: I agree thoroughly. A teacher-centered computer lab is absurd. The professor has to learn to relinquish total control when the students are in the lab.

SK: "Most of us were trained with the idea that the whole point of mathematics is to understand precisely why things work. To make the point more strongly, this attitude is what sets us apart from laboratory scientists."

Reaction: I agree and disagree. Throughout our courseware are two recurring themes:

1. One of the goals of mathematics is to explain why things work out the way they do.

2. There are no accidents in mathematics!

On the other hand, good mathematical research has always been a laboratory science. Top quality mathematical research invariably feeds off examples and special cases. With the computer, students can get lots of examples and begin to formulate what mathematical truth might be then go on to explain why— thereby engaging in the whole mathematical process. Students in lecture classes miss out on this opportunity. The lecturer handles all of this for them.

SK: "One of the highest and best uses of computer in mathematics instruction is as the basis for laboratory work."

Reaction: I agree thoroughly. But I hasten to add that not all laboratory work is good. Weekly lab sections tacked onto an otherwise traditional course are of dubious value. The way the calculator is used in high schools to prepare for the AP exam is far from optimal.

SK: "Using the quadratic formula is easy. Analyzing a word problem is hard. A person who cannot do the first will probably not be able to do the second—with or without the aid of a machine."

Reaction: Disagree. One does not follow from the other, as many students in Calculus&`Mathematica` "BioCalc" sections at Illinois have proved.

SK: "If you choose a poor text, you will have to pay for it through the semester."

Reaction: Too many professors are not allowed to choose their texts. And how many really good texts are there out there?

SK: "...the respect a teacher must show his audience."

Reaction: A really good point. But I am uncomfortable with the characterization of students as audience. Audiences are usually faceless and rarely participate; they just watch. The view of students as audience is one of the major defects of the lecture method.

5. Suggestions for Further Reading

Here are some sources for those who want to re-examine their ideas about teaching of mathematics:

a) Ralph Boas's article "Can We Make Mathematics More Intelligible?" (*Amer. Math. Monthly* 88 (1981), 727–731) is a provocative short complement and counterpoint to SK's revision. Samples are included in the text above. This and a number of Boas's other essays were reprinted in the book "Lion Hunting & Other Mathematical Pursuits" (MAA,1995). Enlightening reading!

b) Gian-Carlo Rota's book, *Indiscrete Thoughts* (Birkhäuser,1997). Rota raises the right issues in the way only Rota can. A sample: "One must guard . . . against confusing the presentation of mathematics with the content of mathematics. An axiomatic presentation differs from the fact that is being presented as medicine differs from food. . . . Understanding mathematics means being able to forget the medicine and enjoy the food." If you read this book, you will not forget the experience!

c) Reuben Hersch's book, *What is Mathematics Really?*, (Oxford, 1997) A sample: "A philosophy of mathematics that obscures the teachability of mathematics is unacceptable."

d) H. Poincaré's books *Science and Hypothesis*, *The Value of Science*, and *Science and Method: Some Running Themes*: the value of intuition, arguing against reduction of mathematics to algebra à la Weierstrass; verification is not enough. Poincaré was one of the first to point out that mathematics teaching is not all it could be.

e) Henri Lebesgue's essays, *Measure and Integral* (Kenneth O. May editor, Holden-Day 1964). This books consists of pedagogical essays written by Henri Lebesgue and assembled by Kenneth O. May. The essays are very heavy on pedagogical and mathematical content with a passionate plea for better acceptance of decimal numbers in the mathematics classroom. Two samples: "There is a real hypocrisy, quite frequent in the teaching of mathematics. The teacher takes verbal precautions, which are valid in the sense he gives them, but that the students most assuredly will not understand the same way."

And:

"Unfortunately competitive examinations often encourage [an educational] deception. The teachers must train their students to answer little fragmentary questions quite well, and they give them model answers that are often veritable masterpieces and that leave no room for criticism. To achieve this, the teachers isolate each question from the whole of mathematics and create for this question alone a perfect language without bothering about its relationships to other questions. Mathematics is no longer a monument but a heap."

The Joy of Lecturing—with a Critique of the Romantic Tradition in Education Writing

H. Wu

University of California at Berkeley

1. A Sage on the Stage Speaks His Mind

A cornerstone of the current mathematics education reform is the recommendation that teachers should cease being "the sage on the stage", but should instead assume the role of "a guide on the side".[1] Lecturing is discouraged; direct instruction is passé. Students should be working in small cooperative groups to discover the mathematics for themselves, and the instructor should be merely providing guidance on the side. Indeed, "students frequently working together in small cooperative groups" is second among what an eminent educator considered to be the five preeminent characteristics of the present reform effort ([6, p. 105]).

It appears to me that this rejection of the sage-on-the-stage method of instruction is unjustified. There are situations where lectures are very effective, and in fact there are even circumstances which make this method of instruction mandatory. Furthermore, in recommending the guide-on-the-side strategy, educators should have been more forthcoming about its limitations so that teachers can better decide for themselves whether or not to follow such a recommendation. The purpose of this appendix is to amplify on these remarks. Although there are many alternative methods of instruction other than lectures, I shall limit the present discussion to the guide-on-the-side format on account of its favored status in the current reform.

It may be assumed that a person who rises to the defense of lectures must be someone who has never taught any other way; it would not be unnatural to go even further and conclude that the only way he can teach is by giving lectures. Not so in this case. Although I have been lecturing in the classroom for all thirty-three years of my teaching life, outside of the classroom I rarely give lectures in the sense of systematically presenting a body of knowledge. When undergraduate students come to my office with questions, for example, I do not believe a short lecture by me giving complete answers would do any good in an overwhelming majority of the cases. Instead I try to engage them in a dialogue and employ the Socratic method to expose for their own benefit the gap in their understanding that led to their questions.[2] In other one-on-one situations, I also do not lecture. I have given reading courses to undergraduates, and in such cases, I make clear that learning can only be achieved by the student and that all I can do is to nudge him in the right direction and offer help when absolutely necessary. The student must do all the work and my contribution is essentially limited to asking key questions when we meet. It is the same when I find myself

---

[1]Although this "sage-on-stage and guide-on-side" dictum has been in existence since the late eighties, it appears difficult to find a precise reference to it in the literature. One place where it is mentioned unambiguously is footnote 15 on p. 17 of [2].

[2]Unhappily, most students are only interested in getting simple answers and getting out of my office as fast as they can. My attempt at fostering genuine education only result in bad student evaluations for my "unfriendliness".

tutoring high school students on occasions. No lectures. The only thing I insist on is that no time limit be placed on any of the tutoring sessions: Once we start on a topic, we stay on course until it is finished regardless of how long it takes. For a later purpose, let me describe one specific example of my tutoring experience.

I once had to teach someone the division algorithm for polynomials. I started by remarking that it was just a glorified version of the same algorithm for integers. I asked if he knew the latter (yes), and if he could prove it (no). So I suggested a proof by induction of the division algorithm for given positive integers a and b in the form $a = qb + r$, $0 \leq r < b$. It took a while for him to decide, with some help from me, that one could attempt an induction on a, but in due course he succeeded in writing down a complete proof. Next came the polynomial version $f = qh + r$, $0 \leq \deg r < \deg h$. I asked him whether he could imitate the case of integers. It took some time for him to realize—again with some help from me—that $\deg f$ could be used for induction. However, he immediately saw that the usual induction step of "$P_{n-1} \Rightarrow P_n$" was of no use in this situation. At that point, it was time for me to step in to teach him about complete induction in the form of "$P_1, P_2, \cdots, P_{n-1} \Rightarrow P_n$". Then I let him figure out how to use it to prove the algorithm. Getting an appropriate q to start off the induction argument was not easy for him. While I saw the frustration, I left him alone because the frustration has to be part of the learning process. Finally he got it done. The whole session took something like two hours. I had no doubt that he really learned the algorithm through this tortuous process, and it is likely that for most students this is the only way to learn it. But could I teach in any fashion remotely resembling this in the usual junior level introductory algebra course? Absolutely not. In such a course, the polynomial algorithm merits a discussion of about 25 minutes. If I spent two hours to teach it, I would be fired for pedagogical turpitude,[3] and rightly so.

In an ideal world, I would like to teach all my classes the same way I teach my students in a one-on-one situation. But this is a dream largely unrealized except during the extra problem sessions I offer my students in upper-divisional courses. With a sparse attendance and little time pressure, I can afford to let the students dictate the pace and the direction of the discourse half of the time. Otherwise, I find the obstacle of time-constraint almost impossible to overcome, and this constraint will be a recurring theme of this article.

2. The Hows and Whys of Lecturing

No matter who says what, lecturing is *an* effective way of teaching in a university—and for that matter in grades 7-12—so long as our education system stays the way it is. Such a bald statement requires a careful description of its underlying assumptions, and I will proceed to do that. I assume that:

(i) The instructor is mathematically and pedagogically competent,

(ii) Only 12 years are devoted to public school educations and 4 years

---

[3]In my contract with the University of California, there is a clause that says I could be fired for moral turpitude.

to college education,

(iii) After 12 years of school education students should be competent enough to function as useful citizens in society, and after 4 years of college students should be competent enough to start graduate work in their chosen disciplines, and

(iv) Our education system continues to be one for the masses rather than for a select few, so that each teacher or professor must teach *many* students in each course.

Whatever I say below will apply equally well to teaching in grades 7–12, but for the purpose at hand I will specifically discuss only college teaching. The sage-on-stage style of lecturing has come under attack in the current mathematics education reform, but the attack seems to show no awareness of the basic constraints of **(i)**–**(iv)** above. For example, if the amount of material to be covered in a course can be greatly reduced (thereby violating **(iii)**) and students are expected to spend 8 years in college (thereby violating **(ii)**), then we can all safely abandon the lecture format and engage in a wholesale application of the guide-on-the-side philosophy in our teaching. To put this comment in context, let us continue with the above discussion of the polynomial division algorithm by considering it specifically as a topic in a junior level algebra course.

The purpose of a mathematics course is, naturally, to further students' knowledge of mathematics and logical reasoning skills, but there is also a practical aspect along the line of assumptions **(ii)** and **(iii)** above. Thus a junior level algebra course should enable its students to acquire a minimum mastery of the most basic techniques and ideas in algebra: the concepts of generality and abstraction, the concept of mathematical structure, and certainly the basic vocabulary of groups, rings and fields. The details may vary and the broad framework is susceptible to a certain amount of stretching (cf. [12]), but ultimately the course must serve to fulfill assumptions **(ii)**–**(iii)**. Students coming out of such a course should be ready to embark on more advanced courses in mathematics and the sciences, deal with the basic technical issues in industry, or at least be able to look back on the high school materials of polynomials, triangle congruence, or fractions with renewed understanding. Given that such a course would typically meet for only 45 hours (a semester), class time must be used wisely. This is the reason why only half a lecture can be allotted to the explanation of the polynomial division algorithm.

Learning mathematics is a long and arduous process, and no matter how one defines "learning", it is not possible to learn all the required material of any mathematics course in 45 hours of discussion. To make any kind of teaching possible, professors and students must enter into a contract. The contract can take many forms, but the following would certainly be valid: The professor gives an *outline* of what and how much students should learn, and students do the work on their own outside of the 45 hours of class meetings. Lecturing is one way to implement this contract. It is an efficient way for the professor to dictate the pace and convey his vision to the students, on the condition that students would do their share of groping and staggering toward the goal on their own. It should be clear that without this understanding, lectures would be of no value whatsoever to the students, especially to those who expect to come to class to be spoon-fed all the tricks for getting an A in the course. In advocating "guide-on-the-side" over "sage-on-the-stage", did educators weigh carefully the intrinsic merits of the lecturing format against the apathy of the students before putting the blame squarely on the former? Or is this simply a case of expediency over reason, because there are hidden forces at work which the educators did not bring to the table? Have they perhaps re-defined learning without telling us what they *really* have in mind? If so, then this is an illustration of what I call the *romantic tradition* in education writing: Unpleasant details are left to the imagination because they might interfere with the attractiveness of the advocacy in question.

Let us return once again to the polynomial division algorithm for a more detailed discussion. *Lecturing can take many forms.* In the 25 minutes or so allowed for the teaching of this topic in a junior level mathematics course, one way to approach it in the classroom is for the professor to indicate briefly the long process of possible trials and errors in arriving at the correct proof and to discuss the main points of the proof in precise terms. Most of the 25 minutes would therefore be spent on explaining the details of the induction on the degree. In order to understand such a lecture, students will have to retrace *on their own* the steps of the trials and errors omitted in class (see the discussion of the tutoring session in the preceding section). There are other ways to handle the lecture. For instance, if the textbook is reliable and readable, the professor may decide to let the students read the polished final account at home but use the class time to go through, as much as possible, the tedious learning process in the allotted 25 minutes. Or, the allotted 25 minutes of class time could be used to go through the initial segment of the trail and error process and, following which, students are told what they need to do in order to complete the investigation. For this kind of teaching to work, the students would have to be very mathematically mature. There can be other variations too. No matter. The fact remains that if we abandon lecturing and the underlying assumption of the sharing of labor between professor and students, and insist that the whole learning process (guided discovery, trials and errors, etc.) must take place within the 45 hours of class time, then the amount of material that can be covered in each course would be reduced by half if not more. Unless we stretch college education to 8 or 10 years, this is not a realistic option.

Last semester (Spring 1998) I taught a one-semester introductory analysis course, and I volunteered to give two additional problem-solving sessions each week. Attendance in these sessions was optional, and therefore poor.[4] Since in these sessions time pressure was not a serious concern, I could often indulge myself in my teaching method for private tutoring (see the preceding section). I did not insist on any kind of cooperative learning, but I let them decide for themselves if they wanted to discuss with their neighbors. Once I had about seven students, and I asked them to prove that the function $f(x) = \sqrt{x-5}$ is continuous at $x = 10$ by the use of ϵ and δ. Of course this requires a rationaliza-

---

[4] I have volunteered these problem sessions often, and attendance has always been poor. Typically about 15-20%. I wish to let this fact be known.

tion of the expression. The trick in question happened to have been discussed briefly several weeks before in the context of the limit of sequences, but it would appear that none of them had any recollection of it and, in any case, they could not make the connection.

I walked around the room, talking to each of them trying to coax at least one of them to come up with a reasonable plan of attack. After more than ten minutes of futility, it became clear that they had to be told. So I mentioned the word "rationalization", and one student immediately caught on. A few more explicit hints later, all of them knew what to do. The actually doing, needless to say, took a while at that stage of their mathematical development. I looked at each student's work and literally guided the hands of a few of them. After more than 15 minutes, they all got it done. Then I asked one of them to go to the board to give a complete exposition, and I followed with some general comments, partly to point out the pitfalls along the way and partly to bring closure. They probably learned something from the experience, but it must be pointed out that it took almost the whole 50 minutes of class meeting. At the risk of harping on the obvious, whatever might be the educational benefits of this way of teaching continuity, an introductory analysis course taught entirely—or just *frequently*—this way could hardly get off the ground.

There is another aspect to lecturing that deserves to be discussed. Lecturing allows the professor to share his insight with students beyond what is found in the textbooks. Again allow me to offer an example from my personal experience. Each time I teach calculus, I go through what I have come to call the "catechism of π". Most students believe they know what π is. To my question of "what is π?" usually comes the reply "circumference divided by diameter". So I ask "what is circumference?", the quick rejoinder of "2π times radius" is usually followed by nervous tittering. They know they have been had. Sometimes the catechism replaces "circumference" by "area", but the result is of course the same. Some years ago, I decided to address this issue directly by defining for them, right after the discussion of arclength, the number π as half the circumference of the unit circle:

$$\pi \equiv \frac{1}{2}\left(2\int_{-1}^{1}\frac{1}{\sqrt{1-x^2}}dx\right),$$

and then proceeded to *prove* for them that with this definition of π, the area of the unit disc is actually equal to π, *i.e.*,

$$\int_{-1}^{1}\frac{1}{\sqrt{1-x^2}}dx = 2\int_{-1}^{1}\sqrt{1-x^2}dx.$$

This requires an integration by parts argument. Could I have guided my students on the side to the final conclusion? For 20% of them, maybe—that is just my guess— and only if I had two lectures at my disposal. But I had only half a lecture (25 minutes) to work with, because in the remaining 25 minutes, I also showed that the circumference of a circle of radius r is $2\pi r$, defined the *radian* measure of an angle correctly for the first time, and computed the arclength of a segment of a parabola together with an explanation of why Archimedes could find the area enclosed by the segment but not its arclength (he had no logarithm function, only polynomials). Such a pace is normal for a calculus course at Berkeley.

Now I do not wish to give the impression that each time my colleagues and I give a calculus lecture we always have this kind of interesting information to offer. How can this be true when so much spade work must be done in a basic course of this nature? Nevertheless, it is no stretch of the imagination to say that most of these lectures routinely carry elementary insights about mathematics which comes only with years of working at it. I see no reason why students cannot profit from such insight by making an effort to understand the lectures instead of adamantly resisting them by coming to class unprepared, or for that matter, leaving it without making an effort to understand it later. The teaching of calculus varies in each university, but to any conscientious lecturer, the preceding account must resonate to some extent. We hear that lectures are a relic of the past. Does this mean that it is improper for students to pick up essential information because they *have to* "discover" it by themselves? Or is it the case that, since students no longer want to put in the strenuous effort to learn, the universities must henceforth resolve to teach either the most simplistic aspects of mathematics or only the smallest possible amount consistent with the guiding-on-the-side philosophy?

For advanced (upper-division) courses in mathematics, the professor's understanding and vision of the subject are even more important in providing proper guidance to students—recall that we are assuming a large amount of material is supposed to be covered in each course (assumptions **(ii)** and **(iii)** at the beginning of this section). This is especially true in view of how textbooks are written these days (cf. [12] again). Lecturing is not the only way to provide this guidance but, until there is data to prove otherwise, it is one way of doing it. However, since the lecturing format is most heavily criticized in the context of the teaching of calculus, let me continue with the example of π and discuss calculus lectures. It has been said that the typical calculus lecture is *worthless, because it has virtually no conceptual development in it, is boring, and focuses mostly on techniques.* If this a judgment on the average calculus instructor's pedagogical or mathematical deficiency, then it has no bearing on our present discussion of lectures *per se*. On the other hand, learning about a correct definition of π and understanding for the first time how π enters into the circumference or area formula certainly gives a good account of the conceptual development in mathematics, makes interesting topics for students, and convincingly demonstrates how technique is inseparable from conceptual understanding in mathematics. Such being the case, it is clear that we have not even begun to exhaust the potential of lecturing. The sage on the stage still has work to do.

3. Time-Compressed Instruction

In the preceding sections, I have repeatedly emphasized the *time-compressed* nature of classroom mathematics instruction. In order to learn what is taught in class, students must be willing to spend two to three times the amount of time by themselves. For example, a survey conducted by the PDP (Professional

Development Program) unit on the Berkeley campus in the 1980's shows that those students who got A's and B's in freshmen calculus spent an average of 10 to 14 hours on the course material per week outside of regular class meetings. To put these figures in context, since all the calculus courses are only 4 units each, conventional wisdom would have students spend only 8 hours per week instead of 10 to 14 hours. As another example, a recent article ([10]) makes the following comparison between the work habits of Japanese and American school students:

> Another after-school activity that occupies the time of adolescents is homework. Great emphasis is placed on homework as the basis of the excellent performance of Japanese adolescents in mathematics and science. In a typical survey, therefore, one might ask high school students how many hours they spend doing homework each day. The answer often given by Japanese students is unexpected: none. Only by pursuing the topic further does the actual state of affairs become clear. Additional discussion and questioning reveals that homework is often not assigned, but high school students are expected to spend several hours a night reviewing the day's lessons and anticipating the lessons for the following day.
>
> It is increasingly common in both middle schools and high schools in the U.S. that homework is done in school and simply represents work that teachers expect to be done before the next class meeting. The apparent lack of homework assignments was lamented by American parents and teachers. Parents questioned how their children could complete their homework during school hours, a practice very different from what they remember of their own school days. Teachers were concerned about the tendency of students to equate homework with studying; if there was no homework assignment, there was no studying.

Would it be fair to conclude from this that the bashing of lectures in the U.S. is a direct consequence of the infusion into our campuses students who are rarely asked to work outside of class all through K–12? In an inspiring address commemorating the centennial of the *American Mathematical Monthly* [11], Herbert Wilf bluntly stated: "In recent years, we have witnessed serious decline in the demands that we make on our students for intensive and solid intellectual achievement in mathematics. When we feed them more baby food every year, we thereby become accomplices to their intellectual softening." Wilf did not concern himself with the abolition of lectures, but he might as well have.

There are at least two special features about mathematics that dictate the time-compressed nature of mathematics instruction in the classroom: It is *cumulative* and it is *precise*. By cumulative, I mean that at any given point of a mathematical exposition, it is virtually impossible to understand what is taking place without first acquiring a thorough understanding of all that has gone on *before* that point. The failure to confront this rather brutal fact—in a mathematics class, once behind, forever behind—may be the single factor most responsible for the undoing of our mathematics students. The precision of mathematics stems from its abstract nature. Whereas even in a rigorous discipline such as physics, a photograph or a measurement by a laboratory equipment can render verbal explanations superfluous, the basic concepts of mathematics reside only in the realm of ideas and therefore must be meticulously described. Students must learn to concentrate fully on *every* word that is used in the description, or they run the risk of missing the point entirely. We all remember how as students we had to struggle with the seemingly innocuous quantifiers "for every" and "there exist" in the definitions of limit and continuity. And that is only for starters. The precision in mathematics *is* unforgiving indeed.

These two features make it difficult for students to learn from mathematics lectures if they are unwilling to also invest time and energy before or after the lecture for this purpose. In a videotape ([1]) made available by TIMSS (Third International Mathematics and Science Study) in 1997, two Japanese teachers were shown to give lessons—unequivocally based on direct instruction—with the rapt attention and active participation of their students.[5] Now that we have the preceding account ([10]) of the kind of preparation Japanese students routinely make before coming to class, we are finally in a position to understand why the teaching of mathematics in Japan achieves such good results and why their students always score so well in international tests. The last time I checked, the slogan of "guide-on-the-side but not sage-on-the-stage" has not been aggressively promoted in Japan.

4. The Importance of Being Honest

The folk wisdom that there is no free lunch in this world seems for some reason to be missing in current education writing, and this fact seems to be the genesis of the *romantic tradition* mentioned in §2. Wonderful new prescriptions for ailments of long standing in education are given on a regular basis with nary a hint of the likely detrimental side effects. On the K–12 level, for instance, "real-world" relevance of mathematics has been trumpeted as the salvation of the school mathematics curriculum *without* the caveat that unless this is done in moderation, the abstract nature of mathematics as well as its internal coherence would be jeopardized. Sure enough, the worst fears were realized in most (if not all) of the recent school mathematics texts, which emphasize "real-world" relevance (cf. p. 1535 of [8] and the references given therein).

The advocacy of the abolition of lectures as we know them is another case of promoting an idea without any explicit warnings of the possible losses and gains. For example, a more balanced approach to the subject of lecturing might begin by listing its strengths, its weaknesses, and the range within which it would be effective. A summary of the preceding discussion would include the following among the strengths of lecturing:

> **(a)** It allows the instructor to set the pace of the course. This is an important consideration if the basic parameters of school and college education as we know them are to remain intact. See assumptions **(ii)** and **(iii)** of §2.

---

[5]In particular, no cooperative learning there.

(b) It allows the instructor to share his insight into the subject with students. If we still believe that education is the process of passing the torch from generation to generation, this too is an important consideration.

A side benefit of the lecturing format, one that has not been discussed thus far, is that it forces students to stretch their concentration spans. In these days of MTV when everything is interactive and instantaneous, such a beneficial effect should not be dismissed lightly. In fact, one can speculate on the possible correlation between the onset of the computer-age and the dissatisfaction with lectures. As to the weaknesses of lecturing, the most serious is that, unless students are willing to do their share of the work outside the class and meet the instructor halfway, lectures are a waste of time. It is possible that this aspect of the lecturing format has never been made explicit to a large percentage of our college students, and the dismal student performance in mathematics courses is the result. The guide-on-the-side advocacy then becomes a facile, one-dimensional response to a multifaceted challenge. One reason why lecturing has been the accepted mode of instruction in most universities for so long is probably the assumption that students are there to work. Are we at the dawn of a new era when even such standard assumptions must be re-examined? Perhaps universities can no longer survive as institutions of higher learning but must transform themselves into "caring, nurturing", glorified high schools.

We now come to the guide-on-the-side method of instruction which, as mentioned, means the guided discovery method *in the context of cooperative learning*. What are its implicit assumptions and what are its strengths and weaknesses? By transferring what used to be activities outside of class into the classroom, the discovery-via-cooperative-learning pedagogy tacitly assumes either that students can no longer be trusted to do their share of the work or that they are incapable of doing it. The great advantage of this method of instruction lies in its seeming ability to make mathematics accessible to a much wider audience than is possible in the lecturing format.[6] The slower students who do not wish to put much energy into a mathematics class would certainly find participating in cooperative learning more congenial than listening to lectures. On the debit side, guided discovery and cooperative learning slow down the pace of a course, at least by half. One may surmise that the authors of some textbooks which advocate this particular pedagogy are well aware of the attendant loss of class time, and therefore deliberately set out to cut down on the more mathematically substantive topics. Thus we find calculus texts which do not even present the proof of something as basic as the Fundamental Theorem of Calculus (cf. [7] and [9]). Another drawback of this particular pedagogy has also been discussed: A guide-on-the-side has fewer occasions to share his vision or insights with the students than a lecturer. Those who would otherwise profit from the knowledge and experience of their instructors end up being short-changed by this pedagogy. If we look past the heroic efforts of a few extremely talented instructors, it would be

---

[6]But by no means to ALL students. I will not reproduce here the oft-repeated anecdotes about how some members of a study group sit and do nothing while one or two members take charge and do all the discoveries for them.

fair to say that students in currently advocated programs of guiding-on-the-side typically learn the details in a small area but not acquire much of a perspective overall.

In this context, an additional comment about the possible omission of topics in a guide-on-the-side classroom may not be out of place. It is a fact that American high school graduates are among the least mathematically knowledgeable compared with their counterparts in nations that did well in TIMSS (cf. [2]–[4]). We can also verify directly from our own collective experience that American students are generically the least prepared among our graduate students. Would it not be fair to say that our undergraduate mathematics curriculum is already down to the bone and has no more fat to be trimmed?

The preceding discussion of lecturing and its common alternative is by no means exhaustive, but even this much critical analysis would have been beneficial to the current debate on the mathematics education reform. For instance, where in this advocacy for guided discovery in the classroom do we find an explicit reference to the underlying assumption about the students' unwillingness or inability to work on their own? (Consult [11] again.) Or is it the case that this assumption is a misapprehension? These issues should have been openly debated long ago so that teachers who opt for one or the other pedagogy would have the benefit of knowing what they are getting into. It would be wrong to say that this advocacy has produced nothing of value thus far. Quite the contrary. Because of this advocacy, some of us who had to struggle to become mathematicians—and have always assumed that all students must know the need to do the trials and errors on their own—have been awakened to the fact that we must *tell* the students of this need or even *demonstrate* to them this need by use of examples. But given the human tendency to oversimplify, the danger of a passionate advocacy in a subject such as pedagogy—which is far from a hard science as of 1998—is that blind acceptance and a reckless pursuit would inevitably follow. The classic dictum that if a little bit is good then a lot must be better unfortunately applies only too well in this situation.

Lest this article sound like a defense of the status quo of the lecture format, let it be said—however briefly—that perhaps the quality of some lectures does raise legitimate concerns. There are lecturers who fail to observe the basic etiquette of lecturing (cf. §§1.6 and 2.13 of [8]), and there are also those who still cling to textbook-writing-on-the-boards as a legitimate form of lecturing in spite of the present super-abundance of adequate textbooks on almost every standard topic. For lecturing to survive, the practitioners of this art must continue to be vigilant (see assumption (i) of §2). Nevertheless, the over-riding fact remains that the current discussion of pedagogy fails to meet the most basic requirements of scholarship: Any advocacy should state clearly its goal, its benefits, and its disadvantages. In this light, the advocacy of the guide-on-the-side pedagogy has been presented more like an info-mercial than a scholarly recommendation. It is all good and nothing bad could possibly come of it.

In the field of medicine, the FDA has made the listing of the precise range of applicability and the side effects of each drug mandatory. Would it be too much to ask that the same consideration of fairness be also extended to teachers so that all education writings are always accompanied by an analysis of the limitations of a particular proposal, including its drawbacks and the conditions under which it would not be applicable?

Acknowledgement. I wish to thank Ralph Raimi, Dick Stanley, and Andre Toom most warmly for their invaluable criticisms of an early draft of this article. In addition, Roger Howe's pithy observations led to significant improvements.

References

[1] The Third International Mathematics and Science Study Eighth-Grade Mathematics Lessons: United States, Japan, and Germany, *videotape*, 1997.

[2] *Mathematics 1: Japanese Grade 10*, K. Kodaira, ed., Amer. Math. Soc., 1996.

[3] *Mathematics 2: Japanese Grade 11*, K. Kodaira, ed., Amer. Math. Soc., 1997.

[4] *New Elementary Mathematics: Syllabus D*, Volumes 1-4, W. H. Yoong, ed., Pan Pacific Publications, Singapore, 1991-4.

[5] Robert B. Davis, What mathematics should students learn, *J. Math. Behavior*, 13 (1994), 3-33.

[6] Robert B. Davis, One very complete view (though only one) of how children learn mathematics, *J. Research in Math. Ed.*, 27 (1996), 100-106.

[7] Deborah Hughes-Hallet et al., *Calculus*, John Wiley & Sons, 1994.

[8] Steven G. Krantz, *How to Teach Mathematics*, 2nd edition, Amer. Math. Soc, 1999.

[9] Donald R. LaTorre et al., *Calculus Concepts*, Preliminary Edition, D.C. Heath and Co., 1997.

[10] Harold W. Stevenson, A Study of Three Cultures: Germany, Japan, and the United States—Am Overview of the TIMSS Case Study Project, *Phi Delta Kappa International, Online article*, http://www.pdkintl.org/kappan/kste9803.htm

[11] Herbert S. Wilf, The wind, the trees, and the flame, *Focus*, 14 (1994), 26-27.

[12] H. Wu, On the education of math majors, in *Issues in Contemporary Mathematics Instruction*, E. Gavosto, S. G. Krantz, and W. G. McCallum, (eds.), Cambridge University Press, to appear.

[13] H. Wu, The mathematician and the mathematics education reform, Notices AMS, 43 (1996), 1531-1537.

Teaching Freshmen to Learn Mathematics

Steven Zucker

Johns Hopkins University

From a mathematics instructor at another university:

"I have just finished reading your article [T.U.L.] in the Notices of the AMS.[1] I absolutely agree [with] every word. I taught some undergraduate courses [here] getting extremely good student evaluations. First I was very surprised, then the next semester the ratings were even higher ... and I still didn't understand.

"Now in the light of your article it makes sense: I taught like this were a high school."

From *The sum of mediocrity*, an article on pre-college education by Pat Wingert, in the December 2, 1996 issue of *Newsweek*:

"We expect less from our students, and they meet our expectations."

One thing that should be happening during the first semester of the freshman year is that the students start to figure out how to learn on their own, and get beyond skimming the surface of the subject. This is the main academic adjustment that most students must make when they get to college. We should therefore insist that they do it. This could involve running our calculus courses in a way that is very different from what they are used to from high school. A simple illustration of this is insisting that the students read the textbook, both for concept and examples; we will get back to this later.

At the beginning of the Spring '97 semester, I was talking to a student from my Fall '96 Calculus II (Physical Science and Engineering) course. I had gotten to know her through visits to my office hours for help in getting started on some of the assigned problems. For instance, at the beginning of the course she was unable to make sense of the problem: Show that when n is a multiple of 4, the sum $1 + 2 + 3 + \ldots + n$ is an even number. Later, she could not figure out how to deal with a sequence defined by a simple two-step recursion. After failing Exam 1, she told me about being panicked during the exam, and she sought my advice. I told her to engage a more secure comprehension of the material, so that it wouldn't get lost under the strain of examination conditions. Her performance on subsequent exams was satisfactory, and her course grade was a C. (I would describe the exams as pretty straightforward, but very thorough.) This is just one example of evidence that the course is being run at a reasonable level for our students.

I felt that my course had helped her substantially, and I sought confirmation from her. What she told me at first I'd heard before: she didn't like what I was doing in the beginning. *After the semester was over*, she said, she had looked back and realized that the way I conducted the course forced her to learn how to

---

[1]The article was titled *Teaching at the University Level*. It appeared in the August 1996 issue of the Notices. Letters to the editor about it were published in the November 1996 issue; a letter of mine appeared in the December 1996 issue. We reprint as an appendix to the present article the appendix from T.U.L., *Academic Orientation* [A.O.], with mild editing. Item #7 from A.O., omitted in the August issue, has been restored and enhanced.

learn. Moreover, I was astonished to hear, mine was the only one of her courses
that did![2]
     In a way, she was a good student. Before continuing with this, I want to
describe the backdrop. The students in the course were almost all freshmen with
A.P. credit for Calculus I. The review of the most pertinent material from Calculus I was left to the students in the first homework assignment. The handling
of integer variables, material that would become relevant later for sequences and
series, was treated. An important thing presented in the first week of lectures
was the mathematical usage of "if/then" and "only if"—a majority of the students seem to think at first that "if" means "if and only if"—so they could read
the textbook correctly. A good way to confront the issue is by showing the two
sentences from conversational English:

  a) *If it stops raining, I'll go to the store.*

  b) *If I win the lottery, I'll buy a new car.*

These have parallel structure but different connotation. It is not hard to convince
the students that we cannot afford such ambiguity in mathematical writing.
     The assignment for the next week consisted of three problems designed for
"consciousness-raising" (in a course where the students had been told to expect
8 hours of work per week outside of class). One of these was the problem about
$1 + 2 + 3 + \ldots + n$ mentioned above. A better one, I think, was

True or false, and explain fully (i.e., verify or give a counterexample):

  a) $f(x)$ *is a rational function only if* $\int f(x)dx$ *is rational.*

  b) *If* $f(x)$ *is a rational function,* $f'(x)$ *is a rational function.*

The answers to these are familiar from Calculus I experience, but few of the
students had ever thought about the processes of differentiation and integration
in the large. To guide them, I went through the verification that the sum of
two rational functions is a rational function. Also, I explained in lecture what
is meant by a counterexample. They were instructed to write up their solutions
carefully. Most of the students discover that they can do such problems if they
persevere. While it seems hard to squeeze them in, problems of similar depth
should be given as the course progresses.
     To return to the student in my office, I asked her to evaluate $1+2+3+\ldots+n$
in order, for $n = 1, 2, 3, 4$, and emphasized the parity of the answers. That already triggered something in her head, and she was prepared to persist with the
problem. A bit later, she reported something very "unusual": As she spent more
and more time with it, she saw her understanding start to grow. I assured her

---

[2]A likely explanation for this is given, more or less, in T.U.L. Untenured faculty are afraid
of incurring complaints about their teaching that could hurt their careers. Tenured faculty
have lost their will to stand up to student indifference and resistance; besides, doing so can
also cost in salary raises, given the way teaching is often evaluated.

that this was very normal. The expectation that the answer is either there or not there is one of the many misconceptions that freshmen have about mathematics.

I have been teaching that Calculus II course every fall semester since 1993. In '94, I changed my attitude toward my role as instructor,[3] substantially increasing the amount of thought during the preparation and energy into the delivery of the lectures, and additional effort in printing up handouts to supplement the lectures as needed. But I also expected the students to match it with increased learning; the aspiration had become command of the material. Though the results were good, there was still a lot of grumbling in the class. Indeed, the libelous review of that course in the student course guide led to the shutting down of that publication. My aspirations have not changed much since 1994, though my understanding of the educational issues has.

One of the students in the 1994 course, who also reported being unhappy at first, wrote of the lectures, "He made the material very easy to understand, if and only if you were doing the work necessary to keep up with the class." I think he was trying to convey the message that a student who was not keeping up would be unable to see that the material was being explained in a clear and helpful manner. It reflects poorly on the current state of affairs that I started to wonder whether that was fair! With the support of my department chair, I decided that it was. I was encouraged further by the dean of the college, now provost (whose academic credentials are in the humanities, by the way).

I remember vividly when, during the second or third week of the 1993 course, a student came to see me, feeling that he was hopelessly lost. I probed with a few questions, after which both of us could see that he was very close to understanding the material. He, like so many other high school graduates, had been trained to absorb mathematics in tiny controlled doses, which are to be memorized and later regurgitated. It is no wonder that the suggestion of learning concept often gets perceived by students as irrelevant theoretical digression, rather than the means to better comprehension that it is.

It struck me just a couple of weeks before Fall '95 began that there was no clear way for the students to figure out what it was, so different from their high school experience, that I wanted them to do. Some students make the transition instinctively,[4] while others simply blame the instructor for teaching poorly. This led me to do some serious orientation for my own students that year. It was a nice coincidence that during the first week of classes, I crossed paths with a senior who had taken Calculus I from me in his freshman year. I started telling him about my ideas on academic orientation. At one point he said to me "You have to remember that they are freshmen, and that they don't realize that whatever they think of their instructor, they'll be learning most of the material on their own anyway." I recall that I came back with *"Why don't they know that?!"* If the high schools and older students aren't communicating that, then it has to come from us.

---

[3]I should thank my colleague W. Stephen Wilson, who was Chair at the time, for pushing me in this direction.

[4]A strong student reported that it took her a couple of weeks to get into stride.

I then pressed hard for academic orientation in the university. For obvious reasons, it is better that the message have the ostensible endorsement of the university: of the math department and the academic deans. I ended up giving a presentation on mathematics during Orientation Week, 1996, a first. At Hopkins, it comes down to communicating to the entering students what most sophomores, and virtually all juniors and seniors, know about education in college. This involves making the new rules explicit (see the appendix Or-2 *Academic Orientation* [A.O.]), and debunking the misconceptions that so many of the freshmen bring to college. Appended as Or-1 is a compilation of such misconceptions. The first five items were discussed first to help soften the impact of the potentially startling A.O., which was circulated about half-way through the presentation.

I am convinced that the level at which my course is now conducted is about right (for its audience). If we assume that this is correct, it is hard to justify on educational grounds running the course at a lower level. My university has 900 to 1000 entering students each year, who are ambitious (at least in the abstract) and "bright" (mean SAT Math score a little over 700). Most of them did not have to work hard in high school; here lies a big part of the problem. One of my 1996 students, who *was* used to having to work hard, reported what another student had said: "I'm so pissed off! I got a *C*–. I couldn't believe it. I never worked in high school, and I always got *A*'s!" And that was *after* orientation! Nobody said that our task is an easy one.

Because some students can't, most—around 85%—of our students were never asked to learn mathematics in high school by reading from a textbook. They grew accustomed to picking up the material from classroom presentation alone, even in A.P. courses. But in fact, the students who attend a good college are capable, with some exertion, of reading the book; we therefore want them to do it. For some, it's a struggle. Even the stronger students will encounter things in the book that they can't, or just don't, figure out. However, there are numerous ways they can have this straightened out, viz., lecture, section meeting, TA office hours, help room, professor's office hours, discussion with classmates, ... Given the advent of extended help room hours, I find even more outrageous the suggestion that the lectures be aimed at those who want to skimp on their effort in learning the material.

When college students say they want a good teacher, some want a good educator, one who will help them to make it *through* the material. They accept that inherent to mathematics is the need to decide what to do to solve a problem, to make distinctions and choices, to reject ideas that are fallacious, and to persist when one's first attempt doesn't work. This flexibility[5] is often absent from high school experience, where mathematics is taught largely by repetition. As such, they were trained to learn the subject *inflexibly*. Many students want it to stay that way in college; they want to be helped *around* the material, in effect, bailed out by the instructor from having to understand it. Pandering to the latter group is slowly but surely eroding the quality of mathematics education in American

---

[5]Intellectual flexibility is a notion that makes sense in *all* disciplines. As such, it is wise to talk about it when we explain to educators in non-scientific fields what we want from our students in mathematics courses.

colleges and universities, and even abroad.

An interesting thing I learned in 1996, when the course was divided into two lectures, with the other run in tandem with mine by a remarkably skillful assistant professor, was that the freshmen here will praise an instructor who runs the course at a high level provided the lectures are well-organized, focused, and "self-contained". However, it does not follow that they learn better; his students and mine performed comparably on exams (cf. misconception #13). All too many students want to come into class cold, expecting to get "taught" by the professor, from the ground up.[6]

What is the point of the instructor's commitment of time and effort to attempt to supply teaching that will be judged to be "better" when it is not getting matched by demonstrated better learning? I would go further. *No style of teaching mathematics can substitute for insisting that the students pick up their share of the work,* unless one is willing to compromise standards. We should stop seeking panaceas that place the burden on the competent instructor; I do not believe in the "Fountain of Education", and it is time to stop looking for it!

How can we fulfill our role as educator? The main theme is that one must aim for the determination of the appropriate level for the course, one that matches the level of the students who take the course *in one's own university.* Of course, I do not mean here the level that we see when the students follow their high school instincts, thinking that they won't be reading the textbook, or equating learning to memorizing a list of formulas, or declaring that spending three or four hours on homework is a lot of work. We must then have the conviction, and ideally the backing of the department and the administration, to hold to that level. However, in doing this we also put a greater obligation upon ourselves, for the threshold requirements[7] for the instructor, in giving a course at a higher level, are correspondingly higher.

Academic orientation is necessary, of course. The students must be told how things are going to be different from high school. This is more likely to succeed if there is a web of support that will block the students from resisting "because their professor has these crazy ideas." As reasonable as we may find the statements of *Academic Orientation,* most freshmen are shocked by them; they even wonder if it's for real. It doesn't help matters if some of our colleagues teach calculus as though it were "grade 13 of high school", an easy way to endear oneself with freshmen.

On the other hand, many students *have* heard in high school such things as "Don't just memorize. Learn concept." However, they found that they could score well on tests by ignoring this advice and behaving as they were advised not to. That must stop in college. Another big problem is that many freshmen wrongly believe that adjustment is needed only for students other than (weaker than) themselves.

---

[6]That would serve to keep the level of the course down, unless the students are expected to pick up a lot more as they read the textbook later. Students here are capable of learning the easier things in the course largely by themselves, and I remind them of that.

[7]This notion is mentioned in misconception #13. I know no algorithm for determining where the threshold is for a given level of aspirations and a given student body.

To put the preceding into effect, we must be free to give the students what they need, not what they say they want. But if we do so, we risk lowering our ratings in the course evaluations,[8] for students who retain their belief that they are entitled to do well without exerting themselves are not going to be happy. This places us in the ironic situation that we might be penalized just for doing our job conscientiously.[9]

I'll summarize what I have been doing toward upgrading the freshmen's expectations.[10] When I gave a forceful presentation in my course in 1995, the Department had circulated J. Martino's *Survival Guide* to all students taking a large lecture course, and that reinforced my message nicely. In Fall 1996, my presentation was on the Orientation Week program, and that "implied" backing by the University. In Fall 1997, my previous experience enabled me to carry out efficient major orientation for my own course. I appealed to something I had ascertained was mentioned by deans of both the College of Arts & Sciences and the Engineering School in their addresses to the freshmen: the amount of work outside of class expected in a college-level course (cf. *Academic Orientation*, #3). After reminding the students of that during the first lecture, I told them, "I don't want to hear any griping about the workload in this course unless you are consistently putting in more than 8 hours a week. And if you are not highly talented, you may decide that you *want* to put in more." I feel that I've told the students this year everything I might want to say in the way of orientation.

Above all, the students should get the message that we intend to help them learn the material, but we are not going to bail them out if they don't. The exams in the course must uphold the level announced for the course. They must be made up so that they (de facto) penalize students who insist on operating on the basis of high school notions that we have declared inappropriate. The problems should be taken from the heart of the material, not from the surface (as students learn to expect in high school). No practice exams to suggest programming. I tell them that doing new problems of sufficient difficulty with books closed comes closest to simulating the exam situation. In lecture this year, I asserted before the first exam, "If you take all of the homework problems, examples done in lecture and the book, problems from past years' exams [available to the students], the problems on the exam will be different from all of them; but they can be done by the same methods."

The messages of academic orientation must be reinforced throughout the semester, and frequently in the first part of the course. Here's a sample of what my class heard or read this year:

— Talent and background will make some difference, but *you are going to have to work in order to succeed*, both in this course and in pursuing your career

---

[8] Actually, my own ratings went up from 1993 to 1994.

[9] I know that some (many?) past students (including, of course, the one mentioned at the beginning of the article) came to realize how much they benefited from my course only *after* the course was over, e.g., while they were taking Calculus III. "The world rewards the appearance of merit oftener than merit itself." –F. de La Rochefoucauld

[10] What one can do in the classroom is based in part on one's nature; it is unreasonable for ourselves or others to regarded us as teaching machines. Also, it may be relevant here that I had control of the entire course.

goals. If you choose to shoot for less, you do so at your own risk.

— The goal is to reduce your dependence on the instructor.

— Think about it after class. You should know by now that you don't *have to* understand it here.

— It's *impossible* for me to explain that to you. Some things you must try to figure out for yourself.

— Mathematics is about concept, attitude and control.

— The purpose of the exam questions is to determine the extent of your command of the material. Though you should be getting the correct answer to the problems if you have good command, it is not the main point. After all, if I only wanted to see the right answers, I'd just do the problems myself!

A strategic point: We shouldn't overlook the power of negative reinforcement. In Calculus II, the following lines of "reasoning" are all too common:

a) Determine whether the series $\sum_{n=1}^{\infty} \frac{1}{n}$ converges. *Well, $\frac{1}{n} \to 0$, so the series converges.*

b) Compute $\lim_{n\to\infty}(1 + \frac{1}{n})^n$. *Well, $1 + \frac{1}{n} \to 1$, and 1 to any power is 1. The limit is 1.*

These are not just silly mistakes; they are fundamental errors that show disrespect for the methods of the course and for the instructor. (Why didn't we teach them the *"easy"* way to do it?) Indeed, it is essential to make the student see that these are wrong.[11] I term **a)** "the ultimate sin", and **(b)** the penultimate; it is announced that committing the penultimate sin in a problem gets an automatic zero credit, and the ultimate sin gets *negative* credit. These sins occur with surprisingly low frequency in my course; the class performs better now with infinite series.

I should mention that the immediate reason for my writing "Teaching at the University Level" was that I felt the calculus reform movement was gaining too much momentum. The notion that students at universities like Harvard and Stanford needed reformed calculus was out of line with my own observations. Moreover, it is unlikely that today's college students are less intelligent than their predecessors; rather, something has happened to their sense of learning. While the transition from high school to college has always been a hurdle for students, today's students find themselves in a weaker position to deal with it. Since I also refuse to believe that they have been irreparably damaged, trying to repair the damage makes far more sense than pretending that the subject needs "genetic

---

[11]Students sometimes object to the idea of showing them that a way of "thinking" is wrong (rather than programming them with the right way to do the problems). I'm quite sure that every math instructor has the experience that students, having been shown the right way, make such mistakes anyway.

engineering". In particular, we don't have to wait for the difficult underlying problem with pre-college mathematics education to get resolved before we can start to remedy the situation in the colleges.

It is not my intention to condemn the calculus reformers outright. But if we are to get serious about resolving the difficulties our students are having in learning mathematics, we must first address the issues that really matter most, namely the low expectations that many of them hold when they arrive in college and their overestimation of the effort that they are putting in. Only then does it make sense to judge the merits of different methods of instruction . . . for a given group of students.

Appendices: Orientation material

Or-1. Misconceptions (and Rejoinders)

A. Common Misconceptions about High School and College

1. *Only a jerk could get less than a B-grade in a course!*

If that's true, then nearly half of the freshman in math courses in previous years are jerks!

2. *I had a good math teacher in high school who taught at my level.*

But most of the freshmen admitted here say they were in the top quarter (say) of their math class, yet agree that a teacher [in high school] is supposed to make sure that even the weakest students in the class learn. Now, whose level was the course run at?

3. *I did well in math, even calculus, in a good high school. I'll have no trouble with math at Hopkins.*

There is a different standard at the college level. The student will have to put in more effort in order to get a good grade (or equivalently, to learn the material sufficiently well by *college* standards).

4. *My AP calculus course in high school was like a calculus course in college.*

The student here is expected to do much more learning outside of the class-room; see also #3 above. (On the other hand, the Advanced Placement Tests are college-level exams.)

5. *College will be like high school, just a little harder.*

(See all of the above)

B. Misconceptions about Learning Mathematics

6. *In a calculus course, theory is irrelevant, for what is really at stake is doing the problems. The lectures should be aimed just at showing you how to do the problems.*

We want you to be able to do all problems—not just particular kinds of problems—to which the methods of the course apply. For that level of command, the student must attain some conceptual understanding and develop judgment. Thus, a certain amount of theory is very relevant, indeed essential. A student who has been trained only to do certain kinds of problems has acquired very limited expertise.

7. *The purpose of the classes and assignments is to prepare the student for the exams.*

The real purpose of the classes and homework is to guide you in achieving the aspiration of the course: *command of the material.* If you have command of the material, you should do well on the exams. On the other hand, some students act as though the exam problems have been decided in advance, and expect the lectures and assignments to be leading up to performance on *those* problems, or ones just like them. The latter would constitute the avoidance of our goal.

8. *The best way to study mathematics is to just memorize everything very carefully.*

As a colleague in the Physics Department once put it, "You can't memorize problem-solving!" Here, problem-solving refers to the ability to take a problem and attempt to carry out whatever methods might be relevant to solve it. This is a skill that grows with experience. (You might keep in mind as an analogy that memorizing the dictionary of a foreign language is not enough to achieve fluency in that language.)

9. *Students learn best when everything they have to know is presented slowly in the classroom.*

If everything the student has to know is presented slowly in the classroom, the total amount of material in the course will be rather little. Thus, students actually learn *least* that way.

10. *It is the teacher's job to cover the material.*

As covering the material is the role of the *textbook*, and the textbook is to be read by the student, the instructor should be doing something else, something that helps the student *grasp* the material. The instructor's role is to guide the students in their learning: to reinforce the essential conceptual points of the subject, and to show the relation between them and the solving of problems (cf. #6).

11. *Since you are supposed to be learning from the book, there's no need to go to the lectures.*

The lectures, the reading, and the homework should combine to produce true comprehension of the material. For most students, reading a math text will not be easy. The lectures should serve to orient the student in learning the material.

12. *A good teacher is one who can eliminate most of the struggle for the student, making the material easy to learn.*

Of course, it is possible to direct the students toward correct ways of thinking, but a certain amount of struggle is inevitable. Experience cannot be taught. Moreover, many topics are inherently difficult so they cannot be understood either passively or quickly. Eliminating the struggle can only be achieved by excising substance from the course (e.g., constricting the scope of the course, or reducing the means for recognizing where the methods of the course apply). Then the fraction of the material that remains could well be easier to learn, but the student will be acquiring diluted skills.

13. *When the students are happy with the instructor's lectures, they learn the material better.*

This statement is wishful thinking. According to the evidence I've seen, once threshold requirements are met the perceived quality of the instructor makes little, if any, difference in learning. What makes a real difference in learning is appropriate effort *by the student*. The best thing that a decent teacher can do, in order to get the students to learn better, is to hold high yet reasonable expectations of them.

Or-2. Academic Orientation

What follows is what an entering freshman should hear about the academic side of university life [in mathematics (and the sciences)]. It is distilled from what I've learned and written concerning the need for academic orientation.

The underlying premise, whose truth is very easy to demonstrate, is that most students who are admitted to a university like JHU were being taught in high school well below their level. The intent here is to reduce the time it takes for the student to appreciate this and to help him or her adjust to the demands of working up to level.

1. **You are no longer in high school.** The great majority of you, not having done so already, will have to discard high school notions of teaching and learning, and replace them by university-level notions. This may be difficult, but it must happen sooner or later, so sooner is better. Our goal is for more than just getting you to reproduce what was told to you in the classroom.

2. Expect to have material covered at *two to three* times the pace of high school. Above that, we aim for greater command of the material, esp. the ability to apply what you have learned to new situations (when relevant).

3. Lecture time is at a premium, so must be used efficiently. You cannot be "taught" everything in the classroom. **It is *your* responsibility to learn the material.** Most of this learning must take place *outside* the classroom. You should willingly put in two hours outside the classroom for each hour of class.

4. The instructor's job is primarily to provide a framework, with *some* of the particulars, to guide you in doing your learning of the concepts and methods that comprise the material of the course. It is not to "program" you with isolated facts and problem types, nor to monitor your progress.

5. You are expected to read the textbook for comprehension. It gives the detailed account of the material of the course. It also contains many examples of problems worked out, and these should be used to supplement those you see in the lecture. The textbook is not a novel, so the reading must often be slow-going and careful. However, there is the clear advantage that you can read it at your own pace. Use pencil and paper to work through the material, and to fill in omitted steps.

6. As for *when* you engage the textbook, you have the following dichotomy:

a) *[recommended for most students]* Read, for the first time, the appropriate section(s) of the book *before* the material is presented in lecture. That is, come prepared for class. Then, the faster-paced college-style lecture will make more sense.

b) If you haven't looked at the book beforehand, try to pick up what you can from the lecture. Though the lecture may seem hard to follow (cf. #2), absorb the general idea and/or take thorough notes, hoping to sort it out later, while studying from the book outside of class.

7. It is the student's responsibility to communicate clearly in writing up solutions of the questions and problems in homework and exams. The rules of language still apply in mathematics, and apply even when symbols are used in formulas, equations, etc. Exams will consist largely of fresh problems that fall within the material that is being tested.

Or-3. Two Telling Tales

1. Analogy: French in high school and college. I knew that the first-year French course (French Elements) covers about the same material as the first two years of high school French. This is typical of first year college language courses. Also, the semesters are shorter here, and one can calculate that the material is covered approximately three times as rapidly here as in high school.

After looking at the catalog description of French Elements, I called the instructor. I felt sure that there was more to it than just the triple speed. "Yes," she replied. "In our course, we aspire for fluency."

I admit that I had four years of French in high school, but no one ever spoke of *fluency*. It should be obvious that most of the work must occur outside of class. You can expect something like the tripling of high school pace, a lot of work outside of class, *plus aiming for the mathematical analogue of fluency* (perhaps *command* is the correct word), in a calculus course here.

2. Analogy: Martial arts. An 18-year-old enters a *tae kwon do* studio, walks up to the instructor, and states proudly, "I want to learn how to put my hand through a stack of bricks!"

The instructor thinks a moment, then replies. "Well, that's very difficult, and will take time. First, you must develop self-control and mental discipline. Then ..."

The youth interrupts, "Don't give me that discipline crap! Just teach me how to put my hand through the bricks!"

The instructor walks away shaking his head, as does the would-be student. One of the regulars of the studio, who teaches math in a local high school, steps up to the instructor. "You know, the young man has a point. All you have to do is make the bricks out of softer material, and crack them a little in advance."

Bibliography

[MP1]   D. Albers and G. Alexanderson, *Mathematical People*, Random House/Birkhäuser, Boston, 1985.

[AMR]   N. Ambady and R. Rosenthal, Half a minute: predicting teacher evaluations from thin slices of nonverbal behavior and physical attractiveness, *Jour. of Personality and Social Psychology* 64(1993), 1–11.

[AND]   D. Anderluh, Proposed math standards divide state's educators, *The Sacramento Bee*, October 26, 1997, p. A23.

[ANDR]   G. Andrews, The irrelevance of calculus reform: Ruminations of a sage-on-the-stage, *UME Trends* 6(1995), 17, 23.

[ANG]   I. Anshel and D. Goldfeld, *Calculus: A Computer Algebra Approach*, International Press Books, Boston, 1996.

[ASI]   A. Asiala, et al, A Framework for Research and Curriculum Development in Undergraduate Mathematics Education. In Research in Collegiate Mathematics Education II, 1996, 1–32.

[ACDS]   M. Asiala, J. Cottrill, E. Dubinsky, and K. Schwingendorf, The development of students' graphical understanding of the derivative, *Journal of Mathematical Behavior* 16(1997).

[BAP]   U. Backlund and L. Persson, Moore's teaching method, preprint.

[BAN]   T. Banchoff, Secrets of My Success, *Focus*, Math. Association of American, Washington, D.C., October, 1996, 26–8.

[BAU]   L. Frank Baum, *The Wizard of Oz*, Bobbs-Merrill Co., Indianapolis, 1899.

[BPA]   E. Beth and J. Piaget, *Mathematical Epistemology and Psychology*, translated from the French by W. Mays, D. Reidel Publishing, Dordrecht, 1966.

[BOA]   R. Boas, Can we make mathematics intelligible?, *Am. Math. Monthly* 88(1981), 727–731.

[BOU]   M. Bouniaev, Stage-by-Stage Development of Mental Actions and Computer Based Instruction, Technology and Teacher Education Annual, 1996, 947–951.

[BRE]   D. Bressoud, Review of *How to Teach Mathematics: A Personal Perspective* by Steven G. Krantz, *UME Trends* 7(1995), 4–5.

[BDDT]   A. Brown, D. DeVries, E. Dubinsky, K. Thomas, Learning binary operations, groups, and subgroups, *Journal of Mathematical Behavior* 16(1997), 187–239.

[CAL]   J. Callahan, et al, *Calculus in Context*, Freeman, New York, 1993.

[CAR]   J. Cargal, The reform calculus debate and the psychology of learning mathematics, preprint.

[CAS]   B. A. Case, ed., *You're the Professor, What's Next?*, Mathematical Association of America, Washington, D.C., 1994.

[CEW]   S. Ceci and W. Williams, Study finds students like a good show, *Science* 278(1997), 229.

[CEN]   J. A. Centra, *Determining Faculty Effectiveness*, Jossey-Bass Publishers, San Francisco, 1979.

[CHC]   "Challenge in the Classroom: the Methods of R. L. Moore", Mathematical Association of America, Washington, D.C., 1966. [*videocassette*]

[CLE]   H. Clemens, Is there a role for mathematicians in math education?, *La Gazette*, 1988, Paris, 41–44.

[COH]   P. A. Cohen, Student ratings of instruction and student achievement: A meta-analysis of multisection validity results, *Review of Educational Research* 41(1981), 511–517.

[CDNS]   J. Cottrill, et al, Understanding the limit concept: Beginning with a coordinated process schema, *Journal of Mathematical Behavior*, 15(1996), 167–192.

[CTUM]   Committee on the Teaching of Undergraduate Mathematics, *College Mathematics: Suggestions on How to Teach it*, Mathematics Association of America, Washington, D.C., 1979.

[COJ]   R. Courant and F. John, *Introduction to Calculus and Analysis*, Springer-Verlag, New York, 1989.

[C4L]   E. Dubinsky and K. Schwingendorf, *Calculus, Concepts, Computers, and Cooperative Learning*, http://www.math.purdue.edu/ ccc/.

[DAV]   B. G. Davis, *Tools for Teaching*, Jossey-Bass Publishers, San Francisco, 1993.

[DIP]   T. Dick and C. M. Patton, *Calculus of a Single Variable*, PWS-Kent, Boston, 1994.

[DZW]   B. W. Dziech and L. Weiner, *The Lecherous Professor*, 2nd. Ed., Univ. of Illinois Press, Urbana, 1990.

[DOU]   R. G. Douglas, *Toward a Lean and Lively Calculus*, Mathematical Association of America, Washington, D.C., 1986.

[DUB]   E. Dubinsky, ISETL: A programming language for learning mathematics, *Comm. Pure and Appl. Math.* 48(1995), 1027–1051.

[DDLZ] E. Dubinsky, J. Dautermann, U. Leron, and R. Zazkis, On learning fundamental concepts of group theory, *Educational Studies in Mathematics* 27(1994), 267–305.

[DUF]   E. Dubinsky, W. Fenton, *Introduction to Discrete Mathematics with* ISETL, New York, Springer, 1996.

[DUL]   E. Dubinsky and U. Leron, *Learning Abstract Algebra with* ISETL, Springer Verlag, New York, 1994.

[DSM]   E. Dubinsky, K. Schwingendorf, D. Mathews, *Applied Calculus, Concepts, and Computers*, 2nd ed., McGraw-Hill, New York, 1995.

[FEL1] K. A. Feldman, Instructional effectiveness of college teachers as judged by teachers themselves: Current and former students, colleagues, administrators, and external (neutral) observers, *Research in Higher Education* 30(1989), 137–194.

[FEL2] K. A. Feldman, The association between student ratings of specific instructional dimensions and student achievement: Refining and extending the synthesis of data from multisection validity studies, *Research in Higher Education* 30(1989), 583–645.

[GKM]   E. A. Gavosto, S. G. Krantz, and W. McCallum, *Contemporary Issues in Mathematics Education*, Cambridge University Press, Cambridge, to appear.

[GS]   I. M. Gelfand, E. G. Glagoleva, and E. E. Shnol, *Functions and Graphs*, translated and edited by Richard Silverman, Gordon and Breach, New York, 1969.

[GGK]   I. M. Gelfand, E. G. Glagoleva, and A. A. Kirillov, *The Method of Coordinates*, Birkhäuser, Boston, 1991.

[GOL]   B. Gold, ed., *Classroom Evaluation Techniques*, MAA Notes, The Mathematical Association of America, to appear.

[HAL]   D. Hughes Hallett, et al, *Calculus*, John Wiley and Sons, New York, 1992.

[HALM]   P. Halmos, *I Want to be a Mathematician*, Springer-Verlag, New York, 1985.

[HEI]   K. Heid, Resequencing Skills and Concepts in Applied Calculus Using the Computer as a Tool, Journal for Research in Mathematics Education 19(1988), 3–25.

[HOF]   D. Hoffman, The Computer-Aided Discovery of New Embedded Minimal Surfaces, *Math. Intelligencer* 9(1987), 8–21.

[HCM]   G. S. Howard, C. G. Conway, and S. E. Maxwell, Construct validity of measures of college teaching effectiveness, *Journal of Educational Psychology* 77(1985), 187–196.

[JAC1]   A. Jackson, The math wars: California battles it out over math education reform (Part I), *Notices* of the AMS 44(1997), 695–702.

[JAC2]   A. Jackson, The math wars: California battles it out over math education reform (Part II), *Notices* of the AMS 44(1997), 817–823.

[KIR]   W. Kirwan, et al, *Moving Beyond Myths*, National Research Council, The National Academy of Sciences, Washington, D.C., 1991.

[KLR]   D. Klein and J. Rosen, Calculus Reform—for the $Millions, *Notices* of the AMS 44(1997), 1324–1325.

[KOB]   N. Koblitz, *Calculus I* and *Calculus II*, at the `ftp` sites `ftp://ftp.math.washington.edu/pub/124notes/k124.tar.gz` and `ftp://ftp.math.washington.edu/pub/125notes/k125.tar.gz`.

[KOW]   S. Kogelman and J. Warren, *Mind over Math: Put Yourself on the road to Success by Freeing Yourself from Math Anxiety*, McGraw-Hill, New York, 1979.

[KRA]   S. G. Krantz, *A Primer of Mathematical Writing*, The American Mathematical Society, Providence, 1997.

[KRT1]   O. Kroeger and J. M. Thuesen, *Type Talk, or, How to Determine your Personality Type and Change Your Life*, Delacorte Press, New York, 1988.

[KRT2]   O. Kroeger and J. M. Thuesen, *Type Talk at Work*, Delacorte Press, New York, 1992.

[KULM]   J. A. Kulik and W. J. McKeachie, The evaluation of teachers in higher education, *Review of Research in Education*, F. N. Kerlinger, ed., Peacock, Itasca, 1975.

[KUM]   Kumon Educational Institute of Chicago, assorted advertising materials, 112 Arlington Heights, Illinois, 1992.

[LET] J. R. C. Leitzel and A. C. Tucker, Eds., *Assessing Calculus Reform Efforts*, The Mathematical Association of America, Washington, D.C., 1994.

[MAR] J. Martino, Dr. Jekyll or Professor Hyde?, preprint.

[MLA] Modern Language Association, *The MLA's Handbook for Writers of Research Papers*, The Modern Language Association, New York, 1984.

[MOO] D. S. Moore, The craft of teaching, *Focus* 15(1995), 5–8.

[MUM] D. Mumford, Calculus Reform—for the Millions, *Notices* of the AMS 44(1997), 559–563.

[NCE] National Commission on Excellence in Education, *A Nation at Risk: The Imperative for Educational Reform*, U. S. Government Printing Office, Washington, D.C., 1983.

[NRC] National Research Council, *Everybody Counts: A Report to the Nation on the Future of Mathematics Education*, National Academy Press, Washington, D.C., 1989.

[OSZ] A. Ostobee and P. Zorn, *Calculus from the Graphical, Numerical, and Symbolic Points of View*, Holt, Rinehart, and Winston, 1995.

[PTR] K. Park and K. J. Travers, A comparative study of a computer-based and a standard college first-year calculus course, *CBMS Issues in Mathematics Education*, 6(1996), 155–176.

[REZ] B. Reznick, *Chalking it Up*, Random House/Birkhäuser, Boston, 1988.

[ROB] A. W. Roberts, *Calculus The Dynamics of Change*, The Mathematical Association of America, Washington, D.C., 1995.

[ROSG] A. Rosenberg, et al, *Suggestions on the Teaching of College Mathematics*, Report of the Committee on the Undergraduate Program in Mathematics, Mathematics Association of America, Washington, D.C., 1972.

[RUD] W. Rudin, *Principles of Mathematical Analysis*, 3rd. Ed., McGraw-Hill Publishing, New York, 1976.

[SMM] D. A. Smith and L. C. Moore, *Calculus: Modeling and Applications*, Houghton-Mifflin College, Boston, 1996.

[SPI] M. Spivak, *Calculus*, Benjamin, New York, 1967.

[STK] G. M. A. Stanic and J. Kilpatrick, Mathematics curriculum reform in the United States: A historical perspective, *Int. J. Educ. Res.* 17(1992), 407–417.

[STE1] L. Steen, *Calculus for a New Century: A Pump, Not a Filter*, Mathematical Association of America, Washington, D.C., 1987.

[STE2]   L. Steen, et al, *Everybody Counts*, National Research Council, The National Academy of Sciences, Washington, D. C., 1989.

[STEW]   J. Stewart, *Calculus: Concepts and Contexts, Single Variable*, Brook/Cole, Pacific Grove, 1997.

[SWJ]   H. Swann and J. Johnson, *E. McSquared's Original, Fantastic, and Highly Edifying Calculus Primer*, W. Kaufman, Los Altos, 1975.

[SYK]   C. Sykes, *Profscam*, St. Martin's Press, New York, 1988.

[THU]   W. Thurston, Mathematical Education, *Notices of the A.M.S.* 37(1990), 844–850.

[TOB]   S. Tobias, *Overcoming Math Anxiety*, Norton, New York, 1978.

[TBJ]   R Traylor, W. Bane, and M. Jones, *Creative Teaching: The Heritage of R. L. Moore*, University of Houston, 1972.

[TRE]   U. Treisman, Studying students studying calculus: a look at the lives of minority mathematics students in college, *The College Mathematics Journal* 23(1992), 362–372.

[TUC]   T. W. Tucker, ed., *Priming the Calculus Pump: Innovations and Resources*, CPUM Subcommittee on Calculus Reform and the First Two Years, The Mathematical Association of America, Washington, D.C., 1990.

[TUR]   S. Turow, *One L*, Warner Books, New York, 1977.

[WAT]   F. Wattenberg, *CALC in a Real and Complex World*, PWS-Kent, Boston, 1995.

[WIE]   K. Wiesenfeld, Making the Grade, *Newsweek*, Jun 17, 1996, 16.

[WIL]   R. L. Wilder, Robert Lee Moore, 1882–1974, *Bulletin of the AMS* 82(1976), 417–427.

[WU1]   H. H. Wu, The mathematician and the mathematics education reform, *Notices of the A.M.S.* 43(1996), 1531–1537.

[WU2]   H. H. Wu, The mathematics education reform: Why you should be concerned and what you can do, *American Math. Monthly* 104(1997), to appear.

[ZUC]   S. Zucker, Teaching at the university level, *Notices of the A.M.S.* 43(1996), 863–865.

Index